THE THEORY AND APPLICATIONS
OF INSTANTON CALCULATIONS

Instantons, or pseudoparticles, are solutions to the equations of motion in classical field theories on a Euclidean spacetime. Instantons are found everywhere in quantum theories as they have many applications in quantum tunnelling. Diverse physical phenomena may be described through quantum tunnelling, for example: the Josephson effect, the decay of meta-stable nuclear states, band formation in tight binding models of crystalline solids, the structure of the gauge theory vacuum, confinement in 2+1 dimensions, and the decay of superheated or supercooled phases. Drawing inspiration from Sidney Coleman's Erice lectures, this volume provides an accessible, detailed introduction to instanton methods, with many applications, making it a valuable resource for graduate students in many areas of physics, from condensed matter, particle and nuclear physics, to string theory.

This title, first published in 2018, has been reissued as an Open Access publication on Cambridge Core.

Manu Paranjape has been a professor at the Université de Montréal for the past 30 years. In this time he has worked on quantum field theory, the Skyrme model, non-commutative geometry, quantum spin tunnelling and conformal gravity. Whilst working on induced fermion numbers, he discovered induced angular momentum on flux tube solitons, and more recently he discovered the existence of negative-mass bubbles in de Sitter space, which merited a prize in the Gravity Research Foundation essay competition.

CAMBRIDGE MONOGRAPHS ON MATHEMATICAL PHYSICS

General Editors: P. V. Landshoff, D. R. Nelson, S. Weinberg

S. J. Aarseth *Gravitational N-Body Simulations: Tools and Algorithms*[†]

J. Ambjørn, B. Durhuus and T. Jonsson *Quantum Geometry: A Statistical Field Theory Approach*[†]

A. M. Anile *Relativistic Fluids and Magneto-fluids: With Applications in Astrophysics and Plasma Physics*

J. A. de Azcárraga and J. M. Izquierdo *Lie Groups, Lie Algebras, Cohomology and Some Applications in Physics*[†]

O. Babelon, D. Bernard and M. Talon *Introduction to Classical Integrable Systems*[†]

F. Bastianelli and P. van Nieuwenhuizen *Path Integrals and Anomalies in Curved Space*[†]

D. Baumann and L. McAllister *Inflation and String Theory*

V. Belinski and M. Henneaux *The Cosmological Singularity*[†]

V. Belinski and E. Verdaguer *Gravitational Solitons*[†]

J. Bernstein *Kinetic Theory in the Expanding Universe*[†]

G. F. Bertsch and R. A. Broglia *Oscillations in Finite Quantum Systems*[†]

N. D. Birrell and P. C. W. Davies *Quantum Fields in Curved Space*[†]

K. Bolejko, A. Krasiński, C. Hellaby and M-N. Célérier *Structures in the Universe by Exact Methods: Formation, Evolution, Interactions*

D. M. Brink *Semi-Classical Methods for Nucleus-Nucleus Scattering*[†]

M. Burgess *Classical Covariant Fields*[†]

E. A. Calzetta and B.-L. B. Hu *Nonequilibrium Quantum Field Theory*

S. Carlip *Quantum Gravity in 2+1 Dimensions*[†]

P. Cartier and C. DeWitt-Morette *Functional Integration: Action and Symmetries*[†]

J. C. Collins *Renormalization: An Introduction to Renormalization, the Renormalization Group and the Operator-Product Expansion*[†]

P. D. B. Collins *An Introduction to Regge Theory and High Energy Physics*[†]

M. Creutz *Quarks, Gluons and Lattices*[†]

P. D. D'Eath *Supersymmetric Quantum Cosmology*[†]

J. Dereziński and C. Gérard *Mathematics of Quantization and Quantum Fields*

F. de Felice and D. Bini *Classical Measurements in Curved Space-Times*

F. de Felice and C. J. S Clarke *Relativity on Curved Manifolds*[†]

B. DeWitt *Supermanifolds, 2nd edition*[†]

P. G. O. Freund *Introduction to Supersymmetry*[†]

F. G. Friedlander *The Wave Equation on a Curved Space-Time*[†]

J. L. Friedman and N. Stergioulas *Rotating Relativistic Stars*

Y. Frishman and J. Sonnenschein *Non-Perturbative Field Theory: From Two Dimensional Conformal Field Theory to QCD in Four Dimensions*

J. A. Fuchs *Affine Lie Algebras and Quantum Groups: An Introduction, with Applications in Conformal Field Theory*[†]

J. Fuchs and C. Schweigert *Symmetries, Lie Algebras and Representations: A Graduate Course for Physicists*[†]

Y. Fujii and K. Maeda *The Scalar-Tensor Theory of Gravitation*[†]

J. A. H. Futterman, F. A. Handler, R. A. Matzner *Scattering from Black Holes*[†]

A. S. Galperin, E. A. Ivanov, V. I. Ogievetsky and E. S. Sokatchev *Harmonic Superspace*[†]

R. Gambini and J. Pullin *Loops, Knots, Gauge Theories and Quantum Gravity*[†]

T. Gannon *Moonshine beyond the Monster: The Bridge Connecting Algebra, Modular Forms and Physics*[†]

A. García-Díaz *Exact Solutions in Three-Dimensional Gravity*

M. Göckeler and T. Schücker *Differential Geometry, Gauge Theories, and Gravity*[†]

C. Gómez, M. Ruiz-Altaba and G. Sierra *Quantum Groups in Two-Dimensional Physics*[†]

M. B. Green, J. H. Schwarz and E. Witten *Superstring Theory Volume 1: Introduction*

M. B. Green, J. H. Schwarz and E. Witten *Superstring Theory Volume 2: Loop Amplitudes, Anomalies and Phenomenology*

V. N. Gribov *The Theory of Complex Angular Momenta: Gribov Lectures on Theoretical Physics*[†]

J. B. Griffiths and J. Podolský *Exact Space-Times in Einstein's General Relativity*[†]

[†] Available in paperback

The Theory and Applications of Instanton Calculations

MANU PARANJAPE

Université de Montréal

CAMBRIDGE
UNIVERSITY PRESS

Shaftesbury Road, Cambridge CB2 8EA, United Kingdom

One Liberty Plaza, 20th Floor, New York, NY 10006, USA

477 Williamstown Road, Port Melbourne, VIC 3207, Australia

314–321, 3rd Floor, Plot 3, Splendor Forum, Jasola District Centre, New Delhi – 110025, India

103 Penang Road, #05–06/07, Visioncrest Commercial, Singapore 238467

Cambridge University Press is part of Cambridge University Press & Assessment,
a department of the University of Cambridge.

We share the University's mission to contribute to society through the pursuit of
education, learning and research at the highest international levels of excellence.

www.cambridge.org
Information on this title: www.cambridge.org/9781009291262

DOI: 10.1017/9781009291248

First published 2018
Reissued as OA 2022

A catalogue record for this publication is available from the British Library.

ISBN 978-1-009-29126-2 Hardback
ISBN 978-1-009-29123-1 Paperback

Contents

Preface

This book is based on a graduate course taught four times, once in French at the Université de Montréal and then three times in English at the Institut für Theoretische Physik, in Innsbruck, Austria, at the Center for Quantum Spacetime, Department of Physics, Sogang University, Seoul, Korea, and most recently, a part of it at the African Institute for Mathematical Sciences (AIMS), Cape Town, South Africa.

The course covered the contents of the magnificent Erice lectures of Coleman [31], "The Uses of Instantons", in addition to several chapters based on independent research papers. However, it might be more properly entitled, "The Uses of Instantons for Dummies". I met Sidney Coleman a few times, more than 30 years ago, and although I am sure that he was less impressed with the meetings than I was and probably relegated them to the dustbin of the memory, my debt to him is enormous. Without his lecture notes I cannot imagine how I would ever have been able to understand what the uses of instantons actually were. However, in his lecture notes, one finds that he also thanks and expresses gratitude to a multitude of eminent and great theoretical physicists of the era, indeed thanking them for "patiently explaining large portions of the subject" to him. Unfortunately, we cannot all be so lucky. Coleman's lecture notes are a work of art; it is clear when one reads them that one is enjoying a master impressionist painter's review of a subject, a review that transmits, as he says, the "awe and joy" of the beauty of the "wonderful things brought back from far places". But then the hard work begins.

Hence, through diligent, fastidious and brute force work, I have been able, I hope, to produce what I believe is a well-rounded, detailed monograph, essentially explaining in a manner accessible to first- and second-year graduate students the beauty and the depth of what is contained in Coleman's lectures and in some elaborations of the whole field itself.

I am indebted to many, but I will thank explicitly Luc Vinet for impelling me to first give this course when I started out at the Université de Montréal; Gebhard Grübl for the opportunity to teach the course at the Universität Innsbruck in Innsbruck, Austria; Bum-Hoon Lee for the same honour at Sogang University in Seoul, Korea; and Fritz Hahne for the opportunity to give the lectures at the African Institute for Mathematical Sciences, Cape Town, South Africa. I thank the many students who took my course and suggested corrections to my lectures. I thank Nick Manton, Chris Dobson, and Duncan Dormor, respectively, Fellow, Master and President of St John's College, University of Cambridge in 2015, for making available to me the many assets of the College that made it possible to work uninterrupted and in a pleasant ambiance on this book, during my stay as an Overseas Visiting Scholar. I also thank my many colleagues and friends who have helped me through discussions and advice; these include Ian Affleck, Richard MacKenzie, Éric Dupuis, Jacques Hurtubise, Keshav Dasgupta and Gordon Semenoff.

I especially thank my wife Suneeti Phadke, who started the typing of my lectures in TeX and effectively typed more than half the book while caring for a six-month-old baby. This was no easy feat for someone with a background in Russian literature, devoid of the intricacies of mathematical typesetting. This book would not have come to fruition had it not been for her monumental efforts.

I also thank my children Kiran and Meghana, whose very existence makes it a joy and a wonder to be alive.

1

Introduction

This book covers the methods by which we can use instantons. What is an instanton? A straightforward definition is the following. Given a quantum system, an instanton is a solution of the equations of motion of the corresponding classical system; however, not for ordinary time, but for the analytically continued classical system in imaginary time. This means that we replace t with $-i\tau$ in the classical equations of motion. Such solutions are alternatively called the solutions of the Euclidean equations of motion.

This type of classical solution can be important in the semi-classical limit $\hbar \to 0$. The Feynman path integral, which we will study in its Euclideanized form in great detail in this book, gives the matrix element corresponding to the amplitude for an initial state at $t = t_i$ to be found in a final state at $t = t_f$ as a "path integral"

$$\langle \text{final}, t_f | \text{initial}, t_f \rangle = \langle \text{final}, t_f | e^{-\frac{i}{\hbar}(t_f - t_i)\hat{h}(\hat{q},\hat{p})} | \text{initial}, t_i \rangle$$

$$= \int_{\text{initial}, t_i}^{\text{final}, t_f} \mathcal{D}p\mathcal{D}q \, e^{\frac{i}{\hbar} \int dt(p\dot{q} - h(p,q))} \tag{1.1}$$

where $\hat{h}(\hat{q},\hat{p})$ is the quantum Hamiltonian and $h(q,p)$ is the corresponding classical Hamiltonian of the dynamical system. The "path integral" and integration measure $\mathcal{D}p\mathcal{D}q$ defines an integration over all classical "paths" which satisfy the boundary conditions corresponding to the initial state at t_i and to the final state at t_f. It is intuitively evident, or certainly from the approximation method of stationary phase, that the dominant contribution, as $\hbar \to 0$, should come from the neighbourhood of the classical path which corresponds to a stationary (critical) point of the exponent, since the contributions from non-stationary points of the exponent become suppressed as the regions around them cause wild, self-annihilating variations of the exponential.

However, the situation can occur where the particle (or quantum system in general) is classically forbidden from entering some parts of the configuration

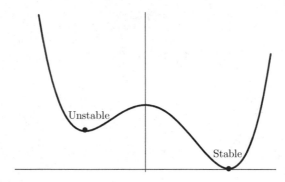

Figure 1.1. A system trapped in the false vacuum will tunnel through the barrier to the state of lower energy

space. In this case we are, generally speaking, considering tunnelling through a barrier, as depicted in Figure 1.1. Classically the particle is not allowed to enter the space where the potential energy is greater than the total energy of the particle. Indeed, if the energy of a particle is given by

$$E = T + V = \frac{\dot{q}^2}{2} + V(q) \tag{1.2}$$

then for a classically fixed energy, regions where $E < V(q)$ require that $T = \frac{\dot{q}^2}{2} < 0$, which means that the kinetic energy has to be negative, and such regions are classically forbidden. Then what takes the role of the dominant contribution in the limit $\hbar \to 0$, since no classical path can interpolate between the initial and final states?

Heuristically such a region is attainable if t becomes imaginary. Indeed, if $t \to -i\tau$ then $\left(\frac{dq}{dt}\right)^2 \to \left(i\frac{dq}{d\tau}\right)^2 = -\left(\frac{dq}{d\tau}\right)^2$, T becomes negative and then perhaps such regions are accessible. Hence it could be interesting to see what happens if we analytically continue to imaginary time, equivalent to continuing from Minkowski spacetime to Euclidean space, which is exactly what we will do. In fact, we will be able to obtain many results of the usual semi-classical WKB (Wentzel, Kramers and Brillouin) approximation [119, 77, 22], using the Euclidean space path integral. The amplitudes that we can calculate, although valid for the small $\hbar$ limit, are not normally attainable in any order in perturbation theory; they behave like $\sim K e^{-S_0/\hbar}(1 + o(\hbar))$. Such a behaviour actually corresponds to an essential singularity at $\hbar = 0$.

The importance of being able to do this is manifold. Indeed, it is interesting to be able to reproduce the results that can be obtained by the standard WKB method for quantum mechanics using a technique that seems to have absolutely no connection with that method. Additionally, the methods that we will enunciate here can be generalized rather easily to essentially any quantum system, especially to the case of quantum field theory. Tunnelling phenomena in quantum field theory are extremely important. The structure of the quantum chromodynamics (QCD) vacuum and its low-energy excitations is intimately connected to tunnelling. Various properties of the phases of quantum field theories are dramatically altered by the existence of tunelling. The decay of the false vacuum and the escape from inflation is also a tunnelling effect that is of paramount importance to cosmology, especially the early universe. The method of instantons lets us study all of these phenomena in one general framework.

1.1 A Note on Notation

We will use the following notation throughout this book:

$$\text{metric} \quad \eta_{\mu\nu} = (1, -1, -1, -1) \tag{1.3}$$

$$\text{Minkowsi time} \quad t \tag{1.4}$$

$$\text{Euclidean time} \quad \tau \tag{1.5}$$

$$\text{Analytic continuation of time} \quad t \to -i\tau \tag{1.6}$$

2

Quantum Mechanics and the Path Integral

2.1 Schrödinger Equation and Probability

Our starting point will be single-particle quantum mechanics as defined by the Schrödinger equation

$$i\hbar\frac{d}{dt}\Psi(x,t) = \hat{h}\left(x, -i\hbar\frac{d}{dx}\right)\Psi(x,t). \tag{2.1}$$

Here $\hat{h}(x, -i\hbar\frac{d}{dx})$ is a self-adjoint operator, the Hamiltonian on the space of wavefunctions $\Psi(x,t)$, where x stands for any number of spatial degrees of freedom. The connection to physics of $\Psi(x,t)$ comes from the interpretation of $\Psi(x,t)$ as the amplitude of the probability to find the particle between x and $x + dx$ at time t; hence, the probability density is given by

$$\mathcal{P}[x, x+dx] = \Psi^*(x,t)\Psi(x,t). \tag{2.2}$$

Correspondingly, the probability of finding the particle in a volume V is given by

$$\mathcal{P}[V] = \int_V dx\,\Psi^*(x,t)\Psi(x,t). \tag{2.3}$$

The state of the system is completely described by the wave function $\Psi(x,t)$. It is the content of a standard course on quantum mechanics to find $\Psi(x,t)$ for a given $\hat{h}(x, -i\hbar\frac{d}{dx})$.

2.2 Position and Momentum Eigenstates

For our purposes, we introduce the set of (improper) states $|x\rangle$ which diagonalize the position operator $\hat{X}$, with

$$\hat{X}|x\rangle = x|x\rangle \tag{2.4}$$

and

$$\int dx\,|x\rangle\langle x| = \mathbb{I}. \tag{2.5}$$

We are in principle working in d dimensions, but we suppress the explicit dependence on the number of coordinates. The states are improper in the sense that the normalization is

$$\langle x|y\rangle = \delta(x-y), \tag{2.6}$$

where $\delta(x-y)$ is the Dirac delta function. We also introduce the set of (improper) states $|p\rangle$ which diagonalize the momentum operator $\hat{P}$

$$\hat{P}|p\rangle = p|p\rangle \tag{2.7}$$

with

$$\int dp|p\rangle\langle p| = 1 \tag{2.8}$$

but as with the position eigenstates

$$\langle p|p'\rangle = \delta(p-p'), \tag{2.9}$$

where $\delta(p-p')$ is the Dirac delta function in momentum space. The improper states $|x\rangle$ and $|p\rangle$ are not vectors in the Hilbert space of states, they have infinite norm. They actually define vector-valued distributions, linear maps from the space of the square integrable functions of x or p or some suitable set of test functions usually taken to be of compact support, to actual vectors in the Hilbert space,

$$|x\rangle: \quad f(x) \to |f\rangle \sim \int dx f(x)|x\rangle, \tag{2.10}$$

where the $\sim$ should be interpreted as "loosely defined by". For a more rigorous definition, see the book by Reed and Simon [107] or Glimm and Jaffe [55].

The operators $\hat{X}$ and $\hat{P}$ must satisfy the canonical commutation relation

$$[\hat{X}, \hat{P}] = i\hbar. \tag{2.11}$$

The algebraic relation Equation (2.11) is not adequate to determine $\hat{P}$ completely; there are infinitely many representations of the commutator Equation (2.11) in which $\hat{X}$ is diagonal. Taking the matrix element of Equation (2.11) between position eigenstates gives

$$(x-y)\langle x|\hat{P}|y\rangle = \langle x|[\hat{X}, \hat{P}]|y\rangle = i\hbar\langle x|y\rangle = i\hbar\delta(x-y). \tag{2.12}$$

For the more mathematically inclined, this expression does not make good sense, since the position and momentum operators are unbounded, though self-adjoint operators. They may only act on their respective domains and, correspondingly, the product of two unbounded operators requires proper analysis of the domains and ranges of the operators concerned and similar other difficulties can exist. We leave these subtleties out in what follows, and refer the interested reader to the book on functional analysis by Reed and Simon [107]. We find the solution for $\langle x|\hat{P}|y\rangle$ as

$$\langle x|\hat{P}|y\rangle = -i\hbar\frac{d}{dx}\delta(x-y) + c\delta(x-y)$$
$$= -i\hbar\frac{d}{dx}\langle x|y\rangle + c\delta(x-y), \tag{2.13}$$

where c is an arbitrary constant, using the property of the δ function that $(x-y)\delta(x-y) \equiv 0$. We will call the x representation the one in which the momentum operator is represented by a simple derivative, *i.e.* $c=0$,

$$\langle x|\hat{P}|y\rangle = -i\hbar\frac{d}{dx}\langle x|y\rangle. \tag{2.14}$$

In this representation,

$$\langle x|\hat{P}|p\rangle = \int dy \langle x|\hat{P}|y\rangle\langle y|p\rangle = \int dy \left(-i\hbar\frac{d}{dx}\langle x|y\rangle\right)\langle y|p\rangle$$
$$= -i\hbar\frac{d}{dx}\langle x|p\rangle. \tag{2.15}$$

Acting to the right directly in the left-hand side of Equation (2.15) gives

$$\langle x|\hat{P}|p\rangle = p\langle x|p\rangle = -i\hbar\frac{d}{dx}\langle x|p\rangle. \tag{2.16}$$

The appropriately normalized solution of the resulting differential equation is

$$\langle x|p\rangle = \frac{1}{(2\pi\hbar)^{\frac{d}{2}}}e^{i\frac{p\cdot x}{\hbar}}, \tag{2.17}$$

where d is the number of spatial dimensions.

2.3　Energy Eigenstates and Semi-Classical States

We can write the eigenstate of the Hamiltonian in the form $|\Psi_E\rangle$,

$$\hat{h}(\hat{X},\hat{P})|\Psi_E\rangle = E|\Psi_E\rangle, \tag{2.18}$$

where $\hat{h}(\hat{X},\hat{P})$ is defined such that

$$\langle x|\hat{h}(\hat{X},\hat{P})|f\rangle = \hat{h}\left(x, -i\hbar\frac{d}{dx}\right)\langle x|f\rangle \tag{2.19}$$

for any vector $|f\rangle$ in the Hilbert space. Then

$$\langle x|\hat{h}(\hat{X},\hat{P})|\Psi_E\rangle = \hat{h}\left(x, -i\hbar\frac{d}{dx}\right)\langle x|\Psi_E\rangle = E\langle x|\Psi_E\rangle, \tag{2.20}$$

which implies the energy eigenfunctions are given by

$$\Psi_E(x) = \langle x|\Psi_E\rangle. \tag{2.21}$$

Correspondingly,

$$|\Psi_E\rangle = \int dx |x\rangle\langle x|\Psi_E\rangle = \int dx \Psi_E(x)|x\rangle \qquad (2.22)$$

and

$$\hat{h}(x, -i\hbar\frac{d}{dx})\Psi_E(x) = E\Psi_E(x). \qquad (2.23)$$

A particle described by $\Psi_E(x)$ is most likely to be found in the region where $\Psi_E(x)$ is peaked. The time-dependent solution of the Schrödinger equation for static Hamiltonians is given by $\Psi_E(x,t) = \Psi_E(x)e^{-\frac{i}{\hbar}Et}$, and the most general state of the system is a linear superposition

$$\Psi(x,t) = \sum_E A_E \Psi_E(x) e^{-\frac{i}{\hbar}Et} \qquad (2.24)$$

with

$$\sum_E A_E^* A_E = 1. \qquad (2.25)$$

Suppose the Hamiltonian can be modified by adjusting the potential, say, such that $\Psi_E(x)$ approaches a delta function:

$$\Psi_E(x) \to \delta(x - x_0). \qquad (2.26)$$

We would then say that a particle in the energy level E is localized at the point x_0. But in the limit of Equation (2.26) we clearly have

$$|\Psi_E\rangle \to |x_0\rangle \qquad (2.27)$$

from Equation (2.22). Thus the states $|x\rangle$ describe particles localized at the spatial point x. This is conceptually important for the semi-classical limit. Semi-classically we think of particles as localized at points in the configuration space. Thus the states $|x\rangle$ and their generalizations are useful in the description of quantum systems in the semi-classical limit.

2.4 Time Evolution and Transition Amplitudes

Given a particle in a state $|\Psi;0\rangle = |\Psi\rangle$ at $t = 0$, the Schrödinger equation, Equation (2.1), governs the time evolution of the state. The state at $t = T$ is given by

$$|\Psi;T\rangle = e^{-i\frac{T}{\hbar}\hat{h}(\hat{X},\hat{P})}|\Psi\rangle, \qquad (2.28)$$

which satisfies the Schrödinger equation. The exponential of a self-adjoint operator, which accurs on the right-hand side of Equation (2.28), is rigorously defined via the spectral representation [107]. The probability amplitude for finding the particle in a state $|\Phi\rangle$ at $t = T$ is then given by

$$\langle\Phi|\Psi;T\rangle = \langle\Phi|e^{-i\frac{T\hat{h}(\hat{X},\hat{P})}{\hbar}}|\Psi\rangle. \qquad (2.29)$$

We could derive an expression for this matrix element in terms of a "path integral". Such an integral would be defined as an integral over the space of all classical paths starting from the initial state and ending at the final state, and we would find that the function that we would integrate is the exponential of $-i$ times the classical action for each path. This is the standard Feynman path integral [45, 46], which was actually suggested by Dirac [40].

2.5 The Euclidean Path Integral

Rather than the matrix element Equation (2.29), we are more interested in a path-integral representation of the matrix element

$$\langle \Phi | e^{-\frac{\beta \hat{h}(\hat{X}, \hat{P})}{\hbar}} | \Psi \rangle, \tag{2.30}$$

where β can be thought of as imaginary time

$$T \to -i\beta. \tag{2.31}$$

The derivation of the path-integral representation of Equation (2.30) is more rigorous than that for Equation (2.29); however, the derivation which follows can be almost identically taken over to the case of real time. This can be completed by the reader. It is the matrix element of Equation (2.30) that will interest us in future chapters.

First of all, due to the linearity of quantum mechanics, it is sufficient to consider the matrix element

$$\langle y | e^{-\frac{\beta}{\hbar} \hat{h}(\hat{X}, \hat{P})} | x \rangle. \tag{2.32}$$

To obtain Equation (2.30) we just integrate over x and y with appropriate smearing functions as in Equation (2.10). Now we write

$$e^{-\frac{\beta \hat{h}(\hat{X}, \hat{P})}{\hbar}} = \underbrace{e^{-\frac{\epsilon \hat{h}(\hat{X}, \hat{P})}{\hbar}} \cdot e^{-\frac{\epsilon \hat{h}(\hat{X}, \hat{P})}{\hbar}} \cdots e^{-\frac{\epsilon \hat{h}(\hat{X}, \hat{P})}{\hbar}}}_{N+1 \text{ factors}}, \tag{2.33}$$

where we mean $N+1$ factors on the right-hand side and $(N+1)\epsilon = \beta$. Next we insert complete sets of position eigenstates

$$\int dz_i |z_i\rangle\langle z_i| = \mathbb{I}, \tag{2.34}$$

where $\mathbb{I}$ is the identity operator. Between the evolution operators appearing on the right-hand side of Equation (2.33), there will be N such insertions, *i.e.* $i : 1 \to N$. Consider one of the matrix elements

$$\langle z_i | e^{-\frac{\epsilon \hat{h}(\hat{X}, \hat{P})}{\hbar}} | z_{i-1} \rangle \tag{2.35}$$

between position eigenstates $|z_i\rangle$ and $|z_{i-1}\rangle$ for Hamiltonians of the form

$$\hat{h}(\hat{X}, \hat{P}) = \frac{\hat{P}^2}{2} + V(\hat{X}). \tag{2.36}$$

Then

$$
\begin{aligned}
\langle z_i | e^{-\frac{\epsilon \hat{h}(\hat{X},\hat{P})}{\hbar}} | z_{i-1} \rangle &= \langle z_i | 1 - \frac{\epsilon}{\hbar} \left(\hat{P}^2/2 + V(\hat{X}) \right) | z_{i-1} \rangle + \mathrm{o}(\epsilon^2) \\
&= \int dp_i \langle z_i | 1 - \frac{\epsilon}{\hbar} \left(p_i^2/2 + V(z_{i-1}) \right) | p_i \rangle \langle p_i | z_{i-1} \rangle + \mathrm{o}(\epsilon^2) \\
&= \int dp_i \left(1 - \frac{\epsilon}{\hbar} \left(p_i^2/2 + V(z_{i-1}) \right) \right) \langle z_i | p_i \rangle \langle p_i | z_{i-1} \rangle + \mathrm{o}(\epsilon^2) \\
&= \int \frac{dp_i}{(2\pi\hbar)^d} e^{-\frac{\epsilon}{\hbar}\left(\frac{p_i^2}{2} + V(z_{i-1}) - i p_i \frac{(z_i - z_{i-1})}{\epsilon} \right)} + \mathrm{o}(\epsilon^2) \\
&= \left(\int \frac{dp_i}{(2\pi\hbar)^d} e^{-\frac{\epsilon}{\hbar}\left(\frac{p_i^2}{2} - i \frac{p_i(z_i - z_{i-1})}{\epsilon} - \frac{1}{2}\left(\frac{z_i - z_{i-1}}{\epsilon} \right)^2 \right)} \right) \times \\
&\quad \times e^{-\frac{\epsilon}{\hbar}\left(\frac{1}{2}\left(\frac{z_i - z_{i-1}}{\epsilon} \right)^2 + V(z_{i-1}) \right)} + \mathrm{o}(\epsilon^2),
\end{aligned} \tag{2.37}
$$

where in the second step, we have inserted a complete set of momentum eigenstates after letting $V(\hat{X})$ act on the position eigenstate $| z_{i-1} \rangle$. The first factor in the last equality is just a (shifted) Gaussian integral, and can be easily evaluated to give

$$
\mathcal{N}_\epsilon = \int \frac{dp_i}{(2\pi\hbar)^d} e^{\frac{-\epsilon}{2\hbar}\left(p_i - i\frac{(z_i - z_{i-1})}{\epsilon} \right)^2} = \left(\frac{1}{\sqrt{2\pi\hbar\epsilon}} \right)^d. \tag{2.38}
$$

Now we use Equations (2.37) and (2.38) in Equation (2.33), inserting an independent complete set of position eigenstates between each of the factors to yield

$$
\begin{aligned}
\langle y | e^{-\frac{\beta}{\hbar}\hat{h}(\hat{X},\hat{P})} | x \rangle &= \int \frac{dz_1 \cdots dz_N}{(2\pi\hbar\epsilon)^{\frac{Nd}{2}}} \prod_{i=1}^{N+1} e^{-\frac{\epsilon}{\hbar}\left(\frac{1}{2}\left(\frac{z_i - z_{i-1}}{\epsilon} \right)^2 + V(z_{i-1}) \right) + \mathrm{o}(\epsilon^2)} \\
&= \int \frac{dz_1 \cdots dz_N}{(2\pi\hbar\epsilon)^{\frac{Nd}{2}}} e^{-\frac{\epsilon}{\hbar}\sum_{i=1}^{N+1}\left(\frac{1}{2}\left(\frac{z_i - z_{i-1}}{\epsilon} \right)^2 + V(z_{i-1}) \right) + \mathrm{o}(\epsilon^2)},
\end{aligned} \tag{2.39}
$$

where we define $z_0 = x$ and $z_{N+1} = y$. Equation (2.39) is actually as far as one can go rigorously. It expresses the matrix element as a path integral over piecewise straight (N pieces), continuous paths weighted by the exponential of a discretized approximation to the negative Euclidean action. In the limit $N \to \infty$, the $\mathrm{o}(\epsilon^2)$ terms are expected to be negligible. Additionally, in the limit that the path becomes differentiable, which is actually almost never the case,

$$
\epsilon \sum_{i=1}^{N} \left(\frac{1}{2}\left(\frac{z_i - z_{i-1}}{\epsilon} \right)^2 + V(z_{i-1}) \right) \to \int d\tau \left(V(z(\tau)) + \frac{1}{2}\dot{z}(\tau))^2 \right), \tag{2.40}
$$

where $\tau \in [0, \beta]$ parametrizes the path such that $z(0) = x$ and $z(\beta) = y$. Hence the matrix element Equation (2.32) can be formally written as the integral over classical paths,

$$\langle y|e^{-\frac{\beta}{\hbar}\hat{h}(\hat{X},\hat{P})}|x\rangle = \mathcal{N}\int \mathcal{D}z(\tau)e^{-\frac{1}{\hbar}\int_0^\beta d\tau\left(\frac{1}{2}(\dot{z}(\tau))^2+V(z(\tau))\right)}$$

$$= \mathcal{N}\int \mathcal{D}z(\tau)e^{-\frac{S_E[z(\tau)]}{\hbar}}, \tag{2.41}$$

where $S_E[z(\tau)]$ is the classical Euclidean action for each path $z(\tau)$, which starts at x and ends at y. $\mathcal{D}z(\tau)$ is the formal integration measure over the space of all such paths and $\mathcal{N}$ is a formally infinite or ill-defined normalization constant, the limit of $\frac{1}{(2\pi\hbar\epsilon)^{\frac{Nd}{2}}}$ as $N \to \infty$.

There exists a celebrated measure defined on the space of paths, the so-called Wiener measure [121], which was defined in the rigorous study of Brownian motion. One can use it to define the Euclidean path integral rigorously and unambiguously, certainly for quantum mechanics, but also in many instances for quantum field theory. We are not interested in these mathematical details, and we will use and manipulate the path integral as if it were an ordinary integral. We will have to define what we mean by this measure and normalization more carefully, later. The measure actually only makes sense, in any rigorous way, for the discretized version Equation (2.39) including the limit $N \to \infty$; however, strictly speaking the path integral for smooth paths, Equation (2.41), is just a formal expression.

We will record here the corresponding formula in Minkowski time:

$$\langle y|e^{-\frac{iT}{\hbar}\hat{h}(\hat{X},\hat{P})}|x\rangle = \mathcal{N}\int \mathcal{D}z(t)e^{\frac{i}{\hbar}\int_0^T dt\left(\frac{1}{2}(\dot{z}(t))^2-V(z(t))\right)}$$

$$= \mathcal{N}\int \mathcal{D}z(t)e^{\frac{i}{\hbar}S_M[z(t)]}. \tag{2.42}$$

This formula can be proved formally by following each of the steps that we have done for the case of the Euclidean path integral; we leave the details to the reader. However, the Gaussian integral that we encountered at Equation (2.38) becomes

$$\mathcal{N}_\epsilon = \int \frac{dp_i}{(2\pi\hbar)^d}e^{-\frac{i\epsilon}{2\hbar}\left(p_i-i\frac{(z_i-z_{i-1})}{\epsilon}\right)^2}. \tag{2.43}$$

This expression is ill-defined, but it only contributes to an irrelevant normalization constant. We can make it well-defined by adding a small negative imaginary part to the Hamiltonian, which then yields

$$\mathcal{N}_\epsilon = \left(\sqrt{\frac{i}{2\pi\hbar\epsilon}}\right)^d. \tag{2.44}$$

Adding the imaginary part to the Hamiltonian is known in other words as the "i-epsilon" prescription (note this "epsilon" has nothing to do with the ϵ appearing in our formulas above). Such a deformation can be effected in the case at hand by changing the $p_i \to (1 - i\xi)p_i$ in the exponent of Equation (2.43) with infinitesimal ξ (instead of using the usual "epsilon"). It is tantamount to defining the Minkowski path integral by starting with the Euclidean path integral and continuing this back to Minkowski space.

For the remainder of this book, we will be interested in the path-integral representation, Equation (2.39), of the matrix element Equation (2.32). We will apply methods that are standard for ordinary integrals to obtain approximations for the matrix element. We will use the saddle point method for evaluation of the path integral. This involves finding the critical points of the Euclidean action and then expanding about the critical point in a (functional) Taylor expansion. The value of the action at the critical point is a constant as far as the integration is concerned and just comes out of the integral. This term alone already gives much novel information about the matrix element. It is usually non-perturbative in the coupling constant. The first variation of the action vanishes by definition at the critical point. The first non-trivial term, the second-order term in the Taylor expansion, yields a Gaussian path integral. The remaining higher-order terms in the Taylor expansion give perturbative corrections to the Gaussian integral. The Gaussian integral can sometimes be done explicitly, although this too can be prohibitively complicated.

We will work with the formal path integral, Equation (2.41), rather than the exact discretized version, Equation (2.39). First of all it is much easier to find the critical points of the classical Euclidean action rather than its discretized analogue. Secondly, in the limit that $N \to \infty$, the critical points for the discrete action should approach those of the classical action. The actual path integral to be done always remains defined by the discretized version. The critical point of the classical action is only to be used as a centre point about which to perform the path integral Equation (2.39) in the Gaussian approximation and in further perturbative expansion. As stressed by Coleman [31], the set of smooth paths is a negligible fraction of the set of all paths. However, this does not dissuade us from using a particular smooth path, that which is a solution of the classical equations of motion, as a centre point about which to perform the functional integration in a Gaussian approximation. The Gaussian path integral corresponds to integration over all paths, especially including those which are arbitrarily non-smooth, but which are centred on the particular smooth path corresponding to the solution of the equations of motion, with a quadratic approximation to the action (or what is called Gaussian since it leads to an (infinite) product of Gaussian integrals). It actually receives most of its contribution from extremely non-smooth paths. However, the Gaussian path integral can be evaluated in some cases exactly, and in other cases in a perturbative approximation. In this way the exact definition of the formal path integral, Equation (2.41), is not absolutely essential for our further considerations. We will, however, continue to frame our analysis in terms of it, content with the understanding that underlying it a more rigorous expression always exists.

3

The Symmetric Double Well

In this chapter we will consider in detail a simple quantum mechanical system where "instantons", critical points of the classical Euclidean action, can be used to uncover non-perturbative information about the energy levels and matrix elements. We will also explicitly see the use of the particular matrix element (2.27) that we consider. The model we will consider has the classical Euclidean action

$$S_E[z(\tau)] = \int_{-\frac{\beta}{2}}^{\frac{\beta}{2}} d\tau \left(\frac{1}{2}(\dot{z}(\tau))^2 + V(z(\tau)) \right). \qquad (3.1)$$

We choose for convenience the domain $[-\frac{\beta}{2}, \frac{\beta}{2}]$ and we will choose the potential explicitly later. We will always have in mind that $\beta \to \infty$, thus if β is considered finite, it is to be treated as arbitrarily large. The potential, for now, is simply required to be a symmetric double well potential, adjusted so that the energy is equal to zero at the bottom of each well, located at $\pm a$, as depicted in Figure 3.1.

3.1 Classical Critical Points

The critical points of the action, Equation (3.1), are achieved at solutions of the equations of motion

$$\left. \frac{\delta S_E[z(\tau)]}{\delta z(\tau')} \right|_{z(\tau') = \bar{z}(\tau')} = -\ddot{\bar{z}}(\tau') + V'(\bar{z}(\tau')) = 0. \qquad (3.2)$$

We assume $\bar{z}(\tau)$ satisfies Equation (3.2). Then writing $z(\tau) = \bar{z}(\tau) + \delta z(\tau)$ and expanding in a Taylor series, we find

$$S_E[z(\tau)] = S_E[\bar{z}(\tau)] + \frac{1}{2} \int d\tau' d\tau'' \left. \frac{\delta^2 S_E[z(\tau)]}{\delta z(\tau')\delta z(\tau'')} \right|_{z(\tau)=\bar{z}(\tau)} \delta z(\tau')\delta z(\tau'') + \cdots,$$

$$(3.3)$$

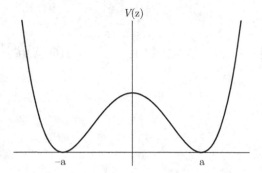

Figure 3.1. A symmetric double well potential with minima at $\pm a$

where we note that the first-order variation is absent as the equations of motion, Equation (3.2), are satisfied. The second-order variation is given by

$$\frac{\delta^2 S_E[z(\tau)]}{\delta z(\tau')\delta z(\tau'')}\bigg|_{z(\tau)=\bar{z}(\tau)} = \left(-\frac{d^2}{d\tau'^2}+V''(\bar{z}(\tau'))\right)\delta(\tau'-\tau''). \qquad (3.4)$$

Then we have

$$S_E[z(\tau)] = S_E[\bar{z}(\tau)] + \frac{1}{2}\int_{-\frac{\beta}{2}}^{\frac{\beta}{2}} d\tau \delta z(\tau)\left(-\frac{d^2}{d\tau^2}+V''(\bar{z}(\tau))\right)\delta z(\tau) + \cdots. \qquad (3.5)$$

We can expand $\delta z(\tau)$ in terms of the complete orthonormal set of eigenfunctions $z_n(\tau)$ of the hermitean operator entering in the second-order term

$$\left(-\frac{d^2}{d\tau^2}+V''(\bar{z}(\tau))\right)z_n(\tau) = \lambda_n z_n(\tau), \quad n=0,1,2,3,\cdots,\infty \qquad (3.6)$$

supplied with the boundary conditions

$$z_n(-\frac{\beta}{2}) = z_n(\frac{\beta}{2}) = 0. \qquad (3.7)$$

Completeness implies

$$\sum_{n=0}^{\infty} z_n(\tau)z_n(\tau') = \delta(\tau-\tau') \qquad (3.8)$$

while orthonormality gives

$$\int_{-\frac{\beta}{2}}^{\frac{\beta}{2}} d\tau\, z_n(\tau)z_m(\tau) = \delta_{nm}. \qquad (3.9)$$

Thus expanding

$$\delta z(\tau) = \sum_{n=0}^{\infty} c_n z_n(\tau) \qquad (3.10)$$

we find

$$S_E[z(\tau)] = S_E[\bar{z}(\tau)] + \frac{1}{2}\sum_{n=0}^{\infty} \lambda_n c_n^2 + \mathrm{o}(c_n^3) \qquad (3.11)$$

using the orthonormality Equation (3.9) of the $z_n(\tau)$'s.

3.2 Analysis of the Euclidean Path Integral

The original matrix element that we wish to study, Equation (2.32), is given by

$$\langle y|e^{-\frac{\beta}{\hbar}\hat{h}(\hat{X},\hat{P})}|x\rangle = \langle \bar{z}(\beta/2)|e^{-\frac{\beta}{\hbar}\hat{h}(\hat{X},\hat{P})}|\bar{z}(-\beta/2)\rangle, \qquad (3.12)$$

as we have not yet picked the boundary conditions on $\bar{z}(\pm\beta/2)$. Then we get

$$\langle \bar{z}(\beta/2)|e^{-\frac{\beta}{\hbar}\hat{h}(\hat{X},\hat{P})}|\bar{z}(-\beta/2)\rangle = \mathcal{N}\int \mathcal{D}z(\tau)e^{-\frac{1}{\hbar}\left(S_E[\bar{z}(\tau)]+\frac{1}{2}\sum_{n=0}^{\infty}\lambda_n c_n^2 + \mathrm{o}(c_n^3)\right)}$$
$$= e^{-\frac{S_E[\bar{z}(\tau)]}{\hbar}}\mathcal{N}\int \mathcal{D}z(\tau)e^{-\frac{1}{\hbar}\left(\sum_{n=0}^{\infty}\frac{1}{2}\lambda_n c_n^2 + \mathrm{o}(c_n^3)\right)}.$$
$$(3.13)$$

Now we will begin to define the path integration measure as

$$\mathcal{D}z(\tau) \to \prod_{n=0}^{\infty}\frac{dc_n}{\sqrt{2\pi\hbar}}, \qquad (3.14)$$

integrating over all possible values of the c_n's as a reasonable way of integrating over all paths. The factor of $\sqrt{2\pi\hbar}$ in the denominator is purely a convention and is done for convenience as we shall see; any difference in the normalization obtained this way can be absorbed into the still undefined normalization constant, $\mathcal{N}$. Then the expression for the matrix element in Equation (3.13) becomes

$$\langle \bar{z}(\beta/2)|e^{-\frac{\beta}{\hbar}\hat{h}(\hat{X},\hat{P})}|\bar{z}(-\beta/2)\rangle = e^{-\frac{S_E[\bar{z}(\tau)]}{\hbar}}\mathcal{N}\prod_{n=0}^{\infty}\int\frac{dc_n}{\sqrt{2\pi\hbar}}e^{-\frac{1}{\hbar}\left(\sum_{n=0}^{\infty}\frac{1}{2}\lambda_n c_n^2 + \mathrm{o}(c_n^3)\right)}.$$
$$(3.15)$$

Scaling $c_n = \tilde{c}_n\sqrt{\hbar}$ gives for the right-hand side

$$= e^{-\frac{S_E[\bar{z}(\tau)]}{\hbar}}\mathcal{N}\prod_{n=0}^{\infty}\int\frac{d\tilde{c}_n}{\sqrt{2\pi}}e^{-\left(\frac{1}{2}\lambda_n \tilde{c}_n^2 + \mathrm{o}(\hbar)\right)}$$
$$= e^{-\frac{S_E[\bar{z}(\tau)]}{\hbar}}\mathcal{N}\prod_{n=0}^{\infty}\left(\frac{1}{\sqrt{\lambda_n}}(1+\mathrm{o}(\hbar))\right). \qquad (3.16)$$

This infinite product of eigenvalues for the operators which arise typically does not converge. We will address and resolve this difficulty later and, assuming that it is so done, we formally write "det" for the product of all the eigenvalues of the operator. This yields the formula

$$\langle \bar{z}(\beta/2)|e^{-\frac{\beta}{\hbar}\hat{h}(\hat{X},\hat{P})}|\bar{z}(-\beta/2)\rangle = e^{-\frac{S_E[\bar{z}(\tau)]}{\hbar}}\left(\mathcal{N}\mathrm{det}^{-\frac{1}{2}}\left[-\frac{d^2}{d\tau^2}+V''(\bar{z}(\tau))\right](1+\mathrm{o}(\hbar))\right).$$
$$(3.17)$$

Thus we see the matrix element has a non-perturbative contribution in $\hbar$ coming from the exponential of the value of the classical action at the critical point divided by $\hbar$, $e^{-\frac{S_E[\bar{z}(\tau)]}{\hbar}}$, multiplying the yet undefined normalization and determinant and an expression which has a perturbative expansion in positive powers of $\hbar$.

3.3 Tunnelling Amplitudes and the Instanton

To proceed further we have to be more specific. We shall consider the following matrix elements

$$\langle \pm a|e^{-\frac{\beta}{\hbar}\hat{h}(\hat{X},\hat{P})}|a\rangle = \langle \mp a|e^{-\frac{\beta}{\hbar}\hat{h}(\hat{X},\hat{P})}|-a\rangle. \tag{3.18}$$

The equality of these matrix elements is easily obtained here by using the assumed parity reflection symmetry of the Hamiltonian,

$$\begin{aligned}
\langle x|e^{-\frac{\beta}{\hbar}\hat{h}(\hat{X},\hat{P})}|y\rangle &= \langle x|\mathfrak{P}\mathfrak{P}e^{-\frac{\beta}{\hbar}\hat{h}(\hat{X},\hat{P})}\mathfrak{P}\mathfrak{P}|y\rangle \\
&= \langle -x|\mathfrak{P}e^{-\frac{\beta}{\hbar}\hat{h}(\hat{X},\hat{P})}\mathfrak{P}|-y\rangle \\
&= \langle -x|e^{-\frac{\beta}{\hbar}\hat{h}(\hat{X},\hat{P})}|-y\rangle,
\end{aligned} \tag{3.19}$$

where $\mathfrak{P}$ is the parity operator which satisfies $\mathfrak{P}^2 = 1$, $\mathfrak{P}|x\rangle = |-x\rangle$ and $[\mathfrak{P}, \hat{h}(\hat{X}, \hat{P})] = 0$.

The equation which $\bar{z}(\tau)$ satisfies is

$$-\ddot{\bar{z}}(\tau) + V'(\bar{z}(\tau)) = 0, \tag{3.20}$$

which is exactly the equation of motion for a particle in real time moving in the reversed potential $-V(z)$, as in Figure 3.2. Because of the matrix elements that we are interested in, Equation (3.18), the corresponding classical solutions are those which start at and return to either $\pm a$ or those that interpolate between

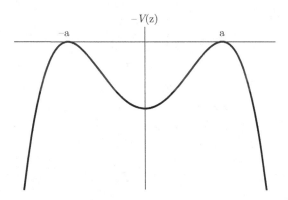

Figure 3.2. Inverted double well potential for $\bar{z}(\tau)$

$\pm a$ and $\mp a$, and each in time β. The trivial solutions

$$\bar{z}(\tau) = \pm a \qquad (3.21)$$

satisfy the first condition while the second condition can be obtained by integrating Equation (3.20). Straightforwardly,

$$\ddot{\bar{z}}(\tau)\dot{\bar{z}}(\tau) = V'(\bar{z}(\tau))\dot{\bar{z}}(\tau), \qquad (3.22)$$

which implies

$$\dot{\bar{z}}(\tau) = \sqrt{2V(\bar{z}(\tau)) + c^2}, \qquad (3.23)$$

where c is an integration constant. Integrating one more time and choosing the solution that interpolates from $-a$ to a, we get

$$\int_{-a}^{\bar{z}(\tau)} \frac{d\bar{z}}{\sqrt{2V(\bar{z}) + c^2}} = \int_{-\frac{\beta}{2}}^{\tau} d\tau = \tau + \frac{\beta}{2} \qquad (3.24)$$

and c is determined by

$$\int_{-a}^{a} \frac{d\bar{z}}{\sqrt{2V(\bar{z}) + c^2}} = \beta. \qquad (3.25)$$

We note that this last Equation (3.25) does not depend on the details of the solution, but only on the fact that it must interpolate from $-a$ to a. Obviously from Equation (3.23), c is the initial velocity. The initial velocity is not arbitrary, the solution must interpolate from $-a$ to a in Euclidean time β, and Equation (3.25) implicitly gives c as a function of β. There is no solution that starts with vanishing initial velocity but interpolates between $\pm a$ in finite time β; vanishing initial velocity requires infinite time.

As $\beta \to \infty$, the only way for the integral in Equation (3.25) to diverge to give an infinite or very large β is for the denominator to vanish. This only occurs for $V(\bar{z}) \to 0$ and for $c \to 0$. $V(\bar{z}) \to 0$ occurs as $\bar{z} \to \pm a$, which is near the start and end of the trajectory. Also, physically, if the particle is to interpolate from $-a$ to a in a longer and longer time, β, then it must start out at $-a$ with a smaller and smaller initial velocity, c. For larger and larger β, c must vanish in an appropriate fashion. Heuristically, for small c, the solution spends most of its time near $\bar{z} = \pm a$ and interpolates from one to the other relatively quickly. Then the major contribution to the integral comes from the region around $\bar{z} = \pm a$. Since the integral diverges logarithmically when $c = 0$, for a typical potential V (which must vanish quadratically at $\bar{z} = \pm a$ as V has a double zero at $\pm a$), the integral must behave as $-\ln c$, *i.e.* $\beta \sim -\ln c$ which is equivalent to $c \sim e^{-\beta}$, which means that c must vanish exponentially with large β. For sufficiently large β we may neglect c altogether.

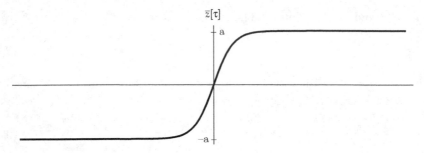

Figure 3.3. Interpolating kink instanton for the symmetric double well

The action for the constant solutions, Equation (3.21), is evidently zero. For the interpolating solution implicitly determined by Equation (3.24), it is

$$S_E[\bar{z}(\tau)] = \int_{-\frac{\beta}{2}}^{\frac{\beta}{2}} d\tau \left(\frac{1}{2}\dot{\bar{z}}^2(\tau) + V(\bar{z}(\tau)) \right)$$

$$= \int_{-\frac{\beta}{2}}^{\frac{\beta}{2}} d\tau \left(\dot{\bar{z}}^2(\tau) - c^2 \right)$$

$$= \left(\int_{-\frac{\beta}{2}}^{\frac{\beta}{2}} \sqrt{2V(\bar{z}(\tau)) + c^2} \frac{d\bar{z}}{d\tau} d\tau \right) - \beta c^2$$

$$= \left(\int_{-a}^{a} d\bar{z}\sqrt{2V(\bar{z}) + c^2} \right) - \beta c^2. \tag{3.26}$$

For large β, we neglect c in the integral for $S_E[\bar{z}(\tau)] \equiv S_0$, and the term $-\beta c^2$, yielding

$$S_0 = \int_{-a}^{a} d\bar{z}\sqrt{2V(\bar{z})}. \tag{3.27}$$

This is exactly the action corresponding to the classical solution for $\beta = \infty$ depicted in Figure 3.3. Such Euclidean time classical solutions are called "instantons".

For large τ the approximate equation satisfied by $\bar{z}(\tau)$ is

$$\frac{d\bar{z}}{d\tau} - \omega(a - \bar{z}), \tag{3.28}$$

obtained by expanding Equation (3.23) as $\bar{z} \to a^-$ from below and where ω^2 is the second derivative of the potential at $\bar{z} = a$. There is a corresponding, equivalent analysis for $\tau \to -\infty$. These have the solution

$$|z(\tau)| = a - Ce^{-\omega|\tau|}. \tag{3.29}$$

Thus the instanton is exponentially close to $\pm a$ for $|\tau| > \frac{1}{\omega}$. Its size is $\frac{1}{\omega}$ which is of order 1, compared with $\hbar$ and β. For large $|\tau|$, the solution is essentially equal to $\pm a$, which is just the trivial solution. The solution is "on" only for an "instant", the relatively short time compared with β, during which it interpolates between $-a$ and $+a$. Hence the name instanton. Reversing the time direction gives another solution which starts at $+a$ and interpolates to $-a$, aptly called an anti-instanton. It clearly has the same action as an instanton.

3.4 The Instanton Contribution to the Path Integral

3.4.1 Translational Invariance Zero Mode

As we have seen, for very large β, the instanton corresponding to infinite β is an arbitrarily close and perfectly good approximation to the true instanton. Evidently with the infinite β instanton, we may choose the time arbitrarily at which the solution crosses over from $-a$ to $+a$. The solution of

$$\int_0^{\bar{z}(\tau)} \frac{dz}{\sqrt{2V(z)}} = \tau - \tau_0 \tag{3.30}$$

corresponds to an instanton which crosses over around $\tau = \tau_0$. Thus the position of the instanton τ_0 gives a one-parameter family of solutions, each with the same classical action. The point is that for large enough β, there exists a one-parameter family of approximate critical points with action arbitrarily close to S_0. The contribution to the path integral from the vicinity of these approximate critical points will be of a slightly modified form, since the first variation of the action about the approximate critical point does not quite vanish. Thus the contribution will be of the form, the exponential of the negative action at the approximate critical point, multiplied by a Gaussian integral with a linear shift, the shift coming from the non-vanishing first variation of the action. The shift will be proportional to some arbitrarily small function $f(\beta)$ as $\beta \to \infty$. The higher-order terms give perturbative corrections in $\hbar$, as in Equation (3.16), and can be dropped. Then, considering a typical Gaussian integral with a small linear shift, as arises in the integration about an approximate critical point, we have

$$\int_{-\infty}^{\infty} \frac{dx}{\sqrt{2\pi}} e^{-\frac{1}{\hbar}\left(\alpha^2 x^2 + 2f(\beta)x\right)} = e^{\frac{f^2(\beta)}{\hbar\alpha^2}} \frac{1}{\alpha}. \tag{3.31}$$

We see that to be able to neglect the effects of the shift, $f(\beta)$ must be so small that $\frac{f^2(\beta)}{\hbar} \ll 1$, given that α, being independent of $\hbar$ and β, is of order 1.

Typically, $f(\beta)$ is exponentially small in β, just as earlier c was found to be. $f(\beta)$ needs to be determined and depends of the details of the dynamics. In any case, β must be greater than a certain value determined by the value of $\hbar$. This is, however, no strong constraint other than imposing that we must consider the limit that β is arbitrarily large while all other constants (especially $\hbar$) are held fixed. Hence, assuming β is sufficiently large, we can neglect the effect of the linear shift and we must include the contribution from these approximate critical points. To do so, we simply integrate over the position of the instanton and perform the Gaussian integral over directions in path space which are orthogonal to the direction corresponding to translations of the instanton.

The easiest way to perform such a constrained Gaussian integral is to use the following observations. In the infinite β limit, the translated instantons become exact critical points and correspondingly the fluctuation directions about a given instanton contain a flat direction. This means that the action does not change to second order for variations along this direction. Precisely, this means that the eigenfrequencies, λ_n, contain a zero mode, $\lambda_0 = 0$. We can explicitly construct this zero mode since

$$\left(-\frac{d^2}{d\tau^2} + V''(\bar{z}(\tau - \tau_1))\right)\frac{d\bar{z}(\tau - \tau_1)}{d\tau_1} = -\frac{d}{d\tau}\left(-\ddot{\bar{z}}(\tau - \tau_1) + V'(\bar{z}(\tau - \tau_1))\right) = 0,$$
(3.32)

the second term vanishing by the equation of motion, Equation (3.20), which is clearly also valid for $\bar{z}(\tau - \tau_1)$. This mode occurs because of the time translation invariance when β is infinite. The corresponding normalized zero mode is

$$z_0(\tau) = \frac{1}{\sqrt{S_0}}\frac{d}{d\tau_1}\bar{z}(\tau - \tau_1).$$
(3.33)

Clearly

$$\int_{-\infty}^{\infty} d\tau \left(\frac{1}{\sqrt{S_0}}\frac{d}{d\tau_1}\bar{z}(\tau - \tau_1)\right)^2 = \frac{1}{S_0}\int_{-\infty}^{\infty} d\tau \left(\frac{1}{2}\dot{\bar{z}}^2(\tau - \tau_1) + V(\bar{z}(\tau - \tau_1))\right) = 1$$
(3.34)

using the equation of motion, Equation (3.23), with $c = 0$ (infinite β).

Integration in the path integral, Equation (3.15), over the coefficient of this mode yields a divergence as the frequency is zero

$$\int \frac{dc_0}{\sqrt{2\pi\hbar}} e^{-\frac{1}{\hbar}\lambda_0 c_0^2} = \int \frac{dc_0}{\sqrt{2\pi\hbar}} 1 = \infty.$$
(3.35)

However, integrating over the position of the instanton is equivalent to integrating over c_0. τ_1 is called a collective coordinate of the instanton corresponding to its position in Euclidean time. Indeed, if $\bar{z}(\tau - \tau_1)$ is an instanton at position τ_1, the change in the path obtained by infinitesimally changing τ_1 is

$$\delta z(\tau) = \frac{d}{d\tau_1}\bar{z}(\tau - \tau_1)d\tau_1 = \sqrt{S_0}z_0(\tau).$$
(3.36)

The change induced by varying c_0 is, however,

$$\delta z(\tau) = z_0(\tau)dc_0.$$
(3.37)

Thus

$$\frac{dc_0}{\sqrt{2\pi\hbar}} = \sqrt{\frac{S_0}{2\pi\hbar}}d\tau_1$$
(3.38)

and when integrating over the position τ_1 we should multiply by the normalizing factor $\sqrt{\frac{S_0}{2\pi\hbar}}$. Clearly for infinite β the integral over τ_1 diverges, reflecting the equivalent infinity obtained when integrating over c_0.

This divergence is not disturbing, since for a positive definite Hamiltonian the infinite β limit of the matrix element, Equation (2.32), is strictly zero, and for large β it is an expression which vanishes exponentially. Thus in the large β limit, the Gaussian integrals in the directions orthogonal to the flat direction must combine to give an expression which indeed vanishes exponentially with β, as we will see. For the time being, for finite β, the integration over the position then gives a factor that is linear in β

$$\sqrt{\frac{S_0}{2\pi\hbar}}\beta. \tag{3.39}$$

Thus, so far the path integral has yielded

$$\langle a|e^{-\frac{\beta}{\hbar}\hat{h}(\hat{X},\hat{P})}|-a\rangle = e^{-\frac{S_0}{\hbar}}\left(\frac{S_0}{2\pi\hbar}\right)^{\frac{1}{2}}\beta\mathcal{N}\left(\mathrm{det}'\left[-\frac{d^2}{d\tau^2}+V''(\bar{z}(\tau))\right]\right)^{-\frac{1}{2}}, \tag{3.40}$$

where det' means the "determinant" excluding the zero eigenvalue. We will leave the evaluation of the determinant for a little later when will show that

$$\mathcal{N}\left(\mathrm{det}'\left[-\frac{d^2}{d\tau^2}+V''(\bar{z}(\tau))\right]\right)^{-\frac{1}{2}} = K\mathcal{N}\left(\mathrm{det}\left[-\frac{d^2}{d\tau^2}+\omega^2\right]\right)^{-\frac{1}{2}}, \tag{3.41}$$

where ω was defined at Equation (3.28), and we will evaluate K, which is, most importantly, independent of $\hbar$ and β.

3.4.2 Multi-instanton Contribution

To proceed further, we must realize that there are also other approximate critical points which give significant contributions to the path integral. These correspond to classical configurations which have, for example, an instanton at τ_1, an anti-instanton at τ_2 and again an instanton at τ_3. If τ_i are well separated within the interval β, these configurations are approximately critical, with an error of the same order as for the approximate critical points previously considered. More generally we can have a string of n pairs of an instanton followed by an anti-instanton, plus a final instanton completing the interpolation from $-a$ to a. We denote such a configuration as $\bar{z}_{2n+1}(\tau)$. The positions are arbitrary except that the order of the instantons and the anti-instantons must be preserved and they must be well separated. The action for $2n+1$ such objects is just $(2n+1)S_0$ to the same degree of accuracy.

One would, at first sight, conclude that this contribution, including the Gaussian integral about these approximate critical points, is exponentially suppressed relative to the contribution from the single instanton sector. Indeed, we would find that the contribution of the $2n+1$-instantons and anti-instantons to the matrix element[1],

$$\langle a|e^{-\frac{\beta}{\hbar}\hat{h}(\hat{X},\hat{P})}|-a\rangle_{2n+1} = e^{-\frac{(2n+1)S_0}{\hbar}}\mathcal{N}\left(\det\left[-\frac{d^2}{d\tau^2}+V''(\bar{z}_{2n+1}(\tau))\right]\right)^{-\frac{1}{2}} \quad (3.42)$$

is suppressed by $e^{-\frac{2nS_0}{\hbar}}$ relative to the one instanton contribution. This is true; however, we must analyse the effects of zero modes.

For $2n+1$ instantons and anti-instantons there are $2n+1$ zero modes corresponding to the independent translation of each object. This is actually only true for infinitely separated objects with β infinite; however, for β large, it is an arbitrarily good approximation. Thus there exist $2n+1$ zero frequencies in the determinant which should not be included in the path integration and, correspondingly, we should integrate over the positions of the $2n+1$ instantons and anti-instantons. This integration is constrained by the condition that their order is preserved. Hence we get the factor

$$\int_{-\frac{\beta}{2}}^{\frac{\beta}{2}}d\tau_1\int_{\tau_1}^{\frac{\beta}{2}}d\tau_2\int_{\tau_2}^{\frac{\beta}{2}}d\tau_3\cdots\int_{\tau_{2n-1}}^{\frac{\beta}{2}}d\tau_{2n}\int_{\tau_{2n}}^{\frac{\beta}{2}}d\tau_{2n+1} = \frac{\beta^{2n+1}}{(2n+1)!}. \quad (3.43)$$

Furthermore, from exactly the same analysis as the integration over the position of the single instanton, the integration is normalized correctly only when each factor is multiplied by $\left(\frac{S_0}{2\pi\hbar}\right)^{\frac{1}{2}}$. Thus we find

$$\begin{aligned}&\left\langle a\left|e^{-\frac{\beta}{\hbar}\hat{h}(\hat{X},\hat{P})}\right|-a\right\rangle_{2n+1}\\&= \left(e^{-\frac{S_0}{\hbar}}\left(\frac{S_0}{2\pi\hbar}\right)^{\frac{1}{2}}\beta\right)^{2n+1}\frac{\mathcal{N}}{(2n+1)!}\left(\det'\left[-\frac{d^2}{d\tau^2}+V''(\bar{z}_{2n+1}(\tau))\right]\right)^{-\frac{1}{2}},\end{aligned} \quad (3.44)$$

where $\det'$ again means the determinant with the $2n+1$ zero modes removed. We will show later that

$$\mathcal{N}\left(\det'\left[-\frac{d^2}{d\tau^2}+V''(\bar{z}_{2n+1}(\tau))\right]\right)^{-\frac{1}{2}} = K^{2n+1}\mathcal{N}\left(\det\left[-\frac{d^2}{d\tau^2}+\omega^2\right]\right)^{-\frac{1}{2}} \quad (3.45)$$

for the same K as in the case of one instanton, as in Equation (3.41).

Now even if $e^{-\frac{S_0}{\hbar}}$ is very small, our whole analysis is done at fixed $\hbar$ with $\beta\to\infty$; the relevant parameter, as can be seen from Equation (3.44), is

$$\delta = \left(\frac{S_0}{2\pi\hbar}\right)^{\frac{1}{2}}e^{-\frac{S_0}{\hbar}}K\beta, \quad (3.46)$$

which is arbitrarily large in this limit. Thus it seems that the contribution from the strings of instanton and anti-instanton pairs is proportional to δ^{2n+1} and

---

[1] Here the subscript $2n+1$ signifies that we are calculating only the contribution to the matrix element from $2n+1$ instantons and anti-instantons.

seems to get larger and larger. However, the denominator contains $(2n+1)!$, which must be taken into account. For large enough n, the denominator always dominates, $\delta^{2n+1} \ll (2n+1)!$, and so renders the contribution small.

We require, however, for the consistency of our approximations that when n is large enough so that this is true, the average space per instanton or anti-instanton, $\frac{\beta}{2n+1}$, is still large compared to the size of these objects $\sim 1/\omega$, which is independent of both $\hbar$ and β. This is satisfied as $\beta \to \infty$. Hence we require n large enough such that

$$\frac{\delta^{2n+1}}{(2n+1)!} \ll 1; \tag{3.47}$$

however, with

$$\frac{\beta}{2n+1} \gg \frac{1}{\omega}. \tag{3.48}$$

Taking the logarithm of Equation (3.47) after multiplying by $(2n+1)!$ yields in the Stirling approximation

$$(2n+1)\ln\delta \ll (2n+1)\ln(2n+1) - (2n+1). \tag{3.49}$$

Neglecting the second term on the right-hand side and combining with Equation (3.48) yields

$$\delta = \left(\left(\frac{S_0}{2\pi\hbar}\right)^{\frac{1}{2}} e^{-\frac{S_0}{\hbar}} K \right) \beta \ll 2n+1 \ll \omega\beta. \tag{3.50}$$

That such an n can exist simply requires $\left(\frac{S_0}{2\pi\hbar}\right)^{\frac{1}{2}} e^{-\frac{S_0}{\hbar}} K \ll \omega$. We will evaluate K explicitly and find that it does not depend on $\hbar$ or β. The inequality then is clearly satisfied for $\hbar \to 0$, which brings into focus that underneath everything we are interested in the semi-classical limit.

A tiny parenthetical remark is in order: in integrating over the positions of the instantons, we should always maintain the constraint that the instantons are well separated. Thus we should not integrate the position of one instanton exactly from that of the preceding one to that of the succeeding one, but we should leave a gap of the order of $\frac{1}{\omega}$ which is the size of the instanton. Such a correction corresponds to a contribution which behaves to leading order as $\frac{1}{\omega}\frac{\beta^{n-1}}{(n-1)!}$, which is negligible in comparison to $\frac{\beta^n}{n!}$ if $\frac{1}{\omega} \ll \beta$.

When the density of instantons and anti-instantons becomes large, all of our approximations break down, and such configurations are no longer even approximately critical. Thus we do not expect any significant contribution to the path integral from the regions of the space of paths which include these configurations. Hence we should actually truncate the series in the number of instantons for some large enough n; however, this is not necessary. We will always assume that we work in the limit that β should be sufficiently large and $\hbar$ sufficiently small so that the contribution from the terms in the series with

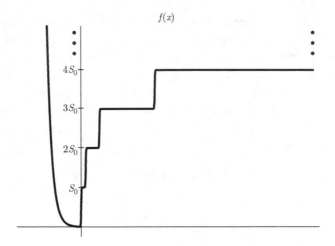

Figure 3.4. A simple function analogous to the action

n greater than some N is already negligible, while there is still a lot of room per instanton, *i.e.* β/N is still large. This should still correspond to a dilute "gas" of instantons and anti-instantons. Then the remaining terms in the series can be maintained, although they do not represent the contribution from any part of path space. It is simply easier to sum the series to infinity, knowing that the contribution added in from n greater than some N makes only a negligible change. The sum to infinity is straightforward. We find

$$\left\langle a \left| e^{-\frac{\beta}{\hbar} \hat{h}(\hat{X}, \hat{P})} \right| -a \right\rangle = \left(\mathcal{N} \left(\det \left[-\frac{d^2}{d\tau^2} + \omega^2 \right] \right)^{-\frac{1}{2}} \right) \sinh \left(\left(\frac{S_0}{2\pi\hbar} \right)^{\frac{1}{2}} e^{-\frac{S_0}{\hbar}} K\beta \right).$$

$$(3.51)$$

3.4.3 Two-dimensional Integral Paradigm

A simple two-dimensional, ordinary integral which serves as a paradigm exhibiting many of the features of the path integral just considered is given by

$$\mathcal{I} = \int dx\, dy\, e^{-\frac{1}{\hbar}(f(x) + \frac{\alpha^2}{2} y^2)} \tag{3.52}$$

where y corresponds to the transverse directions and plays no role. $f(x)$ is a function of the form depicted in Figure 3.4 and increases sharply in steps of S_0, and the length of each plateau is $\frac{\beta^n}{n!}$. In the limit that the steps become sharp, the integral can be done exactly and yields

$$\mathcal{I} = \frac{(2\pi\hbar)^{\frac{1}{2}}}{\alpha} \sum_{n=0}^{\infty} e^{-\frac{nS_0}{\hbar}} \left(\frac{\beta^n}{n!} \right) = \frac{(2\pi\hbar)^{\frac{1}{2}}}{\alpha} e^{\left(\beta e^{-\frac{S_0}{\hbar}} \right)}. \tag{3.53}$$

Obviously this is exactly analogous to the path integral just considered for $\beta \to \infty$ and $\hbar \to 0$. The plateaux correspond to the critical points. Clearly we cannot consider just the lowest critical point since the volume associated with the higher critical points is sufficiently large that their contribution does not damp out until n becomes large enough. In terms of physically intuitive arguments, the volume is like the entropy factor associated with n instantons, $\frac{\beta^n}{n!}$, while the exponential, $e^{-\frac{n S_0}{\hbar}}$, is like the Boltzmann factor. In statistical mechanics, even though the Boltzmann factor is much smaller for higher energy levels, their contribution to the partition function can be significant due to a large enough entropy. We can further model the aspect of approximate critical points by giving the plateaux in Figure 3.4 a very small slope. Clearly the integral is only negligibly modified if the slope is taken to be exponentially small in β.

3.5 Evaluation of the Determinant

Finally, we are left with the evaluation of the determinant. We wish to show for the case of $2n + 1$ instantons and anti-instantons

$$\left(\mathcal{N} \left(\det{}' \left[-\frac{d^2}{d\tau^2} + V''(\bar{z}_{2n+1}(\tau)) \right] \right)^{-\frac{1}{2}} \right) = K^{2n+1} \left(\mathcal{N} \left(\det \left[-\frac{d^2}{d\tau^2} + \omega^2 \right] \right)^{-\frac{1}{2}} \right)$$

(3.54)

and to evaluate K. Physically this means that the effect of each instanton and anti-instanton is simply to multiply the free determinant by a factor of $\frac{1}{K^2}$. Intuitively this is very reasonable, and we expect that for well-separated instantons their effect would be independent of each other.

To obtain the $\det{}'$ we will work in the finite large interval, β, with boundary conditions that the wave function must vanish at the end points. Consider first the case of just one instanton. Because of the finite interval, time translation will not be an exact symmetry and the operator $-\frac{d^2}{d\tau^2} + V''(\bar{z}(\tau))$ will not have an exact zero mode. However, as $\beta \to \infty$ one mode will approach zero. The $\det{}'$ is then obtained by calculating the full determinant on the finite interval, β, and then dividing out by the smallest eigenvalue. There should be a rigorous theorem proving first that the operator in question has a positive definite spectrum on the finite interval, β, for any potential, $V(z)$, of the type considered and the corresponding instanton, $\bar{z}(\tau)$, and secondly as $\beta \to \infty$, one bound state drops to exactly zero; this is reasonable and taken as a hypothesis. Thus we will study the full determinant on the interval β which has the path-integral representation

$$\mathcal{N} \left(\det \left[-\frac{d^2}{d\tau^2} + V''(\bar{z}(\tau - \tau_1)) \right] \right)^{-\frac{1}{2}}$$

$$= \mathcal{N} \int \mathcal{D}z(\tau) e^{-\frac{1}{\hbar} \int_{-\frac{\beta}{2}}^{\frac{\beta}{2}} d\tau \frac{1}{2} \left(\dot{z}^2(\tau) + V''(\bar{z}(\tau - \tau_1)) z^2(\tau) \right)}$$

(3.55)

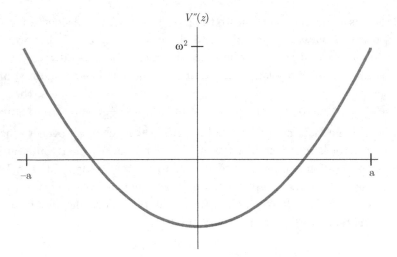

Figure 3.5. The behaviour of $V''(z)$ between $\pm a$

with the boundary conditions that $z(\frac{\beta}{2}) = z(-\frac{\beta}{2}) = 0$ in the path integral. The path integral on the right-hand side is performed in exactly the same manner as in Equation (3.15). This determinant actually corresponds to the matrix element of the Euclidean time evolution operator with a time-dependent Hamiltonian,

$$\langle z = 0 | \mathcal{T}\left(e^{-\frac{1}{\hbar}\int_{-\frac{\beta}{2}}^{\frac{\beta}{2}} d\tau \left(\frac{1}{2}\hat{P}^2 + \frac{V''(\bar{z}(\tau-\tau_1))}{2}\hat{X}^2 \right)} \right) | z = 0 \rangle, \qquad (3.56)$$

where $\mathcal{T}$ denotes the operation of Euclidean time ordering. This time ordering is effectively described by the product representation of Equation (2.33), where the appropriate Hamiltonian is entered into each Euclidean time slice. This can be shown to give the path integral, Equation (3.55), adapting with minimal changes the demonstration in Chapter 2. We leave it to the reader to confirm the details.

Consider first the behaviour of $V''(z)$ which controls the Euclidean time-dependent frequency in the path integral Equation (3.55). $V''(\pm a) = \omega^2$ is the parabolic curvature at the bottom of each well. In between, at $z = 0$, $V''(0)$ will drop to some negative value giving the curvature at the top of the potential hill separating the two wells. We will have a function as depicted in Figure 3.5. Thus $V''(\bar{z}(\tau))$ will start out at ω^2 at $\tau = -\infty$, until $\bar{z}(\tau)$ starts to cross over from $-a$ to a, where it will trace out the potential well of Figure 3.5, and again it will regain the value ω^2 for $\bar{z}(\tau) = a$ at $\tau = \infty$, corresponding to the function of τ as in Figure 3.6. Thus the path integral in Equation (3.55) is exactly equal to the matrix element or "Euclidean persistence amplitude" that a particle at position zero will remain at position zero in Euclidean time β in a quadratic potential with a time-dependent frequency given by $V''(\bar{z}(\tau))$ depicted in Figure 3.6.

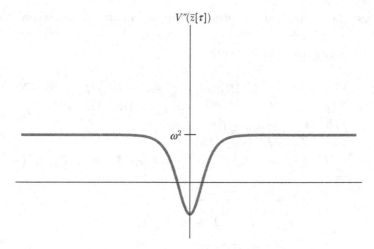

Figure 3.6. The behaviour of $V''(z(\tau))$ between $\tau = \pm\infty$

We will express the matrix element in terms of a Euclidean time evolution operator $\mathcal{U}\left(\frac{\beta}{2}, -\frac{\beta}{2}\right)$ as

$$\mathcal{N} \int \mathcal{D}z(\tau) e^{-\frac{1}{\hbar}\int_{-\frac{\beta}{2}}^{\frac{\beta}{2}} d\tau \frac{1}{2}\left(\dot{z}^2(\tau) + V''(\bar{z}(\tau - \tau_1))z^2(\tau)\right)} \equiv \langle z = 0 | \mathcal{U}\left(\frac{\beta}{2}, -\frac{\beta}{2}\right) | z = 0 \rangle \quad (3.57)$$

with explicitly,

$$\mathcal{U}\left(\frac{\beta}{2}, -\frac{\beta}{2}\right) = \mathcal{T}\left(e^{-\frac{1}{\hbar}\int_{-\frac{\beta}{2}}^{\frac{\beta}{2}} d\tau \left(\frac{1}{2}\hat{P}^2 + \frac{V''(\bar{z}(\tau - \tau_1))}{2}\hat{X}^2\right)}\right). \quad (3.58)$$

Now

$$\begin{aligned}
\mathcal{U}\left(\frac{\beta}{2}, -\frac{\beta}{2}\right) &= \mathcal{U}\left(\frac{\beta}{2}, \tau_1 + \frac{1}{2\omega}\right)\mathcal{U}\left(\tau_1 + \frac{1}{2\omega}, \tau_1 - \frac{1}{2\omega}\right)\mathcal{U}\left(\tau_1 - \frac{1}{2\omega}, -\frac{\beta}{2}\right) \\
&\approx \mathcal{U}^0\left(\frac{\beta}{2}, \tau_1 + \frac{1}{2\omega}\right)\mathcal{U}\left(\tau_1 + \frac{1}{2\omega}, \tau_1 - \frac{1}{2\omega}\right)\mathcal{U}^0\left(\tau_1 - \frac{1}{2\omega}, -\frac{\beta}{2}\right), \quad (3.59)
\end{aligned}$$

where on the intervals $\left[\tau_1 + \frac{1}{2\omega}, \frac{\beta}{2}\right]$ and $\left[-\frac{\beta}{2}, \tau_1 - \frac{1}{2\omega}\right]$ we can replace the full evolution operator with the free evolution operator

$$\mathcal{U}^0(\tau, \tau') = T\left(e^{-\frac{1}{\hbar}\int_{\tau'}^{\tau} d\tau \frac{1}{2}\left(-\hbar^2 \frac{d^2}{dz^2} + \omega^2 z^2\right)}\right) = e^{-\frac{(\tau - \tau')}{\hbar}\hat{h}^0(\hat{X}, \hat{P})} \quad (3.60)$$

as $V''(\bar{z}(\tau))$ is essentially constant and equal to ω^2 on these intervals. Then inserting complete sets of free eigenstates, which are just simple harmonic oscillator states $|E_n\rangle$ for an oscillator of frequency ω, we obtain

$$\mathcal{U}\left(\frac{\beta}{2}, -\frac{\beta}{2}\right) = \sum_{n,m} e^{-\left(\frac{\beta}{2} - \tau_1 - \frac{1}{2\omega}\right)\frac{E_n}{\hbar}} |E_n\rangle\langle E_n | \mathcal{U}\left(\tau_1 + \frac{1}{2\omega}, \tau_1 - \frac{1}{2\omega}\right)|E_m\rangle$$

$$\times \langle E_m | e^{-\left(\tau_1 - \frac{1}{2\omega} + \frac{\beta}{2}\right)\frac{E_m}{\hbar}} \quad (3.61)$$

Now we use the "ground state saturation approximation", *i.e.* when β is huge and the instanton is not near the boundaries, only the ground state contribution is important. Using this twice we obtain

$$
\begin{aligned}
\mathcal{U}\left(\tfrac{\beta}{2},-\tfrac{\beta}{2}\right) &\approx e^{\left(\tfrac{\beta}{2}-\tau_1-\tfrac{1}{2\omega}\right)\tfrac{E_0}{\hbar}} |E_0\rangle \langle E_0| \mathcal{U}\left(\tau_1+\tfrac{1}{2\omega},\tau_1-\tfrac{1}{2\omega}\right)|E_0\rangle \langle E_0| e^{-\left(\tau_1-\tfrac{1}{2\omega}+\tfrac{\beta}{2}\right)\tfrac{E_0}{\hbar}} \\
&= \mathcal{U}^0\left(\tfrac{\beta}{2},\tau_1+\tfrac{1}{2\omega}\right)|E_0\rangle \langle E_0| \mathcal{U}^0\left(\tau_1+\tfrac{1}{2\omega},\tau_1-\tfrac{1}{2\omega}\right)|E_0\rangle \langle E_0| \mathcal{U}^0\left(\tau_1-\tfrac{1}{2\omega},-\tfrac{\beta}{2}\right) \times \\
&\quad \times \frac{\langle E_0| \mathcal{U}\left(\tau_1+\tfrac{1}{2\omega},\tau_1-\tfrac{1}{2\omega}\right)|E_0\rangle}{\langle E_0| \mathcal{U}^0\left(\tau_1+\tfrac{1}{2\omega},\tau_1-\tfrac{1}{2\omega}\right)|E_0\rangle} \\
&\approx \sum_{n,m} \mathcal{U}^0\left(\tfrac{\beta}{2},\tau_1+\tfrac{1}{2\omega}\right)|E_n\rangle \langle E_n| \mathcal{U}^0\left(\tau_1+\tfrac{1}{2\omega},\tau_1-\tfrac{1}{2\omega}\right)|E_m\rangle \langle E_m| \mathcal{U}^0\left(\tau_1-\tfrac{1}{2\omega},-\tfrac{\beta}{2}\right) \times \\
&\quad \times \frac{\langle E_0| \mathcal{U}\left(\tau_1+\tfrac{1}{2\omega},\tau_1-\tfrac{1}{2\omega}\right)|E_0\rangle}{\langle E_0| \mathcal{U}^0\left(\tau_1+\tfrac{1}{2\omega},\tau_1-\tfrac{1}{2\omega}\right)|E_0\rangle} \\
&= \mathcal{U}^0\left(\tfrac{\beta}{2},-\tfrac{\beta}{2}\right) \frac{\langle E_0| \mathcal{U}\left(\tau_1+\tfrac{1}{2\omega},\tau_1-\tfrac{1}{2\omega}\right)|E_0\rangle}{\langle E_0| \mathcal{U}^0\left(\tau_1+\tfrac{1}{2\omega},\tau_1-\tfrac{1}{2\omega}\right)|E_0\rangle} \\
&\equiv \mathcal{U}^0\left(\tfrac{\beta}{2},-\tfrac{\beta}{2}\right)\kappa, \tag{3.62}
\end{aligned}
$$

where κ is the ratio of the two amplitudes over the short time period during which $V''(\bar{z}(\tau))$ is non-trivially time-dependent. κ is surely independent of the position τ_1 of the instanton. The full evolution operator in fact simply does not depend on the position, nor does the denominator. Indeed,

$$
\begin{aligned}
\mathcal{U}\left(\tau_1+\tfrac{1}{2\omega},\tau_1-\tfrac{1}{2\omega}\right) &= T\left(e^{-\tfrac{1}{\hbar}\int_{\tau_1-\tfrac{1}{2\omega}}^{\tau_1+\tfrac{1}{2\omega}} d\tau \tfrac{1}{2}\left(-\hbar^2\tfrac{d^2}{dz^2}+V''(\bar{z}(\tau-\tau_1))z^2\right)}\right) \\
&= T\left(e^{-\tfrac{1}{\hbar}\int_{-\tfrac{1}{2\omega}}^{\tfrac{1}{2\omega}} d\tau' \tfrac{1}{2}\left(-\hbar^2\tfrac{d^2}{dz^2}+V''(\bar{z}(\tau'))z^2\right)}\right), \tag{3.63}
\end{aligned}
$$

since the integration variable is a dummy, thus exhibiting manifest τ_1 independence.

Clearly for n well-separated instantons the result applies also, we simply apply an appropriately adapted version of the same arguments. We convert the determinant into a persistence amplitude for the related quadratic quantum mechanical process, which we then further break up into free evolution in the gaps between the instantons and full evolution during the instanton, use the ground state saturation approximation, giving the result, to leading approximation

$$
\mathcal{N}\left(\det\left[-\frac{d^2}{d\tau^2}+V''(\bar{z}_{2n+1}(\tau))\right]\right)^{-\frac{1}{2}} = \mathcal{N}\left(\det\left[-\frac{d^2}{d\tau^2}+\omega^2\right]\right)^{-\frac{1}{2}}\kappa^{2n+1}. \tag{3.64}
$$

The relationship of κ to the K fixed by Equation (3.41) is obtained by dividing out by the lowest energy eigenvalue, call it λ_0. We will show that this eigenvalue is exponentially small for large β. For $2n+1$ instantons there are $2n+1$ such eigenvalues which are all equal, in first approximation, and we must remove them all giving

$$\mathcal{N}\left(\det{}'\left[-\frac{d^2}{d\tau^2}+V''(\bar{z}_{2n+1}(\tau))\right]\right)^{-\frac{1}{2}}=\mathcal{N}\left(\frac{\det\left[-\frac{d^2}{d\tau^2}+V''(\bar{z}_{2n+1}(\tau))\right]}{\lambda_0^{2n+1}}\right)^{-\frac{1}{2}}$$

$$=\mathcal{N}\left(\det\left[-\frac{d^2}{d\tau^2}+\omega^2\right]\right)^{-\frac{1}{2}}\left(\kappa\lambda_0^{\frac{1}{2}}\right)^{2n+1}\text{(3.65)}$$

Hence

$$K=\kappa\lambda_0^{\frac{1}{2}}.\tag{3.66}$$

It only remains to calculate two things, the free determinant and the correction factor K.

3.5.1 Calculation of the Free Determinant

To calculate the free determinant, we will use the method of Affleck and Coleman [31, 114, 36]. Consider the more general case

$$\det\left[-\frac{d^2}{d\tau^2}+W(\tau)\right],\tag{3.67}$$

where the operator acts on the space of functions which vanish at $\pm\frac{\beta}{2}$. Formally we want to compute the infinite product of the eigenvalues of the eigenvalue problem

$$\left(-\frac{d^2}{d\tau^2}+W(\tau)\right)\psi_{\lambda_n}(\tau)=\lambda_n\psi_{\lambda_n}(\tau),\quad\psi_{\lambda_n}\left(\pm\frac{\beta}{2}\right)=0.\tag{3.68}$$

The eigenvalues generally increase unboundedly, hence the infinite product is actually ill-defined. Consider, nevertheless, an ancillary problem

$$\left(-\frac{d^2}{d\tau^2}+W(\tau)\right)\psi_{\lambda}(\tau)=\lambda\psi_{\lambda}(\tau),\quad\psi_{\lambda}\left(-\frac{\beta}{2}\right)=0,\quad\frac{d}{d\tau}\psi_{\lambda}(\tau)\bigg|_{-\frac{\beta}{2}}=1.\tag{3.69}$$

There exists, in general, a solution for each λ; the second boundary condition can always be satisfied by adjusting the normalization. On the other hand, the equation in λ

$$\psi_{\lambda}\left(\frac{\beta}{2}\right)=0\tag{3.70}$$

has solutions exactly at the eigenvalues $\lambda=\lambda_n$. Affleck and Coleman [31, 114, 36] propose to define the ratio of the determinant for two different potentials as

$$\frac{\det\left[-\frac{d^2}{d\tau^2}+W_1(\tau)-\lambda\right]}{\det\left[-\frac{d^2}{d\tau^2}+W_2(\tau)-\lambda\right]}=\frac{\psi_{\lambda}^1\left(\frac{\beta}{2}\right)}{\psi_{\lambda}^2\left(\frac{\beta}{2}\right)}.\tag{3.71}$$

The left-hand side is defined as the infinite product

$$\prod_{n=1}^{\infty} \frac{(\lambda_n^1 - \lambda)}{(\lambda_n^2 - \lambda)}, \tag{3.72}$$

where the potentials and the labelling of the eigenvalues are assumed to be such that as the eigenvalues become large, they approach each other sufficiently fast,

$$\lim_{n \to \infty} (\lambda_n^1 - \lambda_n^2) = 0 \tag{3.73}$$

so that the infinite product in Equation (3.72) does conceivably converge. To prove Equation (3.71) we observe that the zeros, $\lambda = \lambda_n^1$, and poles, $\lambda = \lambda_n^2$, of the left-hand side are at the same place as those of the right-hand side, as evinced by the solutions of Equation (3.70). Thus the ratio of the two sides

$$\frac{\prod_{n=1}^{\infty} \frac{(\lambda_n^1 - \lambda)}{(\lambda_n^2 - \lambda)}}{\psi_\lambda^1 \left(\frac{\beta}{2}\right) / \psi_\lambda^2 \left(\frac{\beta}{2}\right)} \equiv g(\lambda) \tag{3.74}$$

defines an analytic function $g(\lambda)$ without zeros or poles. Now as $|\lambda| \to \infty$ in all directions except the real axis, the numerator in Equation (3.74) is equal to 1. For the denominator, as $\lambda \to \infty$ the potentials W_1 and W_2 become negligible perturbations compared to the term on the right-hand side of Equation (3.69), which we can consider as a potential $-\lambda$. Neglecting the potentials, clearly $\psi_\lambda^1 \left(\frac{\beta}{2}\right)$ and $\psi_\lambda^2 \left(\frac{\beta}{2}\right)$ approach each other, and hence the denominator also approaches 1 in the same limit. Therefore, $g(\lambda)$ defines an everywhere-analytic function of λ which approaches the constant 1 at infinity, and now in all directions including the real axis, as it does so infinitesimally close to the real axis. By a theorem of complex analysis, a meromorphic function that approaches 1 in all directions at infinity must be equal to 1 everywhere

$$g(\lambda) = 1 \tag{3.75}$$

establishing Equation (3.71). Reorganizing the terms in Equation (3.71), formally we obtain

$$\frac{\det \left[-\frac{d^2}{d\tau^2} + W_1(\tau) - \lambda\right]}{\psi_\lambda^1 \left(\frac{\beta}{2}\right)} = \frac{\det \left[-\frac{d^2}{d\tau^2} + W_2(\tau) - \lambda\right]}{\psi_\lambda^2 \left(\frac{\beta}{2}\right)}, \tag{3.76}$$

where both sides are constants independent of the potentials W_i.

We now finally choose $\mathcal{N}$ by defining

$$\frac{\det \left[-\frac{d^2}{d\tau^2} + W(\tau)\right]}{\psi_0 \left(\frac{\beta}{2}\right)} \equiv 2\pi\hbar\mathcal{N}^2 \tag{3.77}$$

and we will show that this choice is appropriate. Then

$$\mathcal{N}\det{}^{-\frac{1}{2}}\left[-\frac{d^2}{d\tau^2}+\omega^2\right]=\left(2\pi\hbar\psi_0^0\left(\tfrac{\beta}{2}\right)\right)^{-\frac{1}{2}}, \tag{3.78}$$

where $\psi_0^0(\tau)$ is the solution of Equation (3.69) for the free theory. It is easy to see that this solution is given by

$$\psi_0^0(\tau)=\frac{1}{\omega}\sinh\omega\left(\tau+\frac{\beta}{2}\right) \tag{3.79}$$

giving

$$\mathcal{N}\det{}^{\frac{-1}{2}}\left[-\frac{d^2}{d\tau^2}+\omega^2\right]=\left(2\pi\hbar\left(\frac{e^{\omega\beta}-e^{-\omega\beta}}{2\omega}\right)\right)^{-\frac{1}{2}}\approx\left(\frac{\omega}{\pi\hbar}\right)^{\frac{1}{2}}e^{-\omega\frac{\beta}{2}}. \tag{3.80}$$

We can compare this result with the direct calculation of the Euclidean persistence amplitude of the free harmonic oscillator. We find

$$\mathcal{N}\det{}^{-\frac{1}{2}}\left[-\frac{d^2}{d\tau^2}+\omega^2\right]=\left\langle x=0\left|e^{-\frac{\beta}{\hbar}\left(-\frac{\hbar^2}{2}\frac{d^2}{dx^2}+\frac{1}{2}\omega^2x^2\right)}\right|x=0\right\rangle$$

$$=e^{-\frac{\beta E_0}{\hbar}}\langle x=0|E_0\rangle\langle E_0|x=0\rangle+\cdots, \tag{3.81}$$

where $|E_0\rangle$ is the ground state. Clearly the normalized wave function is

$$\langle x|E_0\rangle=\left(\frac{\omega}{\pi\hbar}\right)^{\frac{1}{4}}e^{-\frac{\omega}{2\hbar}x^2} \tag{3.82}$$

while

$$E_0=\frac{1}{2}\hbar\omega \tag{3.83}$$

giving

$$\langle x=0|E_0\rangle=\left(\frac{\omega}{\pi\hbar}\right)^{\frac{1}{4}}. \tag{3.84}$$

Hence Equation (3.81) yields

$$\mathcal{N}\det{}^{-\frac{1}{2}}\left[-\frac{d^2}{d\tau^2}+\omega^2\right]=\left(\frac{\omega}{\pi\hbar}\right)^{\frac{1}{2}}e^{-\omega\frac{\beta}{2}} \tag{3.85}$$

in agreement with Equation (3.80), and confirming the definition of the normalization $\mathcal{N}$ chosen in Equation (3.77).

3.5.2 Evaluation of K

Finally we must evaluate the factor K. K is given by the ratio

$$\frac{1}{K^2}=\frac{\det'\left[-\frac{d^2}{d\tau^2}+V''(\bar{z}(\tau-\tau_1)\right]}{\det\left[-\frac{d^2}{d\tau^2}+\omega^2\right]} \tag{3.86}$$

from Equations (3.64) and (3.66) for $n = 0$. Thus

$$\frac{1}{K^2} = \left(\frac{\psi_0\left(\frac{\beta}{2}\right)/\lambda_0}{\psi_0^0\left(\frac{\beta}{2}\right)}\right), \tag{3.87}$$

where λ_0 is the smallest eigenvalue in the presence of an instanton. To calculate $\psi_0\left(\frac{\beta}{2}\right)$ and λ_0 approximately we describe again the procedure given in Coleman [31]. First we need to solve

$$\left(-\partial_\tau^2 + V''(\bar{z}(\tau))\right)\psi_0(\tau) = 0 \tag{3.88}$$

with the boundary conditions $\psi_0(-\beta/2) = 0$ and $\partial_\tau \psi_0(-\beta/2) = 1$. We already know one solution of Equation (3.88), albeit one that does not satisfy the boundary conditions: the zero mode of the operator in Equation (3.30) due to time translation invariance, we will call it here $x_1(\tau)$:

$$x_1(\tau) = \frac{1}{\sqrt{S_0}}\frac{d\bar{z}}{d\tau}. \tag{3.89}$$

$x_1(\tau) \to A e^{-\omega|\tau|}$ as $\tau \to \pm\infty$. A is determined by the equation of motion, Equation (3.30), which integrated once corresponds to

$$\dot{\bar{z}}(\tau) = \sqrt{2V(\bar{z}(\tau))}. \tag{3.90}$$

Once we have A we can compute $\psi(\frac{\beta}{2})$ and λ_0.

We know that there must exist a second independent solution of the differential Equation (3.88), $y_1(\tau)$ which we normalize so that the Wronskian

$$x_1\frac{dy_1}{d\tau} - y_1\frac{dx_1}{d\tau} = 2A^2. \tag{3.91}$$

We remind the reader that the Wronskian between two linearly independent solutions of a linear second-order differential equation is non-zero, and with no first derivative term, as in Equation (3.88), is a constant. Then as $\tau \to \pm\infty$ we have

$$\dot{y}_1(\tau) \pm \omega y_1(\tau) = 2A\omega e^{\omega|\tau|} \tag{3.92}$$

using the known behaviour of $x_1(\tau)$. The general solution of Equation (3.92) is any particular solution plus an arbitrary factor times the homogeneous solution

$$y_1(\tau) = \pm A e^{\omega|\tau|} + B e^{\mp\omega|\tau|}, \tag{3.93}$$

where B is an arbitrary constant. Evidently the homogenous solution is a negligible perturbation on the particular solution, and $y_1(\tau) \to \pm A e^{\omega|\tau|}$ as $\tau \to \pm\infty$. Then we construct $\psi_0(\tau)$ as

$$\psi_0(\tau) = \frac{1}{2\omega A}\left(e^{\omega\beta/2}x_1(\tau) + e^{-\omega\beta/2}y_1(\tau)\right), \tag{3.94}$$

verifying

$$\psi_0(-\beta/2) = \frac{1}{2\omega A}\left(e^{\omega\beta/2}x_1(-\beta/2)+e^{-\omega\beta/2}y_1(-\beta/2)\right)$$

$$\approx \frac{1}{2\omega A}\left(e^{\omega\beta/2}Ae^{-\omega\beta/2}+e^{-\omega\beta/2}(-A)e^{\omega\beta/2}\right)=0 \qquad (3.95)$$

while

$$\left.\frac{d\psi_0(-\beta/2)}{d\tau}\right|_{\frac{-\beta}{2}} \approx \frac{1}{2\omega A}\left(e^{\omega\beta/2}\left.\frac{d}{d\tau}Ae^{\omega\tau}\right|_{\frac{-\beta}{2}}+e^{-\omega\beta/2}\left.\frac{d}{d\tau}(-A)e^{-\omega\tau}\right|_{\frac{-\beta}{2}}\right)=1.$$

$$(3.96)$$

Then it is also easy to see

$$\psi_0(\beta/2) = \frac{1}{\omega}, \qquad (3.97)$$

which we will need later.

We also need to calculate the smallest eigenvalue λ_0 of Equation (3.69). To do this we convert the differential equation to an integral equation using the corresponding Green function. The Green function satisfying the appropriate boundary conditions is constructed from $x_1(\tau)$ and $y_1(\tau)$ using standard techniques and is given by

$$G(\tau,\tau') = \begin{cases} \frac{1}{2A^2}\left(-y_1(\tau')x_1(\tau)+x_1(\tau')y_1(\tau)\right) & \tau > \tau' \\ 0 & \tau < \tau' \end{cases}. \qquad (3.98)$$

Then the differential equation is converted to an integral equation

$$\psi_\lambda(\tau) = \psi_0(\tau) + \frac{\lambda}{2A^2}\int_{\frac{-\beta}{2}}^{\tau}d\tau'(x_1(\tau')y_1(\tau)-y_1(\tau')x_1(\tau))\psi_\lambda(\tau')$$

$$\approx \psi_0(\tau) + \frac{\lambda}{2A^2}\int_{\frac{-\beta}{2}}^{\tau}d\tau'(x_1(\tau')y_1(\tau)-y_1(\tau')x_1(\tau))\psi_0(\tau'). \qquad (3.99)$$

This wave function vanishes for the lowest eigenvalue λ_0 (and actually for all eigenvalues λ_n) at $\tau=\beta/2$ by Equation (3.70), thus

$$\psi_0(\beta/2) + \frac{\lambda}{2A^2}\int_{\frac{-\beta}{2}}^{\frac{\beta}{2}}d\tau'(x_1(\tau')y_1(\beta/2)-y_1(\tau')x_1(\beta/2))\psi_0(\tau')$$

$$\approx \frac{1}{\omega} - \frac{\lambda}{2A^2}\int_{\frac{-\beta}{2}}^{\frac{\beta}{2}}d\tau'(x_1(\tau')y_1(\beta/2)-y_1(\tau')x_1(\beta/2))$$

$$\frac{1}{2\omega A}\left(e^{\omega\beta/2}x_1(\tau')+e^{-\omega\beta/2}y_1(\tau')\right)$$

$$\approx \frac{1}{\omega} - \frac{\lambda}{2A^2}\int_{\frac{-\beta}{2}}^{\frac{\beta}{2}}d\tau'(x_1(\tau')e^{\omega\beta/2}-y_1(\tau')e^{-\omega\beta/2})$$

$$\frac{1}{2\omega}\left(e^{\omega\beta/2}x_1(\tau')+e^{-\omega\beta/2}y_1(\tau')\right)$$

$$\approx \frac{1}{\omega} - \frac{\lambda}{2A^2\omega} \int_{\frac{-\beta}{2}}^{\frac{\beta}{2}} d\tau'(x_1^2(\tau')e^{\omega\beta} - y_1^2(\tau')e^{-\omega\beta})$$

$$\approx \frac{1}{\omega} - \frac{\lambda}{4A^2\omega} \int_{\frac{-\beta}{2}}^{\frac{\beta}{2}} d\tau' x_1^2(\tau')e^{\omega\beta} = \frac{1}{\omega} - \frac{\lambda}{4A^2\omega}e^{\omega\beta} = 0.$$

$$(3.100)$$

In the penultimate equation, we can drop the second term because it behaves at most as $\sim \beta$, since $y_1(\tau) \sim e^{\beta/2}$ at the boundaries of the integration domain at $\pm\beta/2$, while the first term behaves as $\sim e^\beta$ since $\int x_1^2(\tau)d\tau$ is normalized to 1. This gives quite simply

$$\lambda_0 \approx 4A^2 e^{-\omega\beta}. \qquad (3.101)$$

Then finally we get

$$K = \left(\frac{\psi_0^0(\beta/2)}{\psi_0(\beta/2)/\lambda_0}\right)^{\frac{1}{2}} = \frac{e^{\omega\beta}/2\omega}{(1/\omega 4A^2 e^{-\omega\beta})} = 2A^2. \qquad (3.102)$$

Thus we have found that the matrix element

$$\langle a|e^{-\beta\hat{h}(\hat{X},\hat{P})/\hbar}| - a\rangle = \sinh\left(\left(\frac{S_0}{2\pi\hbar}\right)^{\frac{1}{2}} e^{-S_0/\hbar}2A^2\beta\right)\left(\frac{\omega}{\pi\hbar}\right)^{\frac{1}{2}} e^{-\omega\frac{\beta}{2}}. \qquad (3.103)$$

To see explicitly see how to compute A, we can consider a convenient, completely integrable example, $V(x) = (\gamma^2/2)(x^2 - a^2)^2$, which has $\omega^2 = V''(\pm a) = (2\gamma a)^2$. Then Equation (3.30) yields

$$\int_0^{\bar{z}(\tau-\tau_1)} \frac{dz}{\gamma(z^2 - a^2)} = \tau - \tau_1 \qquad (3.104)$$

with exact solution

$$\bar{z}(\tau) = a\tanh(a\gamma(\tau - \tau_1)). \qquad (3.105)$$

Thus A is determined by

$$x_1(\tau) = \frac{\dot{\bar{z}}(\tau)}{\sqrt{S_0}} = \frac{a^2\gamma}{\sqrt{S_0}\cosh^2(a\gamma(\tau - \tau_1))}, \qquad (3.106)$$

which behaves as

$$\lim_{\tau\to\pm\infty} x_1(\tau) = \frac{4a^2\gamma}{\sqrt{S_0}}e^{-2a\gamma|\tau|} = \frac{2a\omega}{\sqrt{S_0}}e^{-\omega|\tau|} = Ae^{-\omega|\tau|}. \qquad (3.107)$$

$\sqrt{S_0}$ is calculated from Equation (3.27), giving

$$S_0 = \int_{-a}^a dz\gamma(z^2 - a^2) = \frac{4}{3}\gamma a^3 = \frac{2}{3}\omega a^2. \qquad (3.108)$$

Hence $A = \frac{2a\omega}{\sqrt{(2/3)\omega a^2}} = \sqrt{\frac{6}{\omega}}$, for this example.

3.6 Extracting the Lowest Energy Levels

On the other hand, the matrix element of Equation (3.103) can be evaluated by inserting a complete set of energy eigenstates between the operator and the position eigenstates on the left-hand side, yielding

$$\langle a|e^{-\beta \hat{h}(\hat{X},\hat{P})/\hbar}|-a\rangle = e^{-\beta E_0/\hbar}\langle a|E_0\rangle\langle E_0|-a\rangle + e^{-\beta E_1/\hbar}\langle a|E_1\rangle\langle E_1|-a\rangle + \cdots,$$
(3.109)

where we have explicitly written only the first two terms as we expect that the two classical states, $|\pm a\rangle$, are reorganized due to tunnelling into the two lowest-lying states, $|E_0\rangle$ and $|E_1\rangle$. Indeed, comparing Equation (3.103) and Equation (3.109) we find

$$E_0 = \frac{\hbar}{2}\omega - \hbar\left(\frac{S_0}{2\pi\hbar}\right)^{\frac{1}{2}} e^{-S_0/\hbar}2A^2$$
(3.110)

while

$$E_1 = \frac{\hbar}{2}\omega + \hbar\left(\frac{S_0}{2\pi\hbar}\right)^{\frac{1}{2}} e^{-S_0/\hbar}2A^2.$$
(3.111)

It should be stressed that our calculation is only valid for the energy difference, not for the corrections to the energies directly. Indeed, there are ordinary perturbative corrections to the energy levels which are normally far greater than the non-perturbative, exponentially suppressed correction that we have calculated. However, none of these perturbative corrections can see any tunnelling phenomena. Thus our calculation gives the leading term in the correction due to tunnelling. Thus, the energy splitting which relies on tunnelling is found only through our calculation, and not through perturbative calculations.

We also find the relations

$$\langle a|E_0\rangle\langle E_0|-a\rangle = \left(\frac{\omega}{\pi\hbar}\right)^{\frac{1}{2}}$$
(3.112)

in addition to

$$\langle a|E_1\rangle\langle E_1|-a\rangle = -\left(\frac{\omega}{\pi\hbar}\right)^{\frac{1}{2}}$$
(3.113)

while a simple adaptation of our analysis yields

$$\langle a|E_0\rangle\langle E_0|a\rangle = \left(\frac{\omega}{\pi\hbar}\right)^{\frac{1}{2}}$$
(3.114)

in addition to

$$\langle a|E_1\rangle\langle E_1|a\rangle = \left(\frac{\omega}{\pi\hbar}\right)^{\frac{1}{2}}.$$
(3.115)

These yield $\langle E_0|-a\rangle = \langle E_0|a\rangle$ while $\langle E_1|-a\rangle = -\langle E_1|a\rangle$ which are consistent with $|E_0\rangle$ being an even function, *i.e.* $|E_0\rangle$ being an even superposition of the position eigenstates $|a\rangle$ and $|-a\rangle$ while $|E_1\rangle$ being an odd function and hence an odd superposition of these two position eigenstates.

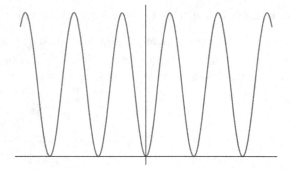

Figure 3.7. A generic periodic potential with minima occurring at na with $n \in \mathbb{Z}$, where a is the distance between neighbouring minima

3.7 Tunnelling in Periodic Potentials

We will end this chapter with an application of the method to periodic potentials. Periodic potentials are very important in condensed matter physics, as crystal lattices are well-approximated by the theory of electrons in a periodic potential furnished by the atomic nuclei. The idea is easiest to enunciate in a one-dimensional example. Consider a potential of the form given in Figure 3.7. A particle in the presence of such a potential with minimal energy will classically, certainly, be localized in the bottom of the wells of the potential. If there is no tunnelling, there would be an infinite number of degenerate states corresponding to the state where the particle is localized in state labelled by integer $n \in \mathbb{Z}$. This could also be a very large, finite number of minima. However, quantum tunnelling will completely change the spectrum. Just as in the case of the double well potential, the states will reorganize so that the most symmetric superposition will correspond to the true ground state, and various other superpositions will give rise to excited states, albeit with excitation energies proportional to the tunnelling amplitude. The tunnelling amplitude is expected to be exponentially small and non-perturbative in the coupling constant.

As in the case of the double well potential, the instanton trajectories will correspond to solutions of the analogous dynamical problem in the inverted potential in Euclidean time (as depicted in Figure 3.8), where the trajectories commence at the top of a potential hill, stay there for a long time, then quickly fall through the minimum of the inverted potential, and then arrive at the top of the adjacent potential hill, and stay there for the remaining positive Euclidean time.

For the simple, real-time Lagrangian

$$L = \frac{1}{2}\dot{x}^2 - V(x), \tag{3.116}$$

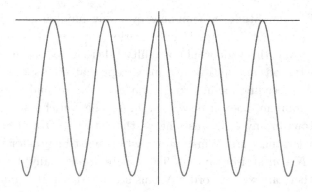

Figure 3.8. The inverted generic periodic potential with maxima occurring at na with $n \in \mathbb{Z}$

where $\dot{x} = \frac{dx(t)}{dt}$, while the Euclidean Lagrangian is simply

$$L = \frac{1}{2}\dot{x}^2 + V(x), \qquad (3.117)$$

where $\dot{x} = \frac{dx(\tau)}{d\tau}$. As $V(na) = 0$ for $n \in \mathbb{Z}$, we impose the boundary conditions $x(\tau = -\infty) = na$ but $x(\tau = \infty) = (n+1)a$ for an instanton and $x(\tau = \infty) = (n-1)a$ for an anti-instanton and look for solutions of the Euclidean equations of motion

$$\frac{d^2x(\tau)}{d\tau^2} - V'(x(\tau)) = 0. \qquad (3.118)$$

This immediately affords a first integral; multiplying by $\dot{x}(\tau)$ and integrating gives

$$\frac{1}{2}\dot{x}^2(\tau) - V(x(\tau)) = 0, \qquad (3.119)$$

where we have fixed the constant with the boundary conditions. This equation admits a solution in general, the instanton, but it does depend on the explicit details of the potential. However, we can find the action of the corresponding instanton, which only depends on an integral of the potential, by first isolating

$$\dot{x} = \sqrt{2V(x)}, \qquad (3.120)$$

and then

$$S_0 = \int_{-\infty}^{\infty} d\tau\, \frac{1}{2}\dot{x}^2 + V(x) = \int_{-\infty}^{\infty} d\tau \left(\frac{1}{2}\dot{x}\sqrt{2V(x)} + \frac{1}{2}\dot{x}\sqrt{2V(x)}\right)$$

$$= \int_{na}^{(n+1)a} dx\, \sqrt{2V(x)}. \qquad (3.121)$$

Although we may naively want to compute the amplitude for tunnelling between neighbouring vacua, it is actually more informative to compute the amplitude for a transition from vacuum n to vacuum $n+m$. Naively we would approximate this amplitude by summing over any number of pairs of widely separated instanton anti-instanton configurations appended by a string of m instantons. However, this logic would be faulty. There is no reason to restrict the order of the instantons and anti-instantons except that they should tunnel from the immediately preceding vacuum to an adjacent vacuum, and finally we should arrive at the minimum indexed by $n + m$. Thus, one can choose the instantons or anti-instantons in any order, as long as they start at n and end at $n + m$. This means that if there are N instantons, which must be greater than m, then there must be $N - m$ anti-instantons. Thus there are as many distinct paths of instantons as there are ways to order N plus signs and $N - m$ minus signs. This gives a degeneracy factor of

$$\frac{(2N - m)!}{N!(N - m)!}. \tag{3.122}$$

Furthermore, when we integrate over the Gaussian fluctuations for each instanton or anti-instanton, we get the usual determinantal factor K for each instanton or anti-instanton, but we do encounter one zero mode corresponding to each one's position, which we omit in the determinant. Then we integrate over the positions of the instantons and anti-instantons, except that the position of each instanton or anti-instanton must occur at the position after the preceding one, as the instantons and anti-instantons correspond to specific tunnelling between specific vacua. This gives the integral

$$\int_{-\beta/2}^{\beta/2} d\tau_1 \int_{\tau_1}^{\beta/2} d\tau_2 \cdots \int_{\tau_{2N-1}}^{\beta/2} d\tau_{2N-m} = \frac{\beta^{2N-m}}{(2N - m)!}. \tag{3.123}$$

As usual, the action for any instanton or anti-instanton is the same and equal to S_0. Thus, for N instantons and $N - m$ anti-instantons we get

$$\langle n + m | e^{-\beta \hat{h}(\hat{X}, \hat{P})/\hbar} | n \rangle = \sum_{N=m}^{\infty} e^{-(2N-m)S_0/\hbar} K^{2N-m} \frac{(2N - m)!}{N!(N - m)!} \frac{\beta^{2N-m}}{(2N - m)!}. \tag{3.124}$$

This sum is unclear for identifying the underlying spectrum and the contribution of each energy eigenstate; however, if we re-write the sum as a double sum over N instantons and M anti-instantons with a constraint $M = N - m$ we have

$$\langle n + m | e^{-\beta \hat{h}(\hat{X}, \hat{P})/\hbar} | n \rangle = \left(\frac{\omega}{\pi \hbar} \right)^{1/2} e^{-\beta \omega/2}$$
$$\times \sum_{N,M=0}^{\infty} e^{-(N+M)S_0/\hbar} K^{N+M} \frac{\beta^{N+M}}{N!M!} \delta_{N-m,M}, \tag{3.125}$$

where $\omega^2 = V''(na)$. Now the Kronecker delta can be expressed via its Fourier series as

$$\delta_{N-m,M} = \int_0^{2\pi} \frac{d\theta}{2\pi} e^{i\theta(N-m-M)} \tag{3.126}$$

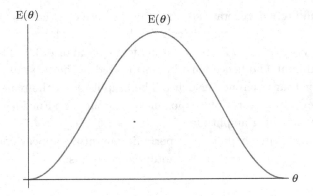

Figure 3.9. The energy band as a function of θ

and so we easily find

$$\langle n+m|e^{-\beta\hat{h}(\hat{X},\hat{P})/\hbar}|n\rangle =$$

$$= \left(\frac{\omega}{\pi\hbar}\right)^{1/2} e^{-\beta\omega/2} \int_0^{2\pi} \frac{d\theta}{2\pi} \sum_{N,M=1}^{\infty} \frac{(K\beta e^{-S_0/\hbar})^{N+M}}{N!M!} e^{i\theta(N-m-M)}$$

$$= \left(\frac{\omega}{\pi\hbar}\right)^{1/2} e^{-\beta\omega/2} \int_0^{2\pi} \frac{d\theta}{2\pi} e^{-im\theta} \sum_{N,M=1}^{\infty} \frac{(K\beta e^{-S_0/\hbar})^{N+M}}{N!M!} e^{i\theta(N-M)}$$

$$= \left(\frac{\omega}{\pi\hbar}\right)^{1/2} e^{-\beta\omega/2} \int_0^{2\pi} \frac{d\theta}{2\pi} e^{-im\theta} e^{\left(K\beta e^{-S_0/\hbar}e^{i\theta}\right)} e^{\left(K\beta e^{-S_0/\hbar}e^{-i\theta}\right)}$$

$$= \left(\frac{\omega}{\pi\hbar}\right)^{1/2} e^{-\beta\omega/2} \int_0^{2\pi} \frac{d\theta}{2\pi} e^{-im\theta} e^{\left(K\beta e^{-S_0/\hbar}\left(e^{i\theta}+e^{-i\theta}\right)\right)}$$

$$= \left(\frac{\omega}{\pi\hbar}\right)^{1/2} e^{-\beta\omega/2} \int_0^{2\pi} \frac{d\theta}{2\pi} e^{-im\theta} e^{\left(2K\beta e^{-S_0/\hbar}\cos\theta\right)}. \tag{3.127}$$

But this expression for the matrix element has a clear interpretation in terms of the spectrum. We see that the spectrum has become a continuum, parametrized by θ. If we write

$$\langle n+m|e^{-\beta\hat{h}(\hat{X},\hat{P})/\hbar}|n\rangle = \int_0^{2\pi} \frac{d\theta}{2\pi} e^{-\beta E(\theta)/\hbar}\langle n+m|E(\theta)\rangle\langle E(\theta)|n\rangle \tag{3.128}$$

we identify

$$E(\theta) = \hbar\omega/2 - 2\hbar K e^{-S_0/\hbar}\cos\theta \tag{3.129}$$

and

$$\langle n+m|E(\theta)\rangle\langle E(\theta)|n\rangle = \left(\frac{\omega}{\pi\hbar}\right)^{1/2} \frac{e^{-im\theta}}{2\pi}, \tag{3.130}$$

which affords the identification

$$\langle n|E(\theta)\rangle = \left(\frac{\omega}{\pi\hbar}\right)^{1/4} \frac{e^{-in\theta}}{\sqrt{2\pi}}. \tag{3.131}$$

Thus our infinitely degenerate spectrum of discrete classical vacua has turned into a continuum of states, what is called a band in condensed matter physics, with an energy that varies as $\cos\theta$, as depicted in Figure 3.9. The states are now in a continuum, and hence must be normalized in the sense of a Dirac delta function rather than a Kronecker delta. The amplitude of the band $2\hbar K e^{-S_0/\hbar}$ contains the tell-tale factor of the exponential of minus the Euclidean action, the hallmark of a tunnelling amplitude.

We will see in future chapters that periodic potentials appear commonly and play an important role in various instanton calculations.

4

Decay of a Meta-stable State

In this chapter we will consider the decays of meta-stable states and calculate
the lifetime for such a state using instanton methods. A meta-stable state arises
due to the existence of a local minimum of the potential, which is not the global
minimum. This corresponds to a potential having the form given in Figure 4.1.
The potential rises steeply to infinity to the left and to the right; after the
potential barrier, it goes down well below the energy of the meta-stable state,
either eventually going to constant or it may even rise to plus infinity again in
order to give an overall stable quantum mechanical problem. However, exactly
what the potential does to the right is considered not to be important; the
behaviour of the potential to the right is assumed to have a negligible effect on
the tunnelling amplitude for a particle initially in the local minimum at $z = 0$
escaping to the right. We have drawn the potential, in Figure 4.1, so that it simply
drops off to the right and we have normalized the potential by adding a constant
such that the local minimum has $V(0) = 0$. Physically we are considering a
potential of the type where a particle is trapped in a local potential well, but
once the particle tunnels out of the well, it is free. The probability that the
initial state is regenerated from the decay products is assumed to be negligible.
This is in contra-distinction to the problem considered in Chapter 3 with two
symmetric wells. Here the tunnelling-back amplitude was sizeable, corresponding
to the anti-instanton, and had to be taken into account.

4.1 Decay Amplitude and Bounce Instantons

In this chapter, we will attempt to calculate the amplitude

$$< z = 0|e^{-\frac{\beta}{\hbar}\hat{h}(\hat{X},\hat{P})}|z = 0 >= \mathcal{N}\int \mathcal{D}z(\tau)e^{-\frac{S_E[z(\tau)]}{\hbar}} = e^{-\frac{\beta E_0}{\hbar}}|\langle E_0 \,|\, z = 0\rangle|^2 +\cdots.$$

$$(4.1)$$

From this amplitude we expect to be able to identify and calculate the energy

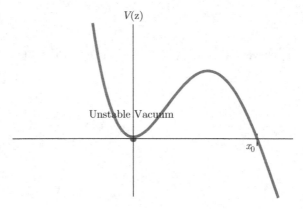

Figure 4.1. A potential with a meta-stable state at $z = 0$ that will decay via tunnelling

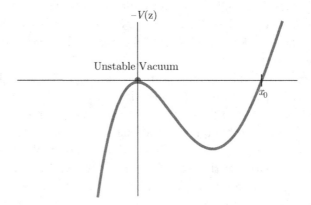

Figure 4.2. The flipped potential for instanton Euclidean classical solution

E_0 for the ground state. For a stable state, localized at $z = 0$, we expect E_0, in first approximation, to correspond to the ground-state energy of the harmonic oscillator appropriate to the well at $z = 0$, and $|\langle E_0 | z = 0 \rangle|$ to be the magnitude of the ground-state wave function at $z = 0$. Now because of tunnelling we imagine that E_0 gains an imaginary part, $E_0 \to E_0 + i\Gamma/2$. We will directly attempt to use the path integral, and calculate it in a Gaussian approximation about an appropriate set of critical points, as in Chapter 3.

The equation of motion corresponds to particle motion in the inverted potential $-V(z)$, as depicted in Figure 4.2, with boundary condition that $z(\pm\frac{\beta}{2}) = 0$. There are two solutions, the trivial one $z(\tau) = 0$ for all τ, and a non-trivial true instanton solution $\bar{z}(\tau)$. Here the particle begins at $\tau = -\frac{\beta}{2}$ with a small positive velocity at $z = 0$, falls through the potential well and rises again to height zero at $z = x_0$, at around $\tau = 0$, and then bounces back, reversing its

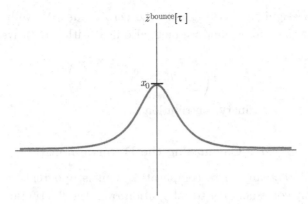

Figure 4.3. The bounce instanton which mediates tunnelling of a meta-stable state

steps and arriving at $z = 0$ again at $\tau = \frac{\beta}{2}$. Clearly from symmetry such a solution exists if β is sufficiently large. We call this instanton, after Coleman, the bounce, $\bar{z}^{\text{bounce}}(\tau)$. The action for the bounce essentially comes from the short time interval during which the particle is significantly away from $z = 0$. One can easily show that the bounce is exponentially close to zero except for a region around $\tau = 0$ of size $\frac{1}{\omega}$, where again $\omega^2 = V''(0)$. We call the action for the bounce $S_0 = S_E[\bar{z}^{\text{bounce}}(\tau)]$ for the case $\beta = \infty$. Due to the time translation invariance in the $\beta = \infty$ case, again, there exists a one-parameter family of configurations, approximate bounces, which correspond to the bounce occurring at any time $\tau_0 \in [-\frac{\beta}{2}, \frac{\beta}{2}]$. The action for these configurations is exponentially close to S_0 and hence the degeneracy is β. Furthermore, approximate critical configurations also exist corresponding to n bounces occurring at widely separated times with action exponentially close to nS_0. The degeneracy of these configurations is $\frac{\beta^n}{n!}$ as they are exactly analogous to identical particles. Thus we expect the matrix element to be expressable as

$$\langle z=0|e^{-\frac{\beta}{\hbar}\hat{h}(\hat{X},\hat{P})}|z=0\rangle = \mathcal{N}\left(\det\left[-\frac{d^2}{d\tau^2}+\omega^2\right]\right)^{-\frac{1}{2}}\sum_{n=0}^{\infty}\left(\left(\frac{S_0}{2\pi\hbar}\right)^{\frac{1}{2}}\beta\right)^n \frac{e^{-\frac{nS_0}{\hbar}}}{n!} \times$$

$$\times \left(\frac{\det'\left[-\frac{d^2}{d\tau^2}+V''(\bar{z}_n^{\text{bounce}}(\tau))\right]}{\det\left[-\frac{d^2}{d\tau^2}+\omega^2\right]}\right)^{-\frac{1}{2}}$$

$$= \left(\frac{\omega}{\pi\hbar}\right)^{\frac{1}{2}}e^{-\omega\beta/2}e^{\beta\sqrt{\frac{S_0}{2\pi\hbar}}Ke^{-S_0}}, \tag{4.2}$$

where

$$K = \frac{\left(\det'\left[-\frac{d^2}{d\tau^2}+V''(\bar{z}^{\text{bounce}}(\tau))\right]\right)^{-\frac{1}{2}}}{\left(\det\left[-\frac{d^2}{d\tau^2}+\omega^2\right]\right)^{-\frac{1}{2}}}. \tag{4.3}$$

Here the prime signifies omitting only the zero mode. We will find that the situation is not that simple, and we must also deal with a negative mode. Then we would find

$$E_0 = \hbar \left(\frac{\omega}{2} + K \left(\frac{S_0}{2\pi\hbar} \right)^{\frac{1}{2}} e^{-\frac{S_0}{\hbar}} \right) \tag{4.4}$$

and we look for an imaginary contribution to K.

4.2 Calculating the Determinant

The situation is actually more complicated than is apparent. K comes from the determinant corresponding to integration over the fluctuations around the critical bounce

$$\left(\det{}' \left[-\frac{d^2}{d\tau^2} + V''(\bar{z}^{\text{bounce}}(\tau)) \right] \right)^{-\frac{1}{2}} = \int \prod_{\substack{n \\ \lambda_n \neq 0}} \frac{dc_n}{\sqrt{2\pi\hbar}} e^{-\frac{1}{\hbar}\frac{1}{2}\sum_n \lambda_n c_n^2} = \prod_{\substack{n \\ \lambda_n \neq 0}} \frac{1}{\sqrt{\lambda_n}}. \tag{4.5}$$

The n's corresponding to vanishing λ_n's are excluded, which is the meaning of the primed determinant. This time, however, the problem is much more serious. One of the λ_n's is actually negative. For this λ_n the integration over the c_n simply does not exist, and hence the determinant, as we wish to calculate it, does not exist. It seems our original idea is doomed. But there is a possible solution: perhaps we can define the integration by analytic continuation. Indeed, analytic continuations of real-valued functions often gain imaginary parts, exactly what we desire. This analytic continuation is in fact possible and we will see how we can perform it appropriately.

4.3 Negative Mode

First we will establish the existence of the negative mode. For $\beta = \infty$ we have an exact zero mode due to time translation invariance

$$\left(-\frac{d^2}{d\tau^2} + V''(\bar{z}^{\text{bounce}}(\tau)) \right) \frac{d}{d\tau} \bar{z}^{\text{bounce}}(\tau)$$

$$= \frac{d}{d\tau} \left(-\frac{d^2}{d\tau^2} \bar{z}^{\text{bounce}}(\tau) + V'(\bar{z}^{\text{bounce}}(\tau)) \right) = 0, \tag{4.6}$$

where the second term vanishes as it is the equation of motion. Since $\bar{z}^{\text{bounce}}(\tau)$ has the increasing and then decreasing form given in Figure 4.3, this implies $\dot{\bar{z}}^{\text{bounce}}(\tau)$ has the form given by Figure 4.4. In contra-distinction to the zero mode of Chapter 3, this zero mode has a node, *i.e.* it has a zero. This is intuitively obvious, the velocity of the particle executing the bounce will vanish exactly when it reverses direction. The analogous quantum mechanical Hamiltonian

$$-\frac{d^2}{d\tau^2} + V''(\bar{z}^{\text{bounce}}(\tau)) \tag{4.7}$$

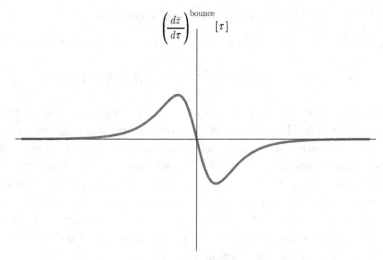

Figure 4.4. The derivative of the bounce $\frac{d\bar{z}^{bounce}(\tau)}{d\tau}$

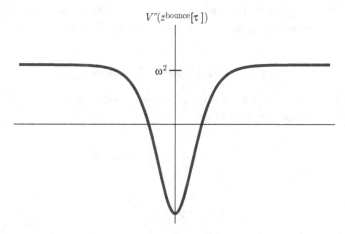

Figure 4.5. The form of the potential $V''(\bar{z}^{bounce}(\tau))$

has the potential given by Figure 4.5. One expects the spectrum to consist of a finite number of bound states and then a continuum beginning at ω^2. The ground-state wave function must have no nodes. The next bound energy level, if it exists, will have one node. We have already found a bound-state wave function with energy exactly zero, but it has one node. Thus there exists exactly one bound-state level, the nodeless ground state, with negative energy. The Gaussian integral in this direction in function space does not exist, and we must only define it through analytic continuation.

4.4 Defining the Analytic Continuation

The original idea that the matrix element has an expansion of the form

$$\langle z=0| e^{-\frac{\beta \hat{h}(\hat{X},\hat{P})}{\hbar}} |z=0\rangle = e^{-\frac{(E+i\Gamma)\beta}{\hbar}} \langle 0| E+i\Gamma \rangle \langle E+i\Gamma| 0\rangle + \cdots \tag{4.8}$$

was ill-conceived. There is no eigenstate of the Hamiltonian corresponding to the meta-stable state. The Hamiltonian is a hermitean operator with all eigenvalues real, an eigenstate with a complex eigenvalue simply does not exist. We can only obtain the imaginary energy of the meta-stable state through analytic continuation. We imagine the analytic continuation in a parameter α which starts at $\alpha = 0$ with a potential with a stable bound state localized at $z = 0$, but yields our original potential at $\alpha = 1$. The energy of the bound state will also be an analytic function of the parameter α. As long as a true bound state exists around $z = 0$, this energy will be a real function of the parameter α. When the parameter is continued to yield our original potential where the bound state becomes meta-stable, we expect that this energy as an analytic function of the parameter α will not remain real and will gain an imaginary part. This imaginary part should correspond to the width of the meta-stable state. These general considerations correspond to a sequence of potentials, as shown in Figures 4.6, 4.7 and 4.8.

4.4.1 An Explicit Example

We will confirm these ideas with an explicit demonstration in a specific solvable potential. The example we consider is

$$V(\alpha, z) = -\left(\alpha - \frac{1}{2}\right) z^4 + \omega^2 z^2 \tag{4.9}$$

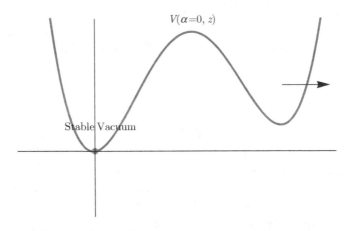

Figure 4.6. The potential with a stable state at $z = 0$ for $\alpha = 0$

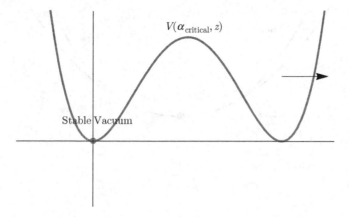

Figure 4.7. The critical potential with a stable state at $z = 0$ for $\alpha_{critical}$

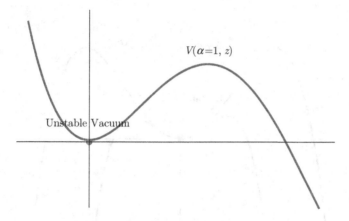

Figure 4.8. The potential with a meta-stable state at $z = 0$ for $\alpha = 1$

and the integral

$$\mathcal{I}(\alpha,\omega) = \int_{-\infty}^{\infty} dz\, e^{-\frac{1}{\hbar}\left(-\left(\alpha-\frac{1}{2}\right)z^4 + \omega^2 z^2\right)}, \tag{4.10}$$

which is analogous to the integral over the direction corresponding to the negative mode in the definition of the determinant, when $\alpha = 1$, as depicted in Figures 4.9 and 4.10. The integral is defined for $\alpha \leq \frac{1}{2}$ and, specifically, it is not defined for $\alpha = 1$. The integral is actually well-defined for complex α, with the condition $\mathfrak{Re}\{\alpha\} \leq \frac{1}{2}$. We can define the analytic function $\mathcal{I}(\alpha,\omega)$ for $\mathfrak{Re}\{\alpha\} > \frac{1}{2}$ by analytic continuation. In this simple case we have no difficulty whatsoever, for $\mathfrak{Re}\{\alpha\} \leq \frac{1}{2}$, the integral is known in terms of special functions,

$$\mathcal{I}(\alpha,\omega) = \frac{1}{2}\sqrt{\frac{\omega^2}{\left(\frac{1}{2}-\alpha\right)}}\, e^{\left(\frac{\omega^4}{8\hbar\left(\frac{1}{2}-\alpha\right)}\right)} K_{\frac{1}{4}}\left(\frac{\omega^4}{8\hbar\left(\frac{1}{2}-\alpha\right)}\right), \tag{4.11}$$

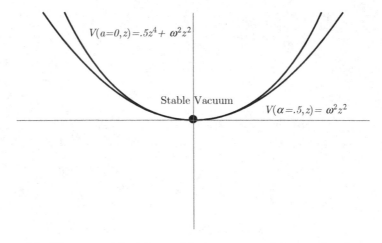

Figure 4.9. The potential with a stable state at $z = 0$ for $\alpha = 0$ and $\alpha = .5$

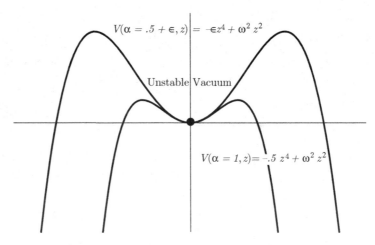

Figure 4.10. The potential with a meta-stable state at $z = 0$ for $\alpha = .5 + \epsilon$ and for $\alpha = 1$

where

$$K_\nu(z) = \frac{\pi i}{2} e^{\frac{\pi}{2}\nu i} \left(J_\nu(iz) + N_\nu(iz)\right) \qquad (4.12)$$

is the modified Bessel function of imaginary argument. The expression in Equation (4.11) has a well-defined analytic continuation throughout the complex α-plane, except on the real α-axis, where starting at $\alpha = \frac{1}{2}$, there is a branch cut.

But in general, we do not have the luxury of knowing the integral exactly. There is, however, a method for performing the analytic continuation more implicitly. Happily, this method allows us to extract the information that we actually seek, the imaginary part of the energy. We apply the method to the specific, exactly

Integration Contours

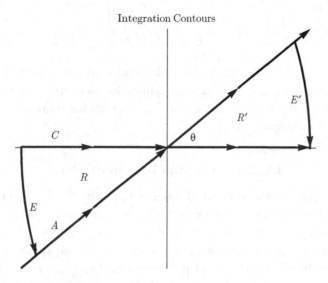

Figure 4.11. The original integration contour C, along the real line and the deformed contour A, a straight line at an angle θ, for the analytic continuation

solvable integral of Equation (4.11) to see in detail how the implicit method works. Indeed, we can obtain the analytic continuation of a function defined by a contour integral, by deforming the integration contour. In our example

$$\mathcal{I}(\alpha,\omega) = \int_{-\infty}^{\infty} dz e^{-\frac{1}{\hbar}\left(-\left(\alpha-\frac{1}{2}\right)z^4+\omega^2 z^2\right)} \quad \text{for} \quad \text{Real}(\alpha) \leq \frac{1}{2} \tag{4.13}$$

corresponds to the integration contour C along the real axis, in Figure 4.11. The integration is defined for $\left|\arg\left(-\left(\alpha-\frac{1}{2}\right)\right)\right| < \frac{\pi}{2}$. We deform the contour to $E + A + E'$ as in Figure 4.11, the integral is invariant since there are no poles in regions R and R' and if the contributions from the circular arcs E, E', vanish for infinite radius, which is assumed to be true, we get

$$\mathcal{I}(\alpha,\omega) = \int_{z=re^{i\theta}} dz e^{-\frac{1}{\hbar}\left(-\left(\alpha-\frac{1}{2}\right)z^4+\omega^2 z^2\right)}. \tag{4.14}$$

But now the integration converges for $\left|\arg\left(-\left(\alpha-\frac{1}{2}\right)\right)+4\theta\right| < \frac{\pi}{2}$ since after replacing $z = re^{i\theta}$ we must have that $-(\alpha-\frac{1}{2})e^{i4\theta}r^4$ has a positive real part. Thus a deformation of the contour defines an analytic continuation of the integral in the parameter α. If we take $\theta = \frac{\pi}{4}$ then $\left|\arg\left(-\left(\alpha-\frac{1}{2}\right)\right)+\pi\right| < \frac{\pi}{2}$. This implies

$$\arg\left(-\left(\alpha-\frac{1}{2}\right)\right) \in \left(\frac{\pi}{2},\frac{3\pi}{2}\right) \tag{4.15}$$

hence the integral is now defined for $\mathfrak{Re}\{-(\alpha-\frac{1}{2})\} < 0$, which is negative. This means $\mathfrak{Re}\{\alpha\} > \frac{1}{2}$. Thus we define, with A corresponding to the contour with

$\theta = \pi/4$,

$$\mathcal{I}\left(\alpha > \frac{1}{2}, \omega\right) = \int_A dz e^{-\frac{1}{\hbar}\left(-\left(\alpha - \frac{1}{2}\right)z^4 + \omega^2 z^2\right)}. \tag{4.16}$$

This is an exact expression for the analytic continuation of the original integral $\mathcal{I}(\alpha < \frac{1}{2}, \omega) \to \mathcal{I}(\alpha > \frac{1}{2}, \omega)$, and there is no question as to its existence. However, we wish to actually evaluate the integral in the approximation as $\hbar \to 0$ and extract only the imaginary part.

4.5 Extracting the Imaginary Part

Consider the part of the contour A from 0 to ∞ in the first quadrant. The other half of the contour clearly gives the same contribution. We will calculate this integral approximately using the method of steepest descent, which is the indicated approximation method in the limit $\hbar \to \infty$. To do this, we further deform the contour from its present path between 0 and $\infty \times e^{i\frac{\pi}{4}}$ to the path of the steepest descent between these points. As there are no poles in the integrand, the integral clearly is invariant under this additional deformation.

4.5.1 A Little Complex Analysis

A contour of the steepest descent for the real part of a complex analytic function keeps the imaginary part constant (and vice versa). We can easily demonstrate this fact. If we have $f(x,y) = R(x,y) + iI(x,y)$ and a curve parametrized by a variable t, $(x(t), y(t))$ with tangent vector $\overrightarrow{(\dot{x}(t), \dot{y}(t))}$, the curve will correspond to the steepest descent of the real part $R(x,y)$ if the tangent vector is anti-parallel to its gradient, as the gradient points in the direction of maximum change. Therefore,

$$\overrightarrow{(\partial_x R(x,y), \partial_y R(x,y))} \times \overrightarrow{(\dot{x}(t), \dot{y}(t))} = \partial_x R(x,y)\dot{y}(t) - \partial_y R(x,y)\dot{x}(t) = 0. \tag{4.17}$$

Due to analyticity, the Cauchy–Riemann equations give

$$\partial_x R(x,y) = \partial_y I(x,y) \quad \text{and} \quad \partial_x I(x,y) = -\partial_y R(x,y) \tag{4.18}$$

thus Equation (4.17) gives

$$\partial_y I(x,y)\dot{y}(t) - (-\partial_x I(x,y))\dot{x}(t) \equiv \frac{d}{dt}I(x,y) = 0. \tag{4.19}$$

But this means $I(t) = \text{constant}$, demonstrating that the imaginary part of the complex analytic function remains constant on the paths of steepest descent of the real part.

In general for an integral of the form

$$\mathcal{I} = \int_a^b dz e^{\lambda f(z)} \tag{4.20}$$

we can describe the process of the method of steepest descent as follows. For the application of the method of steepest descent, a should be a critical point of $f(z)$. We assume $\mathfrak{Re}\{f(a)\} > \mathfrak{Re}\{f(b)\}$ and $\mathfrak{Im}\{f(a)\} > \mathfrak{Im}\{f(b)\}$ and we start from a following the path of steepest descent of the $\mathfrak{Re}\{f(z)\}$ to a' where $\mathfrak{Re}\{f(a')\} = \mathfrak{Re}\{f(b)\}$ and then we append the path of steepest descent of $\mathfrak{Im}\{f(z)\}$ from a' to b along a path where now only the imaginary part of $f(z)$ changes. If $\mathfrak{Im}\{f(a)\} < \mathfrak{Im}\{f(b)\}$, we obviously ascend the appropriate portion. In the limit $\lambda \to \infty$ it is only the first contour which is important, since the second is multiplied by $e^{\lambda \mathfrak{Re}\{f(a')\}} \ll e^{\lambda \mathfrak{Re}\{f(a)\}}$. Finally we perform the integration over the first contour in the Gaussian approximation about $z = a$.

There are two further points to be made. First, we are actually only interested in the imaginary part of the integral, as it is only this part that we believe will have a leading contribution that is non-perturbative in $\hbar$. Second, and this is very important to the first, if the path of steepest descent of the real part of $f(z)$ passes through an ordinary critical point of $f(z)$, it abruptly changes direction by 90°. We can demonstrate this easily. An ordinary critical point of $f(z)$, which requires $f'(z_0) = 0$ and assumes $f''(z_0) \neq 0$, implies the behaviour

$$f(z) = f(z_0) + \frac{1}{2}f''(z_0)(z - z_0)^2 + \cdots . \tag{4.21}$$

Replacing $z - z_0 = x + iy$ we get

$$f(z_0 + x + iy) = f(z_0) + \frac{1}{2}f''(z_0)(x^2 - y^2 + 2ixy) + \cdots . \tag{4.22}$$

Then paths of steepest descent passing through the critical point are paths of the constant imaginary part of $f(z_0 + x + iy)$ passing through $x = y = 0$, i.e. $\mathfrak{Im}\{f(z_0 + x + iy)\} = \mathfrak{Im}\{f(z_0)\}$. Therefore, to lowest non-trivial order, we need paths with $\mathfrak{Im}\{f''(z_0)(x^2 - y^2 + 2ixy)\} = 0$. If $f''(z_0) = r + is$ this gives

$$s(x^2 - y^2) + 2rxy = 0. \tag{4.23}$$

If $s = 0$, the solutions are $x = 0$ or $y = 0$, which are perpendicular horizontal or vertical lines, respectively, hence crossing at 90°. Assuming $s \neq 0$,

$$x^2 + 2\frac{r}{s}xy + \left(\frac{ry}{s}\right)^2 = y^2\left(1 + \left(\frac{r}{s}\right)^2\right). \tag{4.24}$$

This gives

$$\left(x + \frac{ry}{s}\right) = \pm y\left(1 + \left(\frac{r}{s}\right)^2\right)^{\frac{1}{2}} \tag{4.25}$$

that is the curves, which are just the straight lines

$$x = \pm y\left(\left(1 + \left(\frac{r}{s}\right)^2\right)^{\frac{1}{2}} \mp \frac{r}{s}\right) \tag{4.26}$$

with the $\pm$ signs correlated. The tangents at $x = y = 0$ are given by the directions $\left(\pm \left(\left(1 + \left(\frac{r}{s} \right)^2 \right)^{\frac{1}{2}} \mp \frac{r}{s} \right), 1 \right)$. These are clearly orthogonal, as their scalar product vanishes,

$$- \left(\left(1 + \left(\frac{r}{s} \right)^2 \right)^{\frac{1}{2}} - \frac{r}{s} \right) \left(\left(1 + \left(\frac{r}{s} \right)^2 \right)^{\frac{1}{2}} + \frac{r}{s} \right) + 1 = 0. \qquad (4.27)$$

Thus in complete generality, the paths of steepest descent turn abruptly by $90°$ as they pass through an ordinary critical point.

The real or imaginary parts of complex analytic functions are called harmonic functions, which means that they satisfy $\nabla^2 R(x, y) = \nabla^2 I(x, y) = 0$. As should be well known, all critical points of the real or imaginary parts of complex analytic functions are saddle points. Then a path of steepest descent, which descends through an ordinary critical point, must change direction by $90°$, since continuing in the same direction through the critical point would correspond to ascending the other side of the saddle. Turning through $90°$ continues the descent through the saddle point. The above analysis shows that for an ordinary critical point, $f'(z_0) = 0$ but $f''(z_0) \neq 0$, the minimum and maximum directions are at $90°$ to each other.

For our integral Equation (4.16), the real part of the exponent changes from $0 \to -\infty$ as z varies from $0 \to \infty \times e^{i\frac{\pi}{4}}$, while the imaginary part of the exponent is equal to 0 at $z = 0$ but becomes arbitrarily large at $z = \infty \times e^{i\frac{\pi}{4}}$. Thus in this case, along the path of steepest descent of the real part, the imaginary part of the exponent will always be equal to 0, since it must remain constant and it vanishes at the initial point. Such a path will reach a point where $\Re\{f(z_0)\} = -\infty$. Then further following a contour with fixed real part, equal to $-\infty$, but changing imaginary part will be irrelevant since the factor corresponding to the exponential of the real part will already be zero.

Our function actually has three critical points. Indeed,

$$\frac{d}{dz} \left(\left(\alpha - \frac{1}{2} \right) z^4 - \omega^2 z^2 \right) = 4 \left(\alpha - \frac{1}{2} \right) z^3 - 2\omega^2 z = 0 \qquad (4.28)$$

has the solutions $z = 0$ and $z = \frac{\pm \omega}{\sqrt{2\left(\alpha - \frac{1}{2}\right)}}$ for the case at hand, $\alpha > \frac{1}{2}$. Thus the point $z = 0$ happens also to be a critical point, and it is easy to check that the path of steepest descent of the real part from $z = 0$ proceeds along the positive real axis, instead of the contour A, until it reaches the critical point at $z = \frac{\omega}{\sqrt{2\left(\alpha - \frac{1}{2}\right)}}$, and then turns by $90°$ into the complex plane.

The path of steepest descent can be explicitly computed in our special case. The condition that the imaginary part be constant and equal to zero gives, with $z = x + iy$,

$$\Im m \left\{ \left(\alpha - \frac{1}{2} \right) (x^2 - y^2 + 2ixy)^2 - \omega^2 (x^2 - y^2 + 2ixy) \right\} = 0. \qquad (4.29)$$

Path of Steepest Descent

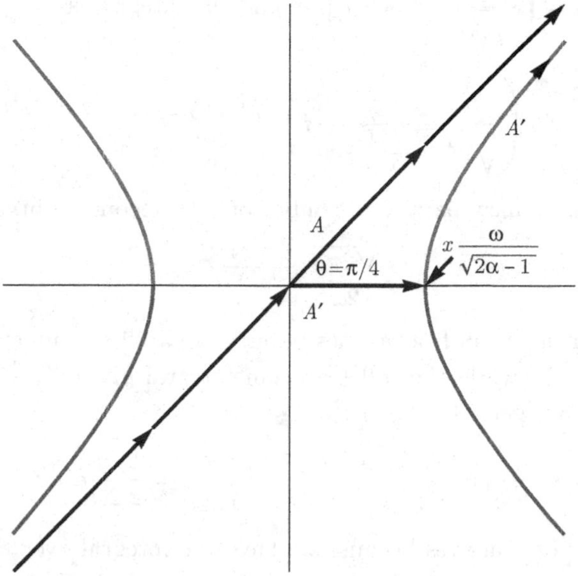

Figure 4.12. The integration contour along the path of steepest descent

Thus

$$\left(4\left(\alpha - \frac{1}{2}\right)(x^2 - y^2) - 2\omega^2\right)xy = 0 \tag{4.30}$$

$$\Rightarrow x = 0, \quad \text{or} \quad y = 0, \quad \text{or} \quad (2\alpha - 1)(x^2 - y^2) = \omega^2 \tag{4.31}$$

The first two solutions simply describe the x and y axes, the third solution corresponds to a hyperbola. Note that all of these curves intersect at 90° as we expect. The path of steepest descent, starting at the origin and going out to infinity at $\infty \times e^{i\frac{\pi}{4}}$, corresponds to the curve A', as depicted in Figure 4.12. Asymptotically the arcs of the hyperbola converge to the lines $y = \pm x$ which is the original contour A. The turn by 90° occurs at the critical point at $z = x = \frac{\omega}{\sqrt{(2\alpha - 1)}}$.

But now, where does the imaginary part to the integral come from? The integrand is always real, and the imaginary part of the function is always zero along the contour of steepest descent of the real part. It can only come from the integration measure dz when the contour follows the hyperbola in the complex plane. The contribution from $z = 0$ to $z = \frac{\omega}{\sqrt{(2\alpha - 1)}}$ along the real axis has no imaginary part, thus we are not interested in it. The integration along the hyperbola we perform in the Gaussian approximation about the critical point at $z = \frac{\omega}{\sqrt{(2\alpha - 1)}}$. We have $x = \sqrt{y^2 + \frac{\omega^2}{(2\alpha - 1)}}$, $dx = \frac{ydy}{\sqrt{y^2 + \frac{\omega^2}{(2\alpha - 1)}}}$, so $dz = dx + idy = \left(\dfrac{y}{\sqrt{y^2 + \frac{\omega^2}{(2\alpha - 1)}}} + i \right) dy$ and the integral is

$$\int_0^\infty dy \left(\frac{y}{\sqrt{y^2 + \frac{\omega^2}{(2\alpha - 1)}}} + i \right) e^{\frac{1}{\hbar} \left(-\frac{\omega^4}{(4\alpha - 2)} - 2\omega^2 y^2 + o(y^4) \right)}. \tag{4.32}$$

Therefore, the imaginary part comes only from the second term, and is given by

$$\frac{i}{2} \frac{\sqrt{2\pi\hbar}}{2\omega} e^{-\frac{\omega^4}{\hbar(4\alpha - 2)}}, \tag{4.33}$$

where the factor of $1/2$ in front comes because we are only integrating over half the Gaussian peak, while the full Gaussian integral gives $\frac{\sqrt{2\pi\hbar}}{2\omega}$. Then for our original integral we get

$$\Im\left\{ \int_{-\infty}^\infty \frac{dz}{\sqrt{2\pi\hbar}} e^{-\frac{1}{\hbar}\left(-\left(\alpha - \frac{1}{2}\right)z^4 + \omega^2 z^2 \right)} \right\}_{\alpha \to 1} = \frac{1}{2} \frac{1}{2\omega} e^{-\frac{\omega^4}{2\hbar}} \times 2, \tag{4.34}$$

where the factor of 2 arrives because we have the integral over the full contour of Figure 4.11, whereas the analysis above was only for half of the contour, the part in the first quadrant. We point out that the imaginary part of the integral simply corresponds to the formal expression of Equation (4.5) with $\lambda_{-1} \to |\lambda_{-1}|$.

4.6 Analysis for the General Case

Now, in the general case, we know what we must do. In order to do the path integral, we parametrize the space of all paths which satisfy the required boundary conditions for $z(\alpha, \tau = -\beta/2)$ and $z(\alpha, \tau = \beta/2)$ (β can be effectively taken to be ∞). We do this parametrization with one special, specific contour $z(\alpha, \tau)$ in the space of all paths, and augmented to this contour, we add the subspace of all paths orthogonal to this contour (which we will label as $z_\perp(\tau)$). To be very clear, a contour is not a path, it is a curve, itself parametrized by α, in the space of paths, where each point along the contour corresponds to a path $z(\alpha, \tau)$. The specific contour will contain two critical points

$$z(\alpha = 0, \tau) = \bar{z}(\tau) = 0, \tag{4.35}$$

which is the "instanton" corresponding to the particle just sitting on top of the unstable initial point in Figure 4.2 and never moving, and the point

$$z(\alpha = 1, \tau) = \bar{z}^{\text{bounce}}(\tau), \tag{4.36}$$

which corresponds to the instanton that we have called the "bounce". This contour is represented pictorially in Figure 4.13 while the corresponding action is represented in Figure 4.14. We will see that the actual paths that the contour passes through are unimportant except for the two critical points. We also insist

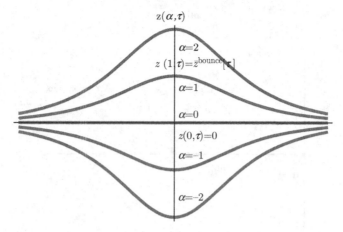

Figure 4.13. The path in function space as a function of α and τ

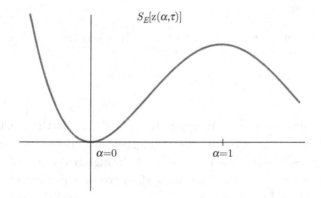

Figure 4.14. The Euclidean action as a function of α

that the "tangent" to the contour at $\alpha = 1$ corresponds to the negative energy mode

$$\frac{d}{d\alpha} z(\alpha, \tau)\big|_{\alpha \to 1} = z_{-1}(\tau). \qquad (4.37)$$

In this way, the orthogonal directions never contain a negative mode and the determinant (path integral over $\mathcal{D}z_\perp$) can be done in principle. We then write the path integral as a nested product of two integrals

$$\mathcal{N} \int \mathcal{D}z(\tau) e^{-\frac{1}{\hbar} S_E[z(\tau)]} = \mathcal{N} \int \frac{d\alpha}{\sqrt{2\pi\hbar}} \mathcal{D}z_\perp(\tau) e^{-\frac{1}{\hbar} S_E[z(\tau)]}. \qquad (4.38)$$

It is important to note that the path integral over the transverse directions is α-dependent. However, we will find that, since we are actually only interested in finding the imaginary part of the full integral, we will need to evaluate this transverse integral only at $\alpha = 1$.

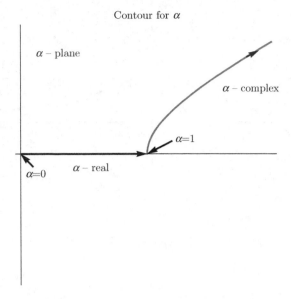

Figure 4.15. The contour for α from the origin, along the real axis and then jutting out into the complex plane at $90°$ at $\alpha = 1$

Now the integral over α is, however, ill-defined due to the existence of the negative mode at $\alpha = 1$. As $\alpha = 0$ is a critical point which is a local minimum, the action increases as we pass from $\alpha = 0$ to $\alpha = 1$ through real values of α. Hence the path in function space is defined as the path of steepest descent of $-S_E[z(\alpha, \tau]$, the exponent (up to the trivial factor of $1/\hbar$) in the integral Equation (4.38). But then we encounter the second critical point at $\alpha = 1$, which is a local maximum of the action, again for real α. The action behaves as depicted in Figure 4.14. Hence continuing the integral past $\alpha = 1$ to $\alpha = \infty$, it fails to converge and give a sensible answer. However, we are actually only trying to find an imaginary component of the original expression. If in fact we could integrate from $\alpha = 1$ on to $\alpha = \infty$, the expression would remain completely real. Thus we can only be content that we must define the integral via analytic continuation, since that is the only possible way that the integral could obtain an imaginary component.

This analytic continuation is expressed as a deformation of the contour of integration into the complex α-plane as we saw in the previous section. From $\alpha = 1$ we must follow along the contour of steepest descent of $-S_E[z(\alpha, \tau)]$. The important point, as we have seen, is that for an ordinary critical point, which is generic and that we assume, this corresponds to a $90°$ turn into the complex plane, as depicted in Figure 4.15. We start at $\alpha = 0$ and go till $\alpha = 1$ on the real α line, then we continue out at $90°$ into the complex α-plane following the line of steepest descent of $-S_E[z(\alpha, \tau)]$.

As before, the imaginary part only comes from the integration measure; the imaginary part of $-S_E[z(\alpha,\tau)]$ on the path of steepest descent is constant and hence always zero. This gives for the imaginary part of the path integral for the fluctuations about one bounce (using the notation $\mathcal{A.C.}$ to mean "analytic continuation"),

$$
\mathfrak{Im}\left\{\mathcal{A.C.}\left(\mathcal{N}\int\frac{d\alpha}{\sqrt{2\pi\hbar}}\mathcal{D}z_\perp(\tau)e^{-\frac{1}{\hbar}S_E[z(\alpha,\tau)]}\right)\right\}=
$$

$$
=\mathfrak{Im}\left\{\mathcal{A.C.}\int\frac{d\alpha}{\sqrt{2\pi\hbar}}e^{-\frac{1}{\hbar}\left(S_E[\bar{z}^{bounce}(\tau)]+\frac{1}{2}\frac{d^2}{d\alpha^2}S_E[z(\alpha,\tau)]|_{\alpha=1}(\alpha-1)^2+\cdots\right)}\right\}\times
$$

$$
\times\mathcal{N}\int\mathcal{D}z_\perp(\tau)e^{-\frac{1}{\hbar}\int d\tau' d\tau''\left(\frac{1}{2}\frac{\delta^2 S_E[z(\tau)]}{\delta z_\perp(\tau')\delta z_\perp(\tau'')}\Big|_{z(\tau)=\bar{z}^{bounce}(\tau)}\delta z_\perp(\tau')\delta z_\perp(\tau'')+\cdots\right)}
$$

$$
=\mathfrak{Im}\left\{\mathcal{A.C.}\int\frac{d\alpha}{\sqrt{2\pi\hbar}}e^{-\frac{1}{\hbar}\left(S_0+\frac{1}{2}\lambda_{-1}(\alpha-1)^2+\cdots\right)}\right\}\times
$$

$$
\times\left(\frac{S_0}{2\pi\hbar}\right)^{\frac{1}{2}}\beta\mathcal{N}\left(\det'\left[-\frac{d^2}{d\tau^2}+V''(\bar{z}^{bounce}(\tau))\right]\right)^{-\frac{1}{2}}
$$

$$
=\frac{1}{2}\times e^{-\frac{S_0}{\hbar}}\times\left(\frac{S_0}{2\pi\hbar}\right)^{\frac{1}{2}}\beta\times\frac{1}{\sqrt{|\lambda_{-1}|}}\times\mathcal{N}\left(\det'\left[-\frac{d^2}{d\tau^2}+V''(\bar{z}^{bounce}(\tau))\right]\right)^{-\frac{1}{2}},
$$

$$
\tag{4.39}
$$

where $\frac{d^2}{d\alpha^2}S_E[z(\alpha,\tau)]|_{\alpha=1}=\lambda_{-1}$ is the negative eigenvalue and $\det'$ now means the determinant is calculated excluding both the zero eigenvalue and the negative eigenvalue. In the last line of Equation (4.39), each factor separated by the $\times$ signs correspond, respectively, to: the factor of one-half since we are integrating over only half of the Gaussian peak, the exponential of minus the action of the bounce divided by $\hbar$, the factor corresponding to the Jacobian of the change of variables and the factor of β when we integrate over the position of the bounce rather than its translational zero mode, the factor of one over the square root of the magnitude of the negative eigenvalue which is the upshot of our analysis of the analytic continuation, and finally the primed determinant over the orthogonal directions in the space of paths where the negative mode and the zero mode are removed. Taking into account the contribution from the multi-bounce sector, the one-bounce contribution, including its imaginary part, just exponentiates as before.

Thus K, as defined in Equation (4.3), changes as $K\rightarrow\mathfrak{Re}\{K\}+i\mathfrak{Im}\{K\}$ and we find

$$
\mathfrak{Im}\{K\}=\frac{1}{2}\frac{1}{\sqrt{|\lambda_{-1}|}}\left(\frac{\det'\left[-\frac{d^2}{d\tau^2}+V''(\bar{z}^{bounce}(\tau))\right]}{\det\left[-\frac{d^2}{d\tau^2}+\omega^2\right]}\right)^{-\frac{1}{2}},
\tag{4.40}
$$

where we now understand the factor of $\frac{1}{2}$ as coming from integrating, in the Gaussian approximation, over just half of the saddle point descending into the

complex α-plane and the primed determinant is now understood to exclude both
the zero mode and the negative mode. Thus the original matrix element that we
wish to calculate, Equation (4.1), is obtained from an analytic continuation

$$\mathcal{A.C.}\left\{\langle z{=}0|e^{-\frac{\beta}{\hbar}\hat{h}(\hat{X},\hat{P})}|_{z=0}\rangle\right\}=\mathcal{A.C.}\left\{\mathcal{N}\int_{z\left(\pm\frac{\beta}{2}\right)=0}\mathcal{D}z(\tau)e^{-\frac{1}{\hbar}S_E[z(\tau)]}\right\}$$

$$=\left(\frac{\omega}{\pi\hbar}\right)^{\frac{1}{2}}e^{-\frac{\beta\omega}{2}}e^{\beta K}\left(\frac{S_0}{2\pi\hbar}\right)^{\frac{1}{2}}e^{-\frac{S_0}{\hbar}}+\cdots$$

$$=e^{-\beta(E_0+i\Gamma/2)/\hbar}\mathcal{A.C.}\left\{|\langle E_0|0\rangle|^2+\cdots\right\},(4.41)$$

where

$$K=\Re e(K)+i\frac{1}{2}\frac{1}{\sqrt{|\lambda_{-1}|}}\left(\frac{\det'\left(-\frac{d^2}{d\tau^2}+V''\left(\bar{z}^{bounce}(\tau)\right)\right)}{\det\left(-\frac{d^2}{d\tau^2}+\omega^2\right)}\right)^{-\frac{1}{2}}. \quad (4.42)$$

This yields the imaginary part to the energy, $i\Gamma/2$, with the width of the state

$$\Gamma=\hbar\left(\frac{S_0}{2\pi\hbar}\right)^{\frac{1}{2}}\frac{e^{-\frac{S_0}{\hbar}}}{\sqrt{|\lambda_{-1}|}}\left(\frac{\det'\left(-\frac{d^2}{d\tau^2}+V''\left(\bar{z}^{bounce}(\tau)\right)\right)}{\det\left(-\frac{d^2}{d\tau^2}+\omega^2\right)}\right)^{-\frac{1}{2}}. \quad (4.43)$$

5

Quantum Field Theory and the Path Integral

5.1 Preliminaries

We will consider the case of a classical scalar field theory and its quantization. Later in this book we will consider both vector and spinor fields. A classical scalar field $\phi(x^\mu)$ is a real-valued function of the coordinates of space and time. The meaning that it is a scalar field is that the value that the function takes is invariant under Lorentz transformations. All inertial observers measure the same value for the field at a given spacetime point.

$$\phi(x^\mu) = \phi'(x'^\mu) \tag{5.1}$$

where

$$x'^\mu = \Lambda^\mu{}_\nu x^\nu \tag{5.2}$$

with the standard notation $x^0 = t$ and x^i, $i = 1, 2, \cdots d$ are the spatial coordinates. The transformation matrix $\Lambda^\mu{}_\nu$ satisfies

$$\Lambda^\mu{}_\nu \eta^{\nu\sigma} \Lambda^\tau{}_\sigma = \eta^{\mu\tau} \tag{5.3}$$

with $\mathrm{diag}\,[\eta^{\nu\sigma}] = (1, -1, -1, \cdots)$ the usual Minkowski space metric, which is the defining condition for a Lorentz transformation. In general, an equation of motion for a classical scalar field is a non-linear partial differential equation. We will restrict ourselves to the case of second-order equations, then Lorentz invariance dictates the form

$$\partial_\nu \partial^\nu \phi(x^\mu) + V'\left(\phi(x^\mu)\right) = 0. \tag{5.4}$$

Written out, this equation is

$$\left(\frac{d^2}{dt^2} - \nabla^2\right) \phi(x^\mu) + V'\left(\phi(x^\mu)\right) = 0. \tag{5.5}$$

Such an equation comes from the variation of an action, $S[\phi(x^\mu)]$, which is a functional of the field $\phi(x^\mu)$, *i.e.*

$$\delta S[\phi(x^\mu)] = \left.\frac{S[\phi(x^\mu) + \epsilon\delta\phi(x^\mu)] - S[\phi(x^\mu)]}{\epsilon}\right|_{\epsilon=0} = 0$$

$$\forall\, \delta\phi(x^\mu) \Rightarrow \partial_\nu\partial^\nu\phi(x^\mu) + V'(\phi(x^\mu)) = 0. \tag{5.6}$$

Then we find the action giving rise to equations of motion, Equation (5.4), is given by

$$S[\phi(x^\mu)] = \int d^dx\left(\frac{1}{2}\partial_\nu\phi(x^\mu)\partial^\nu\phi(x^\mu) - V(\phi(x^\mu))\right) \equiv \int d^dx\mathcal{L} \tag{5.7}$$

where $\mathcal{L}$ is called the Lagrangian density. The kinetic energy is

$$T = \int d^{d-1}x\left(\frac{1}{2}\partial_t\phi(x^\mu)\partial_t\phi(x^\mu)\right) \tag{5.8}$$

while the potential energy is

$$V = \int d^{d-1}x\left(\frac{1}{2}\vec{\nabla}\phi(x^\mu)\cdot\vec{\nabla}\phi(x^\mu) + V(\phi(x^\mu))\right) \tag{5.9}$$

which define the Lagrangian as $L = T - V$ and the action is simply

$$S[\phi(x^\mu)] = \int dt\,(T - V). \tag{5.10}$$

This defines a dynamical system which is an exact analogy to the particle mechanical systems we have been considering in the previous chapters. There are just a few simple conceptual changes. The dynamical variable is a function of space, which evolves through time. For a mechanical system the variables were the positions of particles in space and these positions were evolving through time. Now the spatial coordinates x^i are not the positions of any particle. They are just parameters or labels, and they do not evolve in time. An important point to observe is that a dynamical variable which is a function of space, rather than a point in space, comprises an infinite number of degrees of freedom, in contradistinction to the case of particle mechanics where we typically consider only a finite number of particles. This is easy to make explicit by expanding the scalar field in terms of a fixed orthonormal basis of functions $\phi_n(x^i)$, $n = 0, 1, 2, \cdots$,

$$\phi(x^i, t) = \sum_{n=0}^{\infty} c_n(t)\phi_n(x^i). \tag{5.11}$$

We can thus exchange the dynamical field $\phi(x^i, t)$ for an infinite number of dynamical variables $\{c_n(t)\}_{n=0,1,\cdots,\infty}$.

This difference is the cause of almost all the problems that arise in the quantization of fields. We will proceed with the philosophy that these problems correspond to the extreme ultraviolet or infrared degrees of freedom, this philosophy perhaps to be justified only *a posteriori.* We plead ignorance as to what dynamics actually exist at extremely high energies and simply reject theories where the answers to questions involving processes at only low energies depend on the dynamics at very high energies! Furthermore, we invoke the principles of locality and causality, which stated simply means that configurations at the other end of the universe cannot affect the local dynamics here. In this way we consider only theories which are unaffected by cutting off the infrared degrees of freedom. Thus, effectively, we are interested in theories with an enormous but actually finite number of degrees of freedom, since we can cut the theory off in both the infrared and the ultraviolet. However, this number of degrees of freedom is assumed to be so huge that it is well-approximated by ∞, so long as that limit is sensible.

5.2 Canonical Quantization

5.2.1 Canonical Quantization of Particle Mechanics

The canonical quantization of fields proceeds formally as for particle mechanics. First we briefly review how it works for particle mechanics. We find the classical canonical variables p_i and q_i, $p_i = \frac{\partial L}{\partial \dot{q}_i}$ and the Hamiltonian $H = \sum_i p_i \dot{q}_i - L$. The equations of motion are:

$$\dot{q}_i = \{q_i, h(q_j, p_k)\}$$
$$\dot{p}_i = \{p_i, h(q_j, p_k)\} \tag{5.12}$$

where $\{\cdot, \cdot\}$ is the Poisson bracket,

$$\{A, B\} = \sum_i \frac{\partial A}{\partial q_i}\frac{\partial B}{\partial p_i} - \frac{\partial A}{\partial p_i}\frac{\partial B}{\partial q_i}. \tag{5.13}$$

Quantization proceeds with the replacement

$$\{A, B\} \rightarrow -\frac{i}{\hbar}\left[\hat{A}, \hat{B}\right] \tag{5.14}$$

yielding, for example, the canonical commutation relations:

$$[\hat{q}_i, \hat{p}_j] = i\hbar\delta_{i,j}. \tag{5.15}$$

All dynamical variables become operators, $O \rightarrow \hat{O}$, which act on vectors in a Hilbert space.

5.2.2 Canonical Quantization of Fields

Applying the above to the case of classical fields, we define the conjugate momenta in an analogous way,

$$\Pi(x^i, t) = \frac{\delta L}{\delta \dot{\phi}(x^i, t)}. \tag{5.16}$$

Then

$$H = \int d^{d-1}x \left(\Pi(x^i,t)\dot{\phi}(x^i,t) - \mathcal{L} \right)$$
$$= \int d^{d-1}x \left(\frac{1}{2}\Pi^2(x^i,t) + \frac{1}{2}\vec{\nabla}\phi(x^i,t)\cdot\vec{\nabla}\phi(x^i,t) + V(\phi(x^i,t)) \right). \quad (5.17)$$

The Poisson bracket is now given by (for local functions of $\phi(x^i), \Pi(x^i)$, we can dispense with the functional derivatives and just write partial derivatives, as they give the same answer)

$$\{A,B\} = \int d^{d-1}x \frac{\partial A}{\partial\phi(x^i,t)} \frac{\partial B}{\partial\Pi(x^i,t)} - \frac{\partial A}{\partial\Pi(x^i,t)} \frac{\partial B}{\partial\phi(x^i,t)} \quad (5.18)$$

which includes the fundamental Poisson brackets

$$\{\phi(x^i,t),\Pi(x^j,t)\} = \delta^{d-1}(x^i - x^j). \quad (5.19)$$

We impose the same quantization prescription as in the particle mechanics case, given by Equation (5.14). This yields the celebrated equal time canonical commutation relations

$$\left[\hat{\phi}(x^i,t),\hat{\Pi}(y^i,t) \right] = i\hbar\delta^{d-1}(x^i - y^i). \quad (5.20)$$

The (Heisenberg) equations of motion follow from the commutators:

$$i\hbar\frac{d}{dt}\hat{\phi}(x^i,t) = \left[\hat{\phi}(x^i,t),\hat{H} \right] \quad (5.21)$$

$$i\hbar\frac{d}{dt}\hat{\Pi}(x^i,t) = \left[\hat{\Pi}(x^i,t),\hat{H} \right] \quad (5.22)$$

There is a lot of mathematical subtlety in the definition of the product of the quantum field operators of a one-spacetime point which is required in the definition of the Lagrangian and Hamiltonian. Indeed, the quantum field operators that satisfy Equation (5.20) cannot be simple operators but in fact are operator-valued distributions. The operator products required to define the Lagrangian and the Hamiltonian are not straightforwardly well-defined. Canonical quantization can be made to work reasonably well for the case of linear field theories, for example see [107].

So far we have been considering the quantization in the Heisenberg picture. The variables are dynamical while the states are constant. We can equally well consider the quantization in the Schrödinger picture, with the transformation

$$\hat{\phi}(x^i,t) \to \hat{\phi}^S(x^i) = U(t)\hat{\phi}(x^i,t)U^\dagger(t)$$
$$\hat{\Pi}(x^i,t) \to \hat{\Pi}^S(x^i) = U(t)\hat{\Pi}(x^i,t)U^\dagger(t). \quad (5.23)$$

Then we find,

$$\frac{\partial\hat{\phi}^S(x^i)}{\partial t} = \frac{\partial\hat{\Pi}^S(x^i)}{\partial t} = 0, \quad (5.24)$$

i.e. the fundamental quantum fields in the Schrödinger picture are time-independent, if $U(t)$ satisfies

$$i\hbar\frac{d}{dt}U(t) = \hat{H}U(t). \tag{5.25}$$

The formal solution of this differential equation is $U(t) = e^{-it\hat{H}/\hbar}$. Evidently $\hat{H}$ commutes with $U(t)$. The corresponding transformation of the Hamiltonian yields

$$\hat{H} \to \hat{H}^S = U(t)\hat{H}U^\dagger(t) = \hat{H}. \tag{5.26}$$

This states that the Hamiltonian for time-independent problems does not depend on the representation. If we have an eigenstate of $\hat{H}$,

$$\hat{H}|\Psi\rangle = \mathcal{E}|\Psi\rangle \tag{5.27}$$

then

$$\hat{H}^S U(t)|\Psi\rangle = U(t)\hat{H}U^\dagger(t)U(t)|\Psi\rangle = \mathcal{E}U(t)|\Psi\rangle. \tag{5.28}$$

Thus

$$i\hbar\frac{d}{dt}\left(U(t)|\Psi\rangle\right) = U(t)\hat{H}|\Psi\rangle = \hat{H}^S\left(U(t)|\Psi\rangle\right) = \mathcal{E}\left(U(t)|\Psi\rangle\right) \tag{5.29}$$

which is just the Schrödinger equation.

5.3 Quantization via the Path Integral

Now the path integral for a quantum particle mechanics amplitude in Minkowski time, as given by Equation (2.42), yields

$$\langle y|e^{-\frac{iT\hat{h}(X.P)}{\hbar}}|x\rangle = \mathcal{N}\int_x^y \mathcal{D}z(t)e^{i\frac{S[z(t)]}{\hbar}}. \tag{5.30}$$

This formula was proven assuming nothing of the nature of the space in which x and y took their values. Typically they were coordinates in $\mathbf{R}^n$, but they could have been in any configuration space of unconstrained variables (with constraints additional terms can appear [76]). Actually we have

$$\langle q_f|e^{-\frac{iT\hat{H}(\hat{q}.\hat{p})}{\hbar}}|q_i\rangle = \mathcal{N}\int_{q_i}^{q_f} \mathcal{D}q(t)e^{i\frac{S[q(t)]}{\hbar}}, \tag{5.31}$$

where $q(t)$ could be any generalized coordinate, for example, an angular variable of a rotator or the radius of a bubble which changes its size.

Then, for quantum field theory, we simply let q take values in the space of configurations of a classical field. This gives

$$\langle \phi_f|e^{-\frac{iT\hat{H}(\hat{\phi},\hat{\Pi})}{\hbar}}|\phi_i\rangle = \mathcal{N}\int_{\phi_i}^{\phi_f} \mathcal{D}\phi(x^\mu)e^{-i\frac{S[\phi(x^\mu)]}{\hbar}}. \tag{5.32}$$

The states $|\phi_i\rangle$ and $|\phi_f\rangle$ correspond to a quantum field localized on the configurations $\phi_i(x^\mu)$ and $\phi_f(x^\mu)$, respectively. The states $|\phi\rangle$ are directly analogous to the states $|\vec{x}\rangle$ that we considered earlier in particle quantum mechanics. These were eigenstates of the (Schrödinger) position operator $\hat{\vec{X}}$

$$\hat{\vec{X}}|\vec{x}\rangle = \vec{x}|\vec{x}\rangle . \tag{5.33}$$

In that respect, the states $|\phi\rangle$ are taken to be eigenstates of the field operator

$$\hat{\phi}^S(x^i)|\phi\rangle = \phi(x^i)|\phi\rangle . \tag{5.34}$$

The states $|\phi\rangle$ are also improper vectors, as the states $|\vec{x}\rangle$ were, and true states are obtained by smearing with some profile function

$$|F\rangle = \int \mathcal{D}\phi F(\phi)|\phi\rangle \tag{5.35}$$

where $F(\phi)$ is a functionally square integrable functional. The inner product is defined by

$$\langle F|G\rangle = \int \mathcal{D}\phi F^*(\phi)G(\phi). \tag{5.36}$$

We call the Feynman path integral in this case the functional integral. It is a rather formal object in Minkowski space, but it can be used to generate the usual perturbative expansion of matrix elements, in a rather efficient manner. (Its analogue in Euclidean space, which we will use, can be rigorously defined in some cases.)

5.3.1 The Gaussian Functional Integral

We can essentially perform only one functional integral and that, too, not necessarily in closed form. This is the Gaussian functional integral. However, if we can do the Gaussian functional integral it is sufficient to generate the perturbative expansion. Consider the functional $W[J]$ of some external source field $J(x^\mu)$ defined by

$$W[J] = \mathcal{N} \int \mathcal{D}\phi\, e^{\frac{i}{\hbar}\int d^d x \left(\frac{1}{2}\partial_\mu\phi(x^i,t)\partial^\mu\phi(x^i,t) - \frac{1}{2}m^2\phi^2(x^i,t) - V\left(\phi(x^i,t)\right) + J(x^i,t)\phi(x^i,t)\right)}$$

$$\equiv \sum_{N=0}^{\infty} \frac{i^N}{\hbar^N N!} \int d^d x_1 \cdots d^d x_N J(x_1) \cdots J(x_N) G^N(x_1, \cdots, x_N), \tag{5.37}$$

where the integrations are done over all of spacetime and we impose the boundary conditions on the field the $\phi(x^\mu) \to 0$ as $|x^\mu| \to \infty$. Then the so-called N point Green functions of the theory are obtained via functional differentiation

$$G^N(x_1, \cdots, x_N) = \left(\frac{\hbar}{i}\right)^N \left(\frac{\delta}{\delta J(x_1)} \cdots \frac{\delta}{\delta J(x_N)}\right) W[J]\bigg|_{J=0}. \tag{5.38}$$

Correspondingly, $W[J]$ is called the generating functional since it can be used to generate all the Green functions of the theory. We will show that the $G^N(x_1, \cdots, x_N)$ corresponds in principle to the matrix elements

$$\langle 0| T \left(\hat{\phi}^H(x_1) \cdots \hat{\phi}^H(x_N) \right) |0\rangle \tag{5.39}$$

where the state $|0\rangle$ is the eigenstate of the Schrödinger field operator with eigenvalue $\phi(x^i) = 0$, *i.e.* $\hat{\phi}^S(x^i)|0\rangle = 0$.

For a Hamiltonian that depends on time, $\hat{H}^S(t)$, which is the case here with an arbitrary external source $J(x^\mu)$,

$$\hat{H}^S(t) = \hat{H}^0 + \hat{H}^{int.} \tag{5.40}$$

with

$$\hat{H}^0 = \int d^{d-1}x \left(\frac{1}{2}\hat{\Pi}(x^i)\hat{\Pi}(x^i) + \frac{1}{2}\vec{\nabla}\hat{\phi}^S(x^i) \cdot \vec{\nabla}\hat{\phi}^S(x^i) + V\left(\hat{\phi}^S(x^i)\right) \right) \tag{5.41}$$

and

$$\hat{H}^{int.}(t) = \int d^{d-1}x \left(J(x^i,t)\hat{\phi}^S(x^i) \right), \tag{5.42}$$

one can easily prove that the path integral gives rise to

$$\mathcal{N} \int \mathcal{D}z(t)e^{\frac{i}{\hbar}S[z(t)]} = \lim_{T \to \infty} \langle y| \mathbf{T} \left(e^{-\frac{i}{\hbar}\int_{-T/2}^{T/2} dt \hat{H}^S(t)} \right) |x\rangle \tag{5.43}$$

where $\mathbf{T}(A(t_1)B(t_2)) = \theta(t_1 - t_2)A(t_1)B(t_2) + \theta(t_2 - t_1)B(t_2)A(t_1)$, the usual time-ordered product. The time-ordered product here yields the limiting value of the (infinite) ordered product of infinitesimal unitary time translations over each of N infinitesimal time elements, $\epsilon = T/N$ between $-T/2$ and $T/2$, ordered so that the latest time occurs to the left

$$\mathbf{T} \left(e^{-\frac{i}{\hbar}\int_{-\frac{T}{2}}^{\frac{T}{2}} dt \hat{H}^S(t)} \right)$$
$$= \lim_{N \to \infty} e^{-\frac{i}{\hbar}\epsilon \hat{H}^S(\frac{T}{2})} e^{-\frac{i}{\hbar}\epsilon \hat{H}^S(\frac{T}{2}-\epsilon)} \cdots e^{-\frac{i}{\hbar}\epsilon \hat{H}^S(-\frac{T}{2}+2\epsilon)} e^{-\frac{i}{\hbar}\epsilon \hat{H}^S(-\frac{T}{2}+\epsilon)}. \tag{5.44}$$

The Hamiltonian being time-dependent because of the, in principle, time-dependent external source $J(x^i,t)$. The derivation of the path integral goes through as before by inserting a complete set of states between the infinitesimal unitary transformations. (There is a completely analogous expression for the case of the Euclidean path integral, where the time-ordering is replaced by Euclidean time-ordering, which is sometimes called path-ordering.) Thus we find with

$$W[J] = \lim_{T \to \infty} \langle 0| \mathbf{T} \left(e^{-\frac{i}{\hbar}\int_{-T/2}^{T/2} dt \hat{H}^S(t)} \right) |0\rangle \tag{5.45}$$

then

$$\frac{-i\hbar\delta}{\delta J(x_1)} \cdots \frac{-i\hbar\delta}{\delta J(x_N)} W[J]\bigg|_{J=0} =$$

$$= \frac{-i\hbar\delta}{\delta J(x_1)} \cdots \frac{-i\hbar\delta}{\delta J(x_N)} \langle 0 | \mathbf{T} \left(e^{-\frac{i}{\hbar}\int_{t_1}^{\infty} dt\hat{H}^S(t)} \right)$$

$$\times \mathbf{T} \left(e^{-\frac{i}{\hbar}\int_{t_2}^{t_1} dt\hat{H}^S(t)} \right) \cdots \mathbf{T} \left(e^{-\frac{i}{\hbar}\int_{-\infty}^{t_N} dt\hat{H}^S(t)} \right) | 0 \rangle \bigg|_{J=0}$$

$$\text{for } \quad t_1 > t_2 > \cdots > t_N$$

$$= \langle 0 | \mathbf{T} \left(e^{-\frac{i}{\hbar}\int_{t_1}^{\infty} dt\hat{H}^S(t)} \right) \hat{\phi}^S(x_1^i) \mathbf{T} \left(e^{-\frac{i}{\hbar}\int_{t_2}^{t_1} dt\hat{H}^S(t)} \right) \hat{\phi}^S(x_2^i) \cdots \hat{\phi}^S(x_N^i)$$

$$\times \mathbf{T} \left(e^{-\frac{i}{\hbar}\int_{-\infty}^{t_N} dt\hat{H}^S(t)} \right) | 0 \rangle \bigg|_{J=0}$$

$$= \langle 0 | \mathbf{T} \left(e^{-\frac{i}{\hbar}\int_{-\infty}^{\infty} dt\hat{H}^S(t)} \right) \mathbf{T} \left(e^{\frac{i}{\hbar}\int_{-\infty}^{t_1} dt\hat{H}^S(t)} \right) \hat{\phi}^S(x_1^i) \mathbf{T} \left(e^{-\frac{i}{\hbar}\int_{-\infty}^{t_1} dt\hat{H}^S(t)} \right)$$

$$\times \mathbf{T} \left(e^{\frac{i}{\hbar}\int_{-\infty}^{t_2} dt\hat{H}^S(t)} \right) \hat{\phi}^S(x_2^i) \cdots \hat{\phi}^S(x_N^i) \mathbf{T} \left(e^{-\frac{i}{\hbar}\int_{-\infty}^{t_N} dt\hat{H}^S(t)} \right) | 0 \rangle \bigg|_{J=0}$$

$$= \langle 0 | \mathbf{T} \left(e^{-\frac{i}{\hbar}\int_{-\infty}^{\infty} dt\hat{H}^S(t)} \right) \hat{\phi}^H(x_1^\mu) \hat{\phi}^H(x_2^\mu) \cdots \hat{\phi}^H(x_N^\mu) | 0 \rangle \bigg|_{J=0}$$

$$\to \langle E=0 | \mathbf{T} \left(\hat{\phi}^H(x_1^\mu) \hat{\phi}^H(x_2^\mu) \cdots \hat{\phi}^H(x_N^\mu) \right) | E=0 \rangle \bigg|_{J=0}, \tag{5.46}$$

where we have explicitly written the Heisenberg fields as $\hat{\phi}^H(x^\mu) = \mathbf{T} \left(e^{\frac{i}{\hbar}\int_{-\infty}^{t} dt' \hat{H}^S(t')} \right) \hat{\phi}^S(x^i) \mathbf{T} \left(e^{-\frac{i}{\hbar}\int_{-\infty}^{t} dt' \hat{H}^S(t')} \right)$ while the Schrödinger operators are defined with respect to $t = -\infty$. Here $|0\rangle$ still corresponds to the state with $\phi(x) = 0$ while the state $|E=0\rangle$ corresponds to the true zero-energy vacuum state. However, the last identification in Equation (5.46) requires explanation as it is not exactly the same as Equation (5.39). As we will see, once we define the functional integral more carefully, instead of computing the matrix element in Equation (5.39), the functional integral projects uniformly onto that which corresponds to the matrix element in the state of zero energy, the vacuum state. At the present juncture the definition of the functional integration is extremely formal, and neither the operator-valued matrix element in Equation (5.39) nor its functional integral representation exist.

If we nevertheless continue formally, we find

$$W[J] = \mathcal{N} \int \mathcal{D}\phi \, e^{\frac{-i}{\hbar}\int d^d x V\left(-i\hbar\frac{\delta}{\delta J(x)}\right)} \times$$

$$\times e^{\frac{i}{\hbar}\int d^d x \left(\frac{1}{2}\partial_\mu\phi(x^i,t)\partial^\mu\phi(x^i,t) - \frac{1}{2}m^2\phi^2(x^i,t) + J(x^i,t)\phi(x^i,t)\right)}$$

$$= e^{\frac{-i}{\hbar}\int d^d x V\left(-i\hbar\frac{\delta}{\delta J(x)}\right)} W^0[J]. \tag{5.47}$$

$W^0[J]$ is a Gaussian functional integral, which we can explicitly perform. We use the formula, which as written is only formal but becomes valid if defined via an appropriate analytic continuation

$$\int_{-\infty}^{\infty} \frac{dx}{\sqrt{2\pi}} e^{i\frac{1}{2}(ax^2+2bx)} = \int_{-\infty}^{\infty} \frac{dx}{\sqrt{2\pi}} e^{i\frac{a}{2}\left(x-\frac{b}{a}\right)^2} e^{-i\frac{1}{2}b\left(\frac{1}{a}\right)b}$$

$$= \frac{1}{\sqrt{-ia}} e^{-i\frac{1}{2}b\left(\frac{1}{a}\right)b} \tag{5.48}$$

which generalizes to

$$\int \frac{d^n x}{(2\pi)^{\frac{n}{2}}} e^{i\frac{1}{2}\left((\vec{x},A\cdot\vec{x})+2(\vec{b},\vec{x})\right)} = \left(\det(-iA)\right)^{-\frac{1}{2}} e^{-i\frac{1}{2}\left((\vec{b},A^{-1}\cdot\vec{b})\right)} \tag{5.49}$$

for finite dimensional matrices. Boldly generalizing to the infinite dimensional case, for $W^0[J]$ we find, with $A \to -\left(\partial_\mu \partial^\mu + m^2\right)$ and $b \to J$ (and absorbing an infinite product of i's into the normalization constant),

$$W^0[J] = \frac{\mathcal{N}}{\sqrt{\det\left(\partial_\mu \partial^\mu + m^2\right)}} e^{-\frac{i}{2}\int d^d x\, d^d y (J(x)\,\langle x|\frac{1}{-\left(\partial_\mu\partial^\mu+m^2\right)}|y\rangle\,J(y))} \tag{5.50}$$

5.3.2 The Propagator

It only remains to calculate

$$\langle x| \frac{1}{-\left(\partial_\mu \partial^\mu + m^2\right)} |y\rangle = \int \frac{d^d k}{(2\pi)^d} e^{-ik_\mu(x-y)^\mu} \frac{1}{k_\mu k^\mu - m^2}. \tag{5.51}$$

We seem to be on the right path to defining the functional integral; however, we come up against another problem: this Green function is ambiguous. This problem is only solved via analytic continuation. In the Fourier representation, for example, there are poles in the k_0 integration at $k_0 = \pm\sqrt{|\vec{k}|^2 + m^2}$. We cannot integrate through the poles, we must provide a prescription for integrating around them. Such a prescription translates directly into fixing the asymptotic boundary condition on the solutions of the problem, for ϕ

$$\left(\partial_\mu \partial^\mu + m^2\right)\phi = J. \tag{5.52}$$

Clearly any solution for ϕ is ambiguous up to a solution of the homogeneous equation

$$\left(\partial_\mu \partial^\mu + m^2\right)\phi_0 = 0. \tag{5.53}$$

Correspondingly, the Green function to Equation (5.52) is also ambiguous by the addition of an arbitrary solution of the homogeneous equation. The asymptotic boundary conditions on ϕ fix the Green function. These boundary conditions are equivalent to giving the pole prescription.

5.3.3 Analytic Continuation to Euclidean Time

The existence of homogeneous solutions corresponds to zero modes in the operator $A = -(\partial_\mu \partial^\mu + m^2)$; hence, the original integral was ill-defined. The problem can be traced back to the matrix element

$$W[J] = \lim_{T \to \infty} \langle 0| \mathbf{T} \left(e^{-\frac{i}{\hbar} \int_{-T/2}^{T/2} dt \hat{H}^S(t)} \right) |0\rangle . \tag{5.54}$$

The operator in the matrix element can be written, for an arbitrary future time t,

$$\mathbf{T} \left(e^{-\frac{i}{\hbar} \int_{-T/2}^{t} dt' \hat{H}^S(t')} \right) = e^{-\frac{i}{\hbar} \int_{-T/2}^{t} dt' \hat{H}^0} e^{\frac{i}{\hbar} \int_{-T/2}^{t} dt' \hat{H}^0} \mathbf{T} \left(e^{-\frac{i}{\hbar} \int_{-T/2}^{t} dt' \hat{H}^S(t')} \right)$$

$$\equiv e^{-\frac{i}{\hbar} \int_{-T/2}^{t} dt' \hat{H}^0} U(t, -T/2). \tag{5.55}$$

Then $U(t, -T/2)$ satisfies the differential equation

$$i\hbar \frac{\partial U(t, -T/2)}{\partial t} = e^{-\frac{i}{\hbar} \int_{-T/2}^{t} dt' \hat{H}^0} \hat{H}^{int.}(t) e^{\frac{i}{\hbar} \int_{-T/2}^{t} dt' \hat{H}^0} U(t, -T/2)$$

$$\equiv \hat{H}^I(t) U(t, -T/2) \tag{5.56}$$

where

$$\hat{H}^I(t) = \int d^{d-1}x J(x^i, t) e^{-\frac{i}{\hbar} \int_{-T/2}^{t} dt' \hat{H}^0} \hat{\phi}^S(x^i) e^{\frac{i}{\hbar} \int_{-T/2}^{t} dt' \hat{H}^0}$$

$$\equiv \int d^{d-1}x J(x^i, t) \hat{\phi}^I(x^i, t) \tag{5.57}$$

defines the interaction representation Hamiltonian and the interaction representation field $\hat{\phi}^I(x^i, t)$. The solution of the differential Equation (5.56) is unique with boundary condition $U(-T/2, -T/2) = 1$ and given by

$$U(t, -T/2) = \mathbf{T} \left(e^{-\frac{i}{\hbar} \int_{-T/2}^{t} dt' \hat{H}^I(t')} \right). \tag{5.58}$$

Thus

$$W[J] = \lim_{T \to \infty} \langle 0| e^{-\frac{i}{\hbar} \int_{-T/2}^{T/2} dt' \hat{H}^0} \mathbf{T} \left(e^{-\frac{i}{\hbar} \int_{-T/2}^{T/2} dt' \hat{H}^I(t')} \right) |0\rangle. \tag{5.59}$$

The state $|0\rangle$ corresponds to an eigenstate of the Schrödinger field operator with the eigenvalue zero, and is not an energy eigenstate of the Hamiltonian, hence

$$|0\rangle = \sum_E C_E |E\rangle \tag{5.60}$$

where

$$\hat{H}^0 |E\rangle = E |E\rangle . \tag{5.61}$$

Then the matrix element in Equation (5.59) is given by

$$W[J] = \lim_{T \to \infty} \sum_{E, E'} e^{-\frac{i}{\hbar} TE'} C_{E'}^* C_E \langle E'| \mathbf{T} \left(e^{-\frac{i}{\hbar} \int_{-T/2}^{T/2} dt' \hat{H}^I(t')} \right) |E\rangle \tag{5.62}$$

This expression is generally not well-defined. The infinite phases give an ever-oscillatory contribution which does not exist in the limit $T \to \infty$. We are in fact interested in the matrix element and its various moments which give rise to the Green functions, as $J \to 0$. Even in this limit, we get that $W[J=0]$ is ill-defined; if any of the $C_E \neq 0$ for any $E \neq 0$, then

$$W[J=0] \to \sum_E e^{-\frac{i}{\hbar}(\infty)E}|C_E|^2. \tag{5.63}$$

Thus, somehow we must project onto the ground state, defined to have $E = 0$. This would happen if we can add a negative imaginary part to E. Equivalently, if we rotate

$$t \to \tau = -it \quad d^d x \to -i d^d x \tag{5.64}$$

the action goes to

$$S \to i S^E = i \int d^d x \left(\frac{1}{2}\left(\partial_\mu \phi \partial_\mu \phi + m^2\right) + V(\phi) - J\phi \right), \tag{5.65}$$

and the matrix element is

$$\langle 0|T\left(e^{-\frac{1}{\hbar}\int_{-\infty}^{\infty} dt \hat{H}(t)}\right)|0\rangle \sim \langle E=0|T\left(e^{-\frac{1}{\hbar}\int_{-\infty}^{\infty} dt \hat{H}(t)}\right)|E=0\rangle. \tag{5.66}$$

$|E=0\rangle$ is the zero-energy vacuum state of the theory with $J = 0$. Then the functional integral gives

$$\mathcal{N}'\int \mathcal{D}\phi\, e^{\frac{-S^E}{\hbar}} = \langle E=0|T\left(e^{-\frac{1}{\hbar}\int_{-\infty}^{\infty} dt \hat{H}(t)}\right)|E=0\rangle \tag{5.67}$$

and the Minkowski space functional integral is defined by the analytic continuation of this object to real times.

The rotation $t \to -i\tau$ yields the Euclidean operator $\left(-\partial_\mu \partial_\mu + m^2\right)\phi$ which has no zero modes,

$$\left(-\partial_\mu \partial_\mu + m^2\right)\phi = 0 \Rightarrow \phi = 0. \tag{5.68}$$

Thus

$$\mathcal{N}'\int \mathcal{D}\phi\, e^{\frac{-S^E}{\hbar}}$$

$$= \frac{\mathcal{N}'}{\sqrt{\det\left(-\partial_\mu \partial_\mu + m^2\right)}}\, e^{-\int d^d x V\left(\hbar \frac{\delta}{\delta J(x)}\right)}\, e^{-\int d^d x\, d^d y\left(J(x)\,\langle x|\frac{1}{\left(-\partial_\mu \partial_\mu + m^2\right)}|y\rangle\, J(y)\right)},$$

$$\tag{5.69}$$

where

$$\langle x|\frac{1}{\left(-\partial_\mu \partial_\mu + m^2\right)}|y\rangle = \int \frac{d^d k}{(2\pi)^d} e^{ik_\mu(x-y)_\mu}\frac{1}{(k_\mu k_\mu + m^2)} \tag{5.70}$$

which is now well-defined.

The analytic continuation back to Minkowski space $(x_0 - y_0) \to i(x_0 - y_0)$ gives the Minkowski Green function with the "correct" Feynman prescription at the poles

$$\langle x | \frac{1}{-(\partial_\mu \partial^\mu + m^2)} | y \rangle = \int \frac{d^d k}{(2\pi)^d} e^{-ik_\mu(x-y)^\mu} \frac{1}{(k_\mu k^\mu - m^2 + i\epsilon)}. \tag{5.71}$$

Thus once the Minkowski space functional integral is defined via the analytic continuation back from Euclidean space, it clearly gives the vacuum expectation value

$$W[J] = \langle E = 0 | \mathbf{T} \left(e^{-\frac{i}{\hbar} \int_{-\infty}^{\infty} dt \hat{H}^S(t)} \right) | E = 0 \rangle$$

$$= e^{\frac{-i}{\hbar} \int d^d x V \left(-i\hbar \frac{\delta}{\delta J(x)} \right)} e^{-\frac{i}{2} \int d^d k \frac{\tilde{J}(k)\tilde{J}(-k)}{\left(k_\mu k^\mu - m^2 + i\epsilon \right)}}. \tag{5.72}$$

For example, the Feynman propagator is obtained from

$$\Delta_F(x_1, x_2) = \langle E = 0 | \mathbf{T} \left(\phi(x_1)\phi(x_2) \right) | E = 0 \rangle = \int \frac{d^d k}{(2\pi)^d} \frac{e^{-ik_\mu(x-y)^\mu}}{k_\mu k^\mu - m^2 + i\epsilon}. \tag{5.73}$$

6
Decay of the False Vacuum

In this chapter we give the first example of an application of the methods we have learned so far. We will apply the methods of instantons to the problem of vacuum instability in quantum field theory. We consider a scalar field governed by a Lagrangian of the form

$$L = \int d^3x \, \frac{1}{2} \partial_\mu \phi(x) \partial^\mu \phi(x) - V(\phi(x)). \tag{6.1}$$

The potential $V(\phi(x))$ for $\phi(x) = \phi$, a constant independent of the spacetime coordinates, has the form represented by the graph in Figure 6.1. There are two minima, a global minimum at ϕ_- and a local minimum at ϕ_+. Classically the configurations $\phi(x) = \phi_\pm$ are stable. The energy is given by the functional

$$E = \int d^3x \, \frac{1}{2} \dot{\phi}(x)^2 + \frac{1}{2} \vec{\nabla}\phi(x) \cdot \vec{\nabla}\phi(x) + V(\phi(x)). \tag{6.2}$$

When $\phi(x)$ is a constant the first two terms, which are positive semi-definite, give zero contribution, thus the energy comes solely from the potential term. The potential is minimized and normally adjusted by adding a constant to make it vanish at the global minimum $\phi(x) = \phi_-$, so normally the energy of this classical configuration is zero. At $\phi(x) = \phi_+$ the potential is in a local minimum, however, and then the value of the potential is finite and the total energy is divergent. The divergence is proportional to the volume. However, the physically important quantity is not the total energy but the energy density, which is given directly by the potential. Then the energy density difference between the two classical ground states is finite. ϕ_+ is the false vacuum while ϕ_- is the true vacuum. The false vacuum is unstable while the true vacuum is stable.

We will, however, adjust the zero of the potential not in the normal way but as depicted in Figure 6.1, by adding a constant, so that the energy density of the false vacuum state is zero. Such a redefinition cannot affect the local physics. Then we will calculate the decay of the false vacuum to the true vacuum per

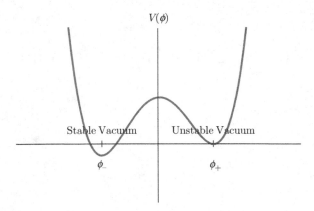

Figure 6.1. The potential giving rise to a false vacuum

unit time and per unit volume, $\frac{\Gamma}{V}$. We will find an expression of the form

$$\frac{\Gamma}{V} = Ae^{-\frac{B}{\hbar}}(1 + 0(\hbar)) \tag{6.3}$$

in the semi-classical limit. This form is exactly that which we have seen for decays via tunnelling. B will correspond to the classical action for a critical configuration while A will come from the quantum considerations. We proceed in an analogous fashion to the problem we considered in quantum mechanics. We wish to define the analytic continuation of the matrix element

$$A.C.\{\langle\phi_+|e^{-\frac{\beta\hat{H}}{\hbar}}|\phi_+\rangle\} \tag{6.4}$$

from a potential for which the vacuum constructed at ϕ_+ is stable to the potential we are considering. As we have seen, the analytic continuation instructs us on how to deal with Gaussian integrals over fluctuations about a critical configuration which correspond to negative frequencies.

6.1 The Bounce Instanton Solution

Otherwise we proceed in the usual way with the semi-classical analysis of the Euclidean functional integral. We look at

$$\mathcal{N}\int\mathcal{D}\phi e^{-\frac{S_E[\phi(x)]}{\hbar}} \tag{6.5}$$

with the boundary conditions $\phi\left(\tau = \pm\frac{\beta}{2}\right) = \phi_+$. Here

$$S_E[\phi(x)] = \int dt d^3x \left(\frac{1}{2}\partial_\mu\phi(x)\partial_\mu\phi(x) + V(\phi(x))\right) \tag{6.6}$$

with the equation of motion corresponding to

$$\frac{\delta S_E[\phi(x)]}{\delta\phi} = -\partial_\mu\partial_\mu\phi(x) + V'(\phi(x)) = 0. \tag{6.7}$$

Here we use the Euclidean metric. This equation is exactly the equation of motion for the scalar field in minus the potential. We take the boundary conditions for the case $\beta = \infty$

$$\lim_{\tau \to \pm \infty} \phi(\vec{x}, \tau) = \phi_+ \tag{6.8}$$

and we add the condition

$$\partial_\tau \phi(\vec{r}, \tau = 0) = 0, \tag{6.9}$$

which determines the Euclidean time at the classical turning point. This time of the classical turning point is completely at our disposal for the case $\beta = \infty$. The condition that the classical action should be finite gives

$$\lim_{|\vec{x}| \to \infty} \phi(\vec{x}, \tau) = \phi_+. \tag{6.10}$$

We assume a form that is $O(4)$-invariant

$$\phi(\vec{x}, \tau) = \phi\left((|\vec{x}|^2 + \tau^2)^{\frac{1}{2}}\right). \tag{6.11}$$

The equation of motion becomes, with $\rho = (|\vec{x}|^2 + \tau^2)^{\frac{1}{2}}$

$$\frac{d^2}{d\rho^2}\phi + \frac{3}{\rho}\frac{d}{d\rho}\phi - V'(\phi) = 0. \tag{6.12}$$

The action is

$$S_E[\phi] = 2\pi^2 \int_0^\infty d\rho \rho^3 \left(\frac{1}{2}\left(\frac{d\phi}{d\rho}\right)^2 + V(\phi)\right) \tag{6.13}$$

with the boundary conditions $\frac{d\phi}{d\rho}\Big|_{\rho=0} = 0$ and $\lim_{\rho \to \infty} \phi(\rho) = \phi_+$. The first condition avoids a singularity at $\rho = 0$ while the second comprises all of the asymptotic boundary conditions.

A rigorous proof of the existence of a solution and that it is the minimum action solution is given by Coleman, Glaser and Martin [34], but we shall be content with the following argument due to Coleman [31]. The equation of motion (6.12) can be interpreted as that for a particle with "position" ϕ moving in "time" ρ. The particle is subject to a force, $-V'(\phi)$, and a frictional force with a "time"-dependent Stokes coefficient of friction $\frac{3}{\rho}$. The equation of motion for a particle in a potential with Stokes coefficient of friction μ is

$$\frac{d^2}{d\rho^2}\phi(\rho) + \mu\frac{d}{d\rho}\phi(\rho) + V'(\phi(\rho)) = 0. \tag{6.14}$$

The solution in the absence of a potential, $V'(\phi(\rho)) = 0$, is simply $\phi(\rho) = a - be^{-\mu\rho}$ for arbitrary constants a, b, with a related to the initial position and b related to the initial velocity. This solution confirms that motion with friction without external forces will come to rest exponentially fast. In the present case μ depends on ρ.

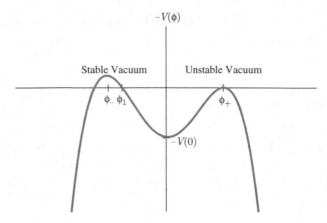

Figure 6.2. The reversed potential and effective dynamical problem

We can prove the existence of a solution satisfying our boundary conditions by the following continuity argument. We must show that there exists an initial point ϕ_0 from which the particle can start at $\rho = 0$ and achieve $\phi = \phi_+$ at $\rho = \infty$. The potential is reversed to give $-V(\phi)$ as depicted in Figure 6.2, and ϕ_1 is defined as the point at which the potential crosses zero. If $\phi_0 > \phi_1$, $\phi(\rho)$ will never reach ϕ_+ even as $\rho \to \infty$ starting with zero velocity. If, however, $\phi_- < \phi_0 < \phi_1$, and ϕ_0 is sufficiently close to ϕ_-, $\phi(\rho)$ will surpass ϕ_+ at some finite time. We can understand this intuitively; if ϕ_0 is arbitrarily close to ϕ_-, the particle will roll off this potential hill arbitrarily slowly. We can make this time so long that the coefficient of friction, $\frac{3}{\rho}$, becomes negligibly small. Then the particle will roll off and eventually climb the hill at ϕ_+ and even surpass ϕ_+ since it is now a conservative system. Indeed, for $\phi(\rho)$ close to ϕ_- we can linearize the equation of motion,

$$\left(\frac{d^2}{d\rho^2}\phi + \frac{3}{\rho}\frac{d}{d\rho}\phi - \omega^2 \right)(\phi(\rho) - \phi_-) = 0, \qquad (6.15)$$

which has the solution

$$(\phi(\rho) - \phi_-) = 2\,(\phi(0) - \phi_-)\,\frac{I_1(\omega\rho)}{\omega\rho} \qquad (6.16)$$

where ω^2 is $V''(\phi_-)$, and $I_1(\omega\rho)$ is the modified Bessel function of the first kind. This implies that for $(\phi(0) - \phi_-)$ sufficiently small, $(\phi(\rho_0) - \phi_-)$ can be kept arbitrarily small such that $\phi(\rho_0) < \phi_1$ so that the potential energy remains positive, where ρ_0 is determined by the condition that the subsequent energy lost to the friction term is negligible. Once the friction term becomes negligible, the system is conservative and, since at ρ_0 the potential energy is positive, the particle will clearly surpass ϕ_+ at finite ρ^f. A measure of the energy lost in the friction is obtained by the integral

$$\int_{\rho_0}^{\rho^f} d\rho \, \frac{3}{\rho} \frac{d}{d\rho} \phi < \frac{3}{\rho_0} \int_{\rho_0}^{\rho^f} d\rho \, \frac{d}{d\rho} \phi \approx \frac{3}{\rho_0} \int_{\phi_-}^{\phi_+} d\phi = \frac{3}{\rho_0} (\phi_+ - \phi_-). \qquad (6.17)$$

Thus we choose ρ_0 large enough so that this energy is negligible in comparison to the energy scales that drive the dynamics, say $V(0)$:

$$\frac{3}{\rho_0} (\phi_+ - \phi_-) \ll V(0). \qquad (6.18)$$

Then finally we conclude that there must exist some intermediate ϕ_0 from which $\phi(\rho)$ will attain ϕ_+ exactly as $\rho \to \infty$. This implies the existence of a solution of the form we desire.

6.2 The Thin-Wall Approximation

We can go much further with the assumption that the energy density difference between the two vacua is small.

$$V(\phi) = U(\phi) + \frac{\epsilon}{2a} (\phi - a) \qquad (6.19)$$

with $U(\phi) = U(-\phi)$, $U'(\pm a) = 0$, $U''(\pm a) = \omega^2$ and ϵ is arbitrarily small, as depicted in Figure 6.3. We can calculate the action for the bounce to first order in ϵ. The reversed potential is given in Figure 6.4. At $\rho = 0$ the field is very close to $-a$, it stays there for a very long "time", and then it rolls relatively quickly through the minimum of the reversed potential, up to the hill at $\phi = +a$ since now the friction is negligible. It achieves $\phi = +a$ only as $\rho \to \infty$. The bounce is like a large four-ball of radius R, in Euclidean space, of true vacuum, separated by a thin wall, from the false vacuum without.

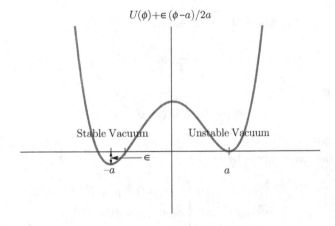

Figure 6.3. The symmetric potential with a small asymmetry

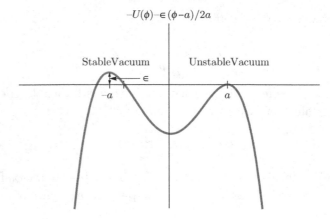

Figure 6.4. The reversed symmetric potential with a small asymmetry

For ρ near R, if we drop the friction term we obtain the equation of motion (to zero order in ϵ)

$$\frac{d^2}{d\rho^2}\phi - U'(\phi) = 0. \tag{6.20}$$

This is exactly the same equation that we have studied in the double-well problem of Chapter (3). The instanton solution interpolates from one well to the other as in Figure 3.3. It is given in this region, which is near the wall, approximately by the equation

$$\rho - R = \int_0^{\tilde{\phi}(\rho)} \frac{d\phi}{\sqrt{2U(\phi(\rho))}}. \tag{6.21}$$

For large $|\rho - R|$, the solution is given by

$$\tilde{\phi}(\rho) = \pm\left(a - \alpha e^{-\omega|\rho - R|}\right). \tag{6.22}$$

For example, for the choice of the potential

$$U(\phi) = \frac{\lambda}{4}\left(\phi^2 - a^2\right)^2 \tag{6.23}$$

the solution is

$$\tilde{\phi}(\rho) = a\tanh\left(\omega\left(\rho - R\right)\right) \tag{6.24}$$

with $\alpha = 2a$ and $\omega^2 = 2\lambda a^2$.

Thus our bounce is given by

$$\phi_{bounce}(\rho) = \begin{cases} -a & 0 < \rho \ll R \\ \tilde{\phi}(\rho) & \rho \approx R \\ a & \rho \gg R \end{cases}. \tag{6.25}$$

To find R we do a variational calculation in R.

$$S_E[\phi_{bounce}] = 2\pi^2 \int_0^{R-\Delta} d\rho\rho^3(-\epsilon) + 2\pi^2 \int_{R-\Delta}^{R+\Delta} d\rho\rho^3 \left(\frac{1}{2}\left(\frac{d\tilde{\phi}(\rho)}{d\rho}\right)^2 + U\left(\tilde{\phi}(\rho)\right)\right) +$$

$$+ 2\pi^2 \int_{R+\Delta}^{\infty} d\rho\rho^3\,(0)$$

$$\approx -\frac{1}{2}\pi^2 R^4\epsilon + 2\pi^2 R^3 S_1, \quad \text{for} \quad R \gg \Delta, \quad (6.26)$$

where S_1 is the action for the one-dimensional instanton $\tilde{\phi}(\rho)$ calculated in Equation (3.27) which is independent of R (we call it S_1 here to emphasize that it is the one-dimensional instanton action),

$$S_1 \approx \int_{-\infty}^{\infty} dx \left(\frac{1}{2}\frac{d^2}{dx^2}\tilde{\phi}(x) + U\left(\tilde{\phi}(x)\right)\right)$$

$$= \int_{-a}^{a} d\phi \sqrt{2U(\phi)}. \quad (6.27)$$

$S_E(R)$ should be stationary under variations of R,

$$\frac{dS_E(R)}{dR} = -2\pi^2 R^2\epsilon + 6\pi^2 R^2 S_1 = 0 \quad (6.28)$$

hence

$$R = \frac{3S_1}{\epsilon}. \quad (6.29)$$

This confirms our expectation that $R \to \infty$ as $\epsilon \to \infty$. Finally, the Euclidean action for the bounce is

$$S_E^{bounce} = \frac{1}{2}\pi^2 \left(\frac{3S_1}{\epsilon}\right)^4 \epsilon + 2\pi^2 \left(\frac{3S_1}{\epsilon}\right)^3 S_1 = \frac{27\pi^2 S_1^4}{2\epsilon}\left(1 + o(\epsilon)^3\right). \quad (6.30)$$

6.3   The Fluctuation Determinant

The calculation of the coefficient A of Equation (6.3) is not so straightforward, even approximately. It is given by the determinant of the operator governing small fluctuations about the bounce.

$$\langle\phi_+|e^{-\frac{\beta\hat{H}}{\hbar}}|\phi_+\rangle = e^{-\frac{S_E^{bounce}}{\hbar}}\mathcal{N}\det^{-\frac{1}{2}}\left(-\frac{d^2}{d\tau^2} - \nabla^2 + V''(\phi_{bounce})\right). \quad (6.31)$$

When we attempt to evaluate the determinant we encounter the same problems that we have already seen in particle quantum mechanics: non-positive frequencies for the spectrum of Gaussian fluctuations.

Zero modes come from invariance of the action under translations. We can translate in space and Euclidean time which gives us four independent zero modes (we write ϕ_{bounce} as ϕ_b for the sake of brevity)

$$\phi_\mu(\vec{x},\tau) = N\frac{\partial}{\partial x^\mu}\phi_b(\vec{x},\tau). \quad (6.32)$$

Zero modes correspond to continuous degeneracies of the critical point of the Euclidean action. Here they correspond to the arbitrariness of the location of the centre of the bounce in Euclidean $\mathbf{R}^4$ which is actually $\mathbb{R}$. We cannot integrate over these directions in the integrations over fluctuations about the bounce; however, we can equivalently integrate over the position of the bounce in $\mathbf{R}^4$ which is actually $\mathbb{R}$. This gives a (divergent) factor of βV and a Jacobian corresponding to the change of integration variable from the fluctuation degree of freedom to the coordinate giving the position of the bounce. The Jacobian factor is of the same type as before, indeed,

$$\delta\phi = \frac{1}{N}\frac{\partial}{\partial x_0^\mu}\phi_b\left((x-x_0)^\nu\right)dc^\mu \tag{6.33}$$

for an infinitesimal change dc_μ of the coefficient of the Gaussian fluctuation along the normalized zero-mode direction, $\frac{1}{N}\frac{\partial}{\partial x_0^\mu}\phi_b\left((x-x_0)^\nu\right)$, while

$$\delta\phi = \frac{\partial}{\partial x_0^\mu}\phi_b\left((x-x_0)^\nu\right)dx_0^\mu. \tag{6.34}$$

Equating the variation in Equations (6.33) and (6.34) gives

$$\frac{dc^\mu}{\sqrt{2\pi\hbar}} = \frac{N}{\sqrt{2\pi\hbar}}dx_0^\mu. \tag{6.35}$$

Now

$$\int d^4x\, \frac{1}{N^2}\frac{\partial}{\partial x_0^\mu}\phi_b\left((x-x_0)^\nu\right)\frac{\partial}{\partial x_0^\nu}\phi_b\left((x-x_0)^\nu\right) = \frac{\delta_{\mu\nu}}{4N^2}\int d^4x\left(\partial_\lambda\phi_b(x)\partial_\lambda\phi_b(x)\right). \tag{6.36}$$

We can evaluate this integral by using the fact that the action S_E is stationary at the bounce.

$$
\begin{aligned}
0 = \frac{d}{d\lambda}S_E\left[\phi_b(\lambda x)\right]\Big|_{\lambda=1} &= \frac{d}{d\lambda}\int d^4x\left(\frac{1}{2}\left(\partial_\mu\phi_b(\lambda x)\partial_\mu\phi_b(\lambda x)\right) + V\left(\phi_b(\lambda x)\right)\right)\Big|_{\lambda=1} \\
&= \frac{d}{d\lambda}\int d^4x\left(\frac{1}{\lambda^2}\frac{1}{2}\left(\partial_\mu\phi_b(x)\partial_\mu\phi_b(x)\right) + \frac{1}{\lambda^4}V\left(\phi_b(x)\right)\right)\Big|_{\lambda=1} \\
&= \int d^4x\left(-2\frac{1}{2}\left(\partial_\mu\phi_b(x)\partial_\mu\phi_b(x)\right) - 4V\left(\phi_b(x)\right)\right) \\
&= -4S_E\left[\phi_b(x)\right] + \int d^4x\left(\partial_\mu\phi_b(x)\partial_\mu\phi_b(x)\right). \tag{6.37}
\end{aligned}
$$

Hence

$$\int d^4x\left(\partial_\mu\phi_b(x)\partial_\mu\phi_b(x)\right) = 4S_E\left[\phi_b(x)\right] \tag{6.38}$$

and finally

$$N = \sqrt{S_E\left[\phi_b(x)\right]} \tag{6.39}$$

exactly as in the one-dimensional case. The Jacobian factor becomes $\left(\sqrt{\frac{S_E[\phi_b(x)]}{2\pi\hbar}}\right)^4$, giving the integration over the position of the bounce

$$\frac{\left(S_E\left[\phi_b(x)\right]\right)^2}{4\pi^2\hbar^2}\beta V. \tag{6.40}$$

We do the same analysis for N well-separated bounces, which are approximate critical points, which gives us

$$\left(\frac{(S_E[\phi_b(x)])^2}{4\pi^2\hbar^2}\right)^N \frac{(\beta V)^N}{N!},$$ (6.41)

where the $N!$ simply indicates that the permutations of the positions of the bounces do not give new configurations. This gives

$$\langle\phi_+|e^{-\frac{\beta\hat{H}}{\hbar}}|\phi_+\rangle = \mathcal{N}\det^{-\frac{1}{2}}(-\partial_\mu\partial_\mu + V''(\phi_+))e^{-\left(\beta V\left(e^{-\frac{S_E[\phi_b(x)]}{\hbar}}\right)\frac{(S_E[\phi_b(x)])^2}{4\pi^2\hbar^2}K\right)},$$ (6.42)

where K is now the ratio

$$K = \left(\frac{\det'(-\partial_\mu\partial_\mu + V''(\phi_b))}{\det(-\partial_\mu\partial_\mu + V''(\phi_+))}\right)^{-\frac{1}{2}}$$ (6.43)

and the prime indicates that the zero modes are removed. The normalization constant $\mathcal{N}$ is defined to exactly cancel the free determinant that appears

$$\mathcal{N}\det^{-\frac{1}{2}}(-\partial_\mu\partial_\mu + V''(\phi_+)) = 1$$ (6.44)

This is, not the whole story, because the operator

$$-\partial_\mu\partial_\mu + V''(\phi_b)$$ (6.45)

has a negative mode. Again our analysis of meta-stable states in quantum mechanics applies directly. Taking into account the factor of $\frac{1}{2}$ which comes from the analytic continuation and deformation of the contour, we find

$$i\frac{\Gamma}{V} = \frac{(S_E[\phi_b(x)])^2}{4\pi^2\hbar^2}\left(e^{-\frac{S_E[\phi_b(x)]}{\hbar}}\right)\left(\frac{\det'(-\partial_\mu\partial_\mu + V''(\phi_b))}{\det(-\partial_\mu\partial_\mu + V''(\phi_+))}\right)^{-\frac{1}{2}}.$$ (6.46)

The prime still indicates that only the zero modes are removed, the square root of the negative eigenvalue reproduces the imaginary nature and the factor of $\frac{1}{2}$ is taken into account because the lifetime is $\frac{1}{2}$ of the imaginary part. Analysis of the negative modes is left for Section 6.5.

6.4 The Fate of the False Vacuum Continued

We continue our analysis of the decay of the false vacuum by considering the evolution of the field after the tunnelling event. We can obtain some intuition from the WKB analysis of tunnelling in particle quantum mechanics. Consider the decay of a nucleus by α-particle emission. A reasonably successful phenomenological potential has the form of a square well of depth extending to less than zero attached to a short-range drop off potential from the top reaching to zero, as depicted in Figure 6.5. The negative energy levels in the well are stable,

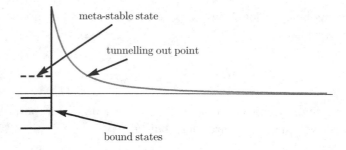

Figure 6.5. A nuclear tunnelling potential

but the positive energy levels are meta-stable and decay by tunnelling. The semi-classical description of the decay process proceeds as follows. The particle stays in the well up to a time, the "transition time", which is a random variable, when it makes a quantum jump to the other side of the barrier. It appears suddenly at the other side at a point, which we call the "tunnelling out point", with the same energy as the meta-stable state within. Subsequently, it continues like a free classical particle until it eventually moves off to infinity.

Quantum mechanics only enters in the calculation of the process of barrier penetration. It allows us to calculate the mean value of the "transition time". In the WKB analysis, the tunnelling out point is the point on the other side of the barrier with equal energy to the energy of the meta-stable state inside, from which, if the particle were released, it would move off to infinity under the classical dynamics. This is the turning point in the usual WKB analysis. We identify this point as the point where all velocities are zero in the bounce solution. We choose this point by the condition

$$\partial_\tau \phi(\vec{x}, \tau)|_{\tau=0} = 0. \tag{6.47}$$

This is satisfied by the $O(4)$ symmetric ansatz that we have taken,

$$\partial_\tau \phi(\rho)|_{\tau=0} = \partial_\rho \phi(\rho) \left(\frac{\tau}{\rho}\right)\bigg|_{\tau=0} = 0. \tag{6.48}$$

The field appears at $\tau = 0$ in the state described by $\phi_b(\vec{x}, \tau = 0)$ and then evolves classically. The WKB analysis should not be taken too literally. It will not be accurate for observations made just after the tunnelling event occurs. It is more correctly an asymptotic description for what happens long after and far away from the tunnelling event.

6.4.1 Minkowski Evolution After the Tunnelling

We continue nevertheless with the initial condition for after the tunnelling event

$$\phi(\vec{x}, t = 0) = \phi_b(\vec{x}, \tau = 0), \quad \partial_\tau \phi(\vec{x}, \tau)|_{\tau=0} = 0 \tag{6.49}$$

and then the field evolves according to the classical, now Minkowskian, equation of motion,

$$\left(\frac{d^2}{dt^2} - \nabla^2\right)\phi(\vec{x},t) + V'(\phi(\vec{x},t)) = 0. \tag{6.50}$$

At $t = 0$, $\phi(\vec{x}, t = 0) = \phi_b(\vec{x}, \tau = 0)$ is exactly a bubble of radius R of true vacuum, separated by a thin wall from the false vacuum without. This is because $\phi_b(\vec{x}, \tau) = \phi_b(\sqrt{|\vec{x}|^2 + \tau^2}) \to \phi(r)$ for $t = 0$ with $r = |\vec{x}|$. We can immediately write down the solution to the classical Minkowskian equation of motion for the subsequent evolution of the bubble. Simply

$$\phi(\vec{x},t) = \phi_b\left(\sqrt{|\vec{x}|^2 - t^2}\right). \tag{6.51}$$

In detail for the Minkowskian signature, with $\tilde{\rho} \equiv \sqrt{|\vec{x}|^2 - t^2} = \sqrt{-x_\mu x^\mu}$,

$$\partial_\mu \partial^\mu \phi(\tilde{\rho}) = \partial_\mu\left(\frac{d}{d\tilde{\rho}}\phi(\tilde{\rho})\partial^\mu\tilde{\rho}\right)$$

$$= \frac{d^2}{d\tilde{\rho}^2}\phi(\tilde{\rho})\partial_\mu\tilde{\rho}\partial^\mu\tilde{\rho} + \frac{d}{d\tilde{\rho}}\phi(\tilde{\rho})\partial_\mu\partial^\mu\tilde{\rho}. \tag{6.52}$$

Using $\partial_\mu\tilde{\rho} = -\frac{x_\mu}{\tilde{\rho}}$, $\partial_\mu\tilde{\rho}\partial^\mu\tilde{\rho} = -1$ and hence $\partial_\mu\partial^\mu\tilde{\rho} = -\frac{3}{\tilde{\rho}}$ we get

$$\partial_\mu\partial^\mu\phi(\tilde{\rho}) = -\left(\frac{d^2}{d\tilde{\rho}^2} + \frac{3}{\tilde{\rho}}\right)\phi(\tilde{\rho}). \tag{6.53}$$

The Euclidean equation satisfied by $\phi_b\left(\sqrt{|\vec{x}|^2 + \tau^2}\right)$ is

$$\left(\frac{d^2}{d\tau^2} + \nabla^2\right)\phi_b\left(\sqrt{|\vec{x}|^2 + \tau^2}\right) - V'\left(\phi_b\left(\sqrt{|\vec{x}|^2 + \tau^2}\right)\right) = 0. \tag{6.54}$$

Since

$$\partial_\mu\partial_\mu\phi(\rho) = \left(\frac{d^2}{d\rho^2} + \frac{3}{\rho}\right)\phi(\rho). \tag{6.55}$$

gives

$$\left(\frac{d^2}{d\rho^2} + \frac{3}{\rho}\right)\phi(\rho) - V'(\phi(\rho)) = 0. \tag{6.56}$$

Thus

$$\left(\frac{d^2}{dt^2} - \nabla^2\right)\phi_b\left(\sqrt{|\vec{x}|^2 - t^2}\right) + V'\left(\phi_b\left(\sqrt{|\vec{x}|^2 - t^2}\right)\right) = 0 \tag{6.57}$$

and it should be noted that this solution is only valid for $|\vec{x}|^2 > t^2$, i.e. for the exterior of the bubble.

Then the $O(4)$ invariance of the Euclidean solution is replaced by the $O(3,1)$ invariance of the Minkowskian regime. This implies that the evolution of the bubble appears the same to all Lorentz observers. When the bubble is nucleated, the wall of the bubble is at $r \approx R$, and then it follows the hyperbola, $\tilde{\rho}^2 = r^2 - t^2 = R^2$. This is because the functional form of $\phi(\tilde{\rho})$ describes the wall for all $\tilde{\rho} \approx R^2$. This means that the bubble grows with a speed which approaches

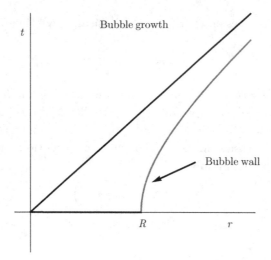

Figure 6.6. The growth of the bubble wall after tunnelling

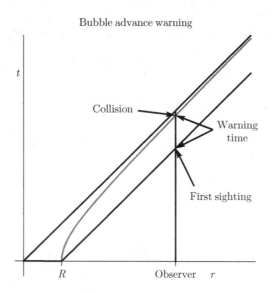

Figure 6.7. Collision warning time with the growth of the bubble wall

the speed of light asymptotically, as depicted in Figure 6.6. How quickly the growth approaches c depends on R. If R is a microscopic number, like $10^{-10} \rightarrow 10^{-30}$ as we would expect, the bubble grows with the speed of light almost instantaneously. If a bubble is coming towards us, the warning time we have is given by the projection of the forward light cone from the creation point to our world line (vertical), as depicted in Figure 6.7. The time this gives us in warning, T, is essentially the time it takes light to travel the distance R, as long as the observer is far from the creation point relative to R. For R micro-physical, T is also microphysical. After the bubble hits us, quoting directly from Coleman [31]: "We are dead. All constants of nature inside the bubble are different. We cannot function biologically or even chemically". But, paraphrasing, as further pointed out by Coleman, this is no cause for concern, since for $R \sim 10^{-15}$ metres, $T \sim 3 \times 10^{-8}$ seconds, this is much less time than the time it takes for a single neuron to fire. If such a bubble is coming towards us, we won't know what hit us.

6.4.2 Energetics

The energy carried by the wall of the bubble is exactly all the energy gained by converting a sphere of radius R of false vacuum into true vacuum. The energy in the wall per unit area is

$$
\begin{aligned}
\mathcal{E} &= \frac{1}{4\pi R^2} \int_{|r| \approx R} d^3x \left(\frac{1}{2} \left(\vec{\nabla}\phi_b \right)^2 + V(\phi_b) \right) \\
&= \frac{1}{R^2} \int_{R-\Delta}^{R+\Delta} dr\, r^2 \left(\frac{1}{2} \left(\vec{\nabla}\phi_b \right)^2 + V(\phi_b) \right) \\
&\approx \int_{-\infty}^{\infty} dr \left(\frac{1}{2} \left(\vec{\nabla}\phi_b \right)^2 + V(\phi_b) \right) = S_1.
\end{aligned}
\tag{6.58}
$$

Now, in time, the wall follows the hyperbola $r^2 - t^2 = R^2$, hence the energy in the wall always stays in the wall. After some time, the element of area will have a velocity v. Energy per unit area just transforms as the zero component of a Lorentz vector,

$$
S_1 \to \frac{S_1}{\sqrt{1-v^2}}.
\tag{6.59}
$$

So at such a time the energy in the wall is

$$
\mathcal{E} = 4\pi r^2 \frac{S_1}{\sqrt{1-v^2}}
\tag{6.60}
$$

with

$$
v = \frac{dr}{dt} = \frac{d}{dt}\sqrt{R^2 + t^2} = \frac{t}{\sqrt{R^2 + t^2}} = \sqrt{\frac{r^2 - R^2}{r^2}} = \sqrt{1 - \frac{R^2}{r^2}}.
\tag{6.61}
$$

Thus $\sqrt{1-v^2} = \sqrt{1 - \left(1 - \frac{R^2}{r^2}\right)} = \frac{R}{r}$, and

$$
\mathcal{E} = 4\pi r^2 S_1 \frac{r}{R} = \frac{4}{3}\pi r^3 \left(\frac{3S_1}{R} \right) = \frac{4}{3}\pi r^3 \epsilon.
\tag{6.62}
$$

(In the thin-wall approximation, we have $R = \frac{3S_1}{\epsilon}$.) This is exactly the energy obtained from the conversion of a ball of radius r of false vacuum into true vacuum. Hence all the energy goes into the wall. Inside the bubble is just the tranquil, true vacuum. There is no boiling, roiling, hot plasma of excitations.

6.5 Technical Details

We complete this chapter with some technical points which we have left unaddressed.

6.5.1 Exactly One Negative Mode

We have assumed that there was exactly one negative energy mode to the operator governing small fluctuations

$$\left(-\partial_\mu \partial_\mu + V''(\phi_b)\right)\phi_n = \lambda_n \phi_n. \tag{6.63}$$

We can prove this in the thin-wall approximation. $O(4)$ invariance means that we can expand in the scalar spherical harmonics in four dimensions

$$\phi_{n,j}(\rho,\Omega) = \frac{1}{\rho^{\frac{3}{2}}} \chi_{n,j}(\rho) Y_{j,m,m'}(\Omega), \tag{6.64}$$

where $Y_{j,m,m'}(\Omega)$ transforms according to the representation D^{jj} of $SO(4) = SO(3) \times SO(3)$, with m and m' independently going from $-j$ to j. These are the eigenfunctions of the transverse Laplacian in four dimensions. Then to zero order in ϵ,

$$\left(-\frac{d^2}{d\rho^2} + \frac{8j(j+1)+3}{4\rho^2} + U''(\phi_b(\rho))\right)\chi_{n,j}(\rho) = \lambda_{n,j}\chi_{n,j}(\rho) \tag{6.65}$$

for the resulting radial equation. This is analogous to the Schrödinger equation for a particle in a radial potential in three dimensions.

The zero modes

$$\frac{1}{\sqrt{S_E^{bounce}}} \partial_\mu \phi_b(\rho) \tag{6.66}$$

transform according to the $j = \frac{1}{2}$ representation. ($\frac{1}{2} + \frac{1}{2} = 1 + 0$ for the three-dimensional rotation subgroup.) Since $\phi_b(\rho)$ is an increasing function, it starts at ϕ_- and increases to ϕ_+ at $\rho = \infty$, the zero modes have no nodes. Hence they are the modes of lowest "energy" for $j = \frac{1}{2}$. For $j > \frac{1}{2}$ the Hamiltonian is simply greater than for $j = \frac{1}{2}$, hence all modes have energy greater than zero. Thus the negative modes can only arise in the sector with $j = 0$. There must be at least one negative mode since the Hamiltonian is simply smaller for $j = 0$. In the thin-wall limit, $U''(\phi_b(\rho))$ has the form given in Figure 6.8 where $\omega^2 = U''(\phi_\pm)$. This is because $\phi_b(\rho)$ starts at ϕ_- at $\rho = 0$ and stays so until about $\rho = R$ where it interpolates relatively quickly to ϕ_+, and then stays essentially constant until $\rho = \infty$. The zero modes, corresponding to derivatives of $\phi_b(\rho)$, hence have support localized at the wall. The negative energy modes must also be localized there. Thus we approximate the equation near $\rho \approx R$ by replacing in the centrifugal term $\rho \to R$. This yields the equation

$$\left(-\frac{d^2}{d\rho^2} + \frac{8j(j+1)+3}{4R^2} + U''(\phi_b(\rho))\right)\chi_{n,j}(\rho) = \lambda_{n,j}\chi_{n,j}(\rho). \tag{6.67}$$

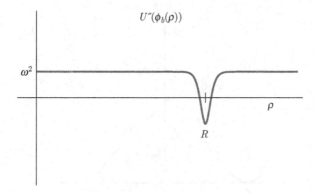

Figure 6.8. The potential for the small fluctuations about a thin-wall bubble

Clearly

$$\lambda_{n,j} = \lambda_n + \frac{8j(j+1)+3}{4R^2} \tag{6.68}$$

with λ_n, ordered to be increasing with n, evidently independent of j. For $R \to \infty$, λ_n are the eigenvalues of the one-dimensional operator

$$\left(-\frac{d^2}{dx^2} + U''(f(x)) \right), \tag{6.69}$$

where $f(x) = \phi_b(x)$ with $x \in [-\infty, \infty]$, *i.e.* we can neglect the effect of the boundary at $\rho = 0$. We already know that for $j = \frac{1}{2}$ the minimum eigenvalue is zero, thus

$$\lambda_0 \to -\frac{8j(j+1)+3}{4R^2}\bigg|_{j=\frac{1}{2}} = -\frac{8 \cdot \frac{1}{2} \cdot \frac{3}{2} + 3}{4R^2} = -\frac{9}{4R^2}. \tag{6.70}$$

This gives

$$\lambda_{0,0} = -\frac{9}{4R^2} + \frac{3}{4R^2} = -\frac{3}{2R^2}, \tag{6.71}$$

which is negative. All other eigenvalues for $j = \frac{1}{2}$ are positive, for all R. This implies that all the other λ_n are greater than zero, since

$$\lim_{R \to \infty} \left(\lambda_n + \frac{8 \cdot \frac{1}{2} \cdot \frac{3}{2} + 3}{4R^2} \right) = \lim_{R \to \infty} (\lambda_n) > 0 \quad \text{for} \quad n > 0. \tag{6.72}$$

Thus also for $j = 0$

$$\lambda_n + \frac{3}{4R^2} > 0, \quad \text{for} \quad n > 0, \tag{6.73}$$

for R large, hence there are no other negative eigenvalues.

In the limit $\epsilon \to 0$ we obtain the double-well potential depicted in Figure 6.9. There are no bounce-type solutions for this potential. Our solution just becomes a ball of true vacuum of infinite radius, $R = \frac{3S_1}{\epsilon} \to \infty$. There exist only the solutions

$$\phi = \phi_- \quad \text{or} \quad \phi = \phi_+ \tag{6.74}$$

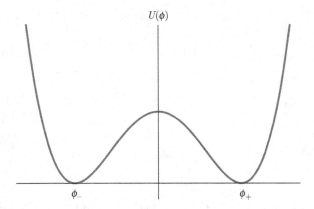

Figure 6.9. The symmetric double-well potential $U(\phi)$

to the Euclidean equation of motion. This is different from the case of particle quantum mechanics, where there are tunnelling-type solutions between the two wells. This difference is completely consistent with our understanding of quantum field theory in a potential with two symmetric wells of the same depth. In such a theory there is spontaneous symmetry breaking. The two vacua, constructed above each well, correspond to inequivalent representations of the quantum field. They cannot exist in the same Hilbert space, and hence there is no tunnelling between them.

6.5.2 Fluctuation Determinant and Renormalization

The determinant that we must compute is

$$\kappa \equiv \det\left(-\partial_\mu\partial_\mu + V''(\phi_b)\right) = e^{\ln\left(\det\left(-\partial_\mu\partial_\mu + V''(\phi_b)\right)\right)} = e^{\operatorname{tr}\ln\left(-\partial_\mu\partial_\mu + V''(\phi_b)\right)}.$$
(6.75)

We expand about $\phi = \phi_+$, then $V''(\phi_+) \approx \omega^2$, then we have

$$\begin{aligned}
\kappa &= e^{\operatorname{tr}\ln\left(-\partial_\mu\partial_\mu + \omega^2 + \left(V''(\phi_b) - \omega^2\right)\right)} \\
&= e^{\operatorname{tr}\ln\left(\left(-\partial_\mu\partial_\mu + \omega^2\right)\left(1 + \left(-\partial_\mu\partial_\mu + \omega^2\right)^{-1}\left(V''(\phi_b) - \omega^2\right)\right)\right)} \\
&= e^{\operatorname{tr}\ln\left(\left(-\partial_\mu\partial_\mu + \omega^2\right) + \operatorname{tr}\ln\left(1 + \left(-\partial_\mu\partial_\mu + \omega^2\right)^{-1}\left(V''(\phi_b) - \omega^2\right)\right)\right)} \\
&= \kappa_0 e^{\operatorname{tr}\left(\left(-\partial_\mu\partial_\mu + \omega^2\right)^{-1}\left(V''(\phi_b) - \omega^2\right) - \frac{1}{2}\left(\left(-\partial_\mu\partial_\mu + \omega^2\right)^{-1}\left(V''(\phi_b) - \omega^2\right)\right)^2 + \cdots\right)}
\end{aligned}$$
(6.76)

where $\kappa_0 = \det\left(-\partial_\mu\partial_\mu + \omega^2\right)$. The free determinant will be absorbed in the definition of the factor $K = (\kappa/\kappa_0)^{-\frac{1}{2}}$ of Equation (6.43).

The first two terms in this expansion are infinite; however, all the rest are finite. $V''(\phi_b) - \omega^2$ is exponentially small for $\rho \gg R$, so we may Fourier transform it to obtain

$$\tilde{f}(k_\mu) = \int \frac{d^4x}{(2\pi)^4} e^{-ik_\mu x_\mu}\left(V''(\phi_b) - \omega^2\right).$$
(6.77)

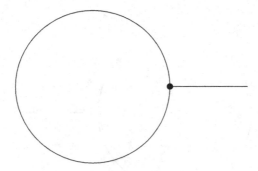

Figure 6.10. Feynman diagram for the first term in the expansion of Equation (6.76)

$\tilde{f}(k_\mu)$ its Fourier transform is then also exponentially small for large k_μ. Then

$$\mathrm{tr}\left(\left(-\partial_\mu\partial_\mu+\omega^2\right)^{-1}\left(V''(\phi_b)-\omega^2\right)\right)$$

$$=\int d^4x d^4y \langle x|\frac{1}{-\partial_\mu\partial_\mu+\omega^2}|y\rangle\langle y|V''(\phi_b)-\omega^2|x\rangle$$

$$=\int d^4x d^4y \int \frac{d^4k}{(2\pi)^4}\frac{e^{ik_\mu(x_\mu-y_\mu)}}{k^2+\omega^2}\left(V''(\phi_b(x))-\omega^2\right)\delta(x-y)$$

$$=\int \frac{d^4k}{(2\pi)^4}\int d^4x d^4y \int d^4q \frac{e^{ik_\mu(x_\mu-y_\mu)}e^{iq_\mu x_\mu}}{k^2+\omega^2}\tilde{f}(q_\mu)\delta(x-y)$$

$$=\int \frac{d^4k}{(2\pi)^4}\int d^4q \int d^4x \frac{e^{iq_\mu x_\mu}\tilde{f}(q_\mu)}{k^2+\omega^2}$$

$$=\int \frac{d^4k}{(2\pi)^4}\frac{1}{k^2+\omega^2}\left(\int d^4q \delta(q_\mu)\tilde{f}(q_\mu)\right). \tag{6.78}$$

The integral over d^4k is divergent, and can be represented by the diagram given in Figure 6.10. The infinity arising here must be absorbed via a non-trivial renormalization of the theory. The next term is

$$\mathrm{tr}\left(\frac{1}{2}\left(\left(-\partial_\mu\partial_\mu+\omega^2\right)^{-1}\left(V''(\phi_b)-\omega^2\right)\right)^2\right)$$

$$=\frac{1}{2}\int d^4x d^4y\langle x|\frac{1}{-\partial_\mu\partial_\mu+\omega^2}|y\rangle\langle y|\frac{1}{-\partial_\mu\partial_\mu+\omega^2}|x\rangle\times$$

$$\left(V''(\phi_b(y))-\omega^2\right)\left(V''(\phi_b(x))-\omega^2\right)$$

$$=\frac{1}{2}\int d^4x d^4y\int\frac{d^4k d^4l d^4p d^4q}{(2\pi)^8}\frac{e^{ik_\mu(x-y)_\mu+il_\mu(y-x)_\mu+iq_\mu y_\mu+ip_\mu x_\mu}}{(k^2+\omega^2)(l^2+\omega^2)}\tilde{f}(q_\mu)\tilde{f}(p_\mu)$$

$$=\frac{1}{2}\int d^4p d^4l\frac{1}{((l-p)^2+\omega^2)}\frac{1}{(l^2+\omega^2)}\tilde{f}(p_\mu)\tilde{f}(-p_\mu), \tag{6.79}$$

where integrating over x and y obtains two delta functions in momentum, and then integrating over k and q eliminates these two variables. The integrals can be represented diagrammatically as depicted in Figure 6.11. The integration over

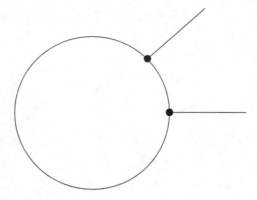

Figure 6.11. Feynman diagram for the second term in the expansion of Equation (6.76)

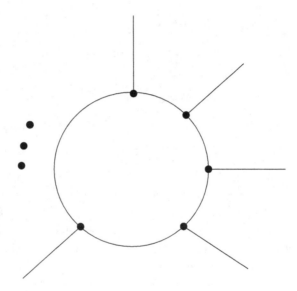

Figure 6.12. General Feynman diagram of the expansion of Equation (6.76)

l is divergent and also requires a non-trivial renormalization of the theory.

In general we get a diagram of the form given in Figure 6.12. It corresponds to the integral

$$\int d^4 l \int \frac{d^4 p_1 \cdots d^4 p_N}{(2\pi)^{4(N-1)}} \frac{\delta(p_1 + p_2 + \cdots + p_N)\tilde{f}(p_{1\mu}) \cdots \tilde{f}(p_{N\mu})}{(l^2 + \omega^2)((l - p_N)^2 + \omega^2) \cdots \left((l - \sum_{i=2}^{N} p_i)^2 + \omega^2\right)}.$$
(6.80)

It is only the integration over l which can cause problems, the $f(p_{i\mu})$ are exponentially decreasing for $p_{i\mu} \to \infty$. For three or more insertions the integral is finite

$$\int d^4l \frac{1}{(l^2+\omega^2)\left((l-p_1)^2+\omega^2\right)\left((l-(p_1+p_2))^2+\omega^2\right)} \sim \int \frac{dl}{l^2}. \qquad (6.81)$$

The solution of the problem of how to remove the divergences is by adding a set of (an infinite number of) counter-terms to the action, which will cancel the infinities arising from the integrations. It is a property of a renormalizable field theory that all such counter-terms can be reabsorbed into a multiplicative redefinition of the coupling constants and fields of the original theory. This means that the counter-terms correspond to terms which are of the same form as those already present.

$$S_{bare}\left(\phi\right) = S^R\left(\phi\right) + \hbar S^1\left(\phi\right) + \cdots, \qquad (6.82)$$

where $S^R\left(\phi\right)$ is finite, but $S^1\left(\phi\right)$ is not and the higher terms are not. This implies a change in the bounce, which will also be of the form

$$\phi_b = \phi_b^R + \hbar\phi^1 + \cdots, \qquad (6.83)$$

where ϕ_b^R is the same function as ϕ_b but now of the renormalized parameters. Now

$$S_{bare}\left(\phi_b^R + \hbar\phi^1 + \cdots\right) = S^R\left(\phi_b^R\right) + \hbar S^1\left(\phi_b^R\right) + \frac{\delta S^R\left(\phi\right)}{\delta\phi}\Big|_{\phi_b^R}\hbar\phi^1 + o(\hbar^2)$$

$$= S^R\left(\phi_b^R\right) + \hbar S^1\left(\phi_b^R\right) + o(\hbar^2), \qquad (6.84)$$

where the third term in the first equality vanishes by the equations of motion. Then

$$\frac{\Gamma}{V} = \frac{\left(S^R\left(\phi_b^R\right)\right)^2}{4\pi^2\hbar^2} e^{-\frac{S^R\left(\phi_b^R\right)+\hbar S^1\left(\phi_b^R\right)+\cdots}{\hbar}} \left(\frac{\det'\left(-\partial^2+V^{R''}\left(\phi_b^R\right)\right)}{\det\left(-\partial^2+V^{R''}\left(\phi_+^R\right)\right)}\right)^{-\frac{1}{2}} \qquad (6.85)$$

with the stipulation that

$$e^{-\frac{\hbar S^1\left(\phi_b^R\right)+\cdots}{\hbar}} \left(\frac{\det'\left(-\partial^2+V^{R''}\left(\phi_b^R\right)\right)}{\det\left(-\partial^2+V^{R''}\left(\phi_+^R\right)\right)}\right)^{-\frac{1}{2}} \qquad (6.86)$$

be finite. We choose $S^1\left(\phi_b\right)$ so that we cancel the two divergent terms in the expansion of the determinant. This can be made even clearer by ensuring that the bare action to $o(\hbar)$ vanish at the renormalized unstable vacuum value ϕ_+^R. This requires

$$S^R\left(\phi_+^R\right) + \hbar S^1\left(\phi_+^R\right) = \hbar S^1\left(\phi_+^R\right) = 0 \qquad (6.87)$$

since by definition $S^R\left(\phi_+^R\right) = 0$. We can achieve this by subtracting the constant $\hbar S^1\left(\phi_+^R\right)$ from the bare action in Equation (6.82), giving

$$S_{bare}\left(\phi\right) = S^R\left(\phi\right) + \hbar\left(S^1\left(\phi\right) - S^1\left(\phi_+^R\right)\right) + \cdots. \qquad (6.88)$$

This change implies the condition that

$$e^{-\frac{\hbar\left(S^1\left(\phi_b^R\right)-S^1\left(\phi_+^R\right)\right)+\cdots}{\hbar}}\left(\frac{\det'\left(-\partial^2+V^{R\prime\prime}\left(\phi_b^R\right)\right)}{\det\left(-\partial^2+V^{R\prime\prime}\left(\phi_+^R\right)\right)}\right)^{-\frac{1}{2}} \tag{6.89}$$

be free of infinities. We see that one factor of counter-terms matches with each determinant, ensuring the independent renormalizability.

In a renormalizable theory, such as ϕ^4 theory, it is possible to prove that it can be done keeping $S^1(\phi)$ of the same form as $S_{bare}(\phi)$. In the general case, it is clear that the infinities can be cancelled; however, it is not clear that it can be done keeping the same functional form of the bare Lagrangian. Continuing the perturbative expansion of the functional integral beyond the Gaussian approximation will yield higher loop corrections and infinities, for which it will be necessary to add further counter-terms, written as $\hbar^2 S^2(\phi)+\cdots$. These again, for a renormalizable theory will be of the same form as the bare Lagrangian. We will not belabour the point any further.

One final avenue for controlling the determinant is to decompose it into angular momentum eigen-sectors using

$$-\partial_j^2+V''(\phi_b)=\frac{d^2}{d\rho^2}+\frac{8j(j+1)+3}{4\rho^2}+V''(\phi_b(\rho)) \tag{6.90}$$

in the angular momentum j sector. The multiplicity of the spherical harmonics of order j is $(2j+1)^2$. Then

$$\frac{\det'\left(-\partial^2+V''(\phi_b)\right)}{\det\left(-\partial^2+\omega^2\right)}=e^{\sum_{j=0,\frac{1}{2},1,\cdots}^{\prime\infty}\left(\operatorname{tr}\ln\left(\frac{-\partial_j^2+V''(\phi_b)}{-\partial_j^2+\omega^2}\right)^{(2j+1)^2}-\text{counter terms}\right)}. \tag{6.91}$$

Each term is a one-dimensional determinant which we know in principle how to calculate. It is finite. The infinities reappear after the summation over j.

6.6 Gravitational Corrections: Coleman–De Luccia

In this section we will consider gravitational corrections to vacuum decay. This is eminently reasonable as the application of these methods will be to situations where gravity is important, such as the evolution of the universe, where we invoke Lorentz invariance. The relevance of gravitational effects to vacuum decay in condensed matter systems may not be so important. However, in cosmological applications, the consideration of gravitational effects is clearly indicated. This analysis was first done by Coleman and De Luccia [33], and we will follow their presentation closely.

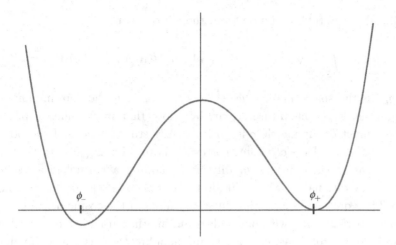

Figure 6.13. The potential with a small asymmetry

For simplicity, we consider a single scalar field with the Euclidean action

$$S_E[\phi] = \int d^4x \left(\left(\frac{1}{2}\partial_\mu \phi\right)^2 + V(\phi) \right), \tag{6.92}$$

which is valid with the absence of gravity. The potential $V(\phi)$ will be as in Figure 6.13, with a true minimum at ϕ_- and a false minimum at ϕ_+; however, we will not assume that the potential is symmetric under reflection $\phi \to -\phi$. We will further assume that the value of the potential at each minimum is very small, proportional to a parameter ϵ. Thus

$$V(\phi) = V_0(\phi) + o(\epsilon), \tag{6.93}$$

where $V_0(\phi_\pm) = 0$.

Adding gravitational corrections may seem pointless at microscopic scales, but for other scales they can be very important. Indeed, if a bubble of radius Λ of false vacuum is converted to a true vacuum, an energy in the amount $E = \epsilon 4\pi \Lambda^3/3$ will be released, and this energy will gravitate in the usual Newton–Einstein fashion. The Schwarzschild radius of the gravitating energy will be $2GE$. This radius will be equal to the radius of the bubble when $\Lambda = 2GE = 2G\epsilon 4\pi\Lambda^3/3$. This gives

$$\Lambda = (8\pi G\epsilon/3)^{-1/2}. \tag{6.94}$$

For energy densities of the order of $\epsilon \approx (1\,GeV)^4$ this gives a radius of about 0.8 kilometres. Thus the gravitational effects of vacuum decay occur at scales which are neither microscopic nor cosmological, but right in the scales of planetary and terrestrial physics. It might well be that gravitational effects in vacuum decay are very relevant.

Adding the gravitational interaction, the action changes to

$$S_E[\phi, g_{\mu\nu}] = \int d^4x \sqrt{g} \mathcal{L}_E = \int d^4x \sqrt{g} \left(\frac{1}{2} g^{\mu\nu} \partial_\mu \phi \partial_\nu \phi + V(\phi) + \frac{1}{16\pi G} R \right),$$

(6.95)

where $g_{\mu\nu}$ is the spacetime metric, $g^{\mu\nu}$ its inverse, g is the determinant of the metric and R is the curvature scalar. We note that in Euclidean spacetime, the determinant of the metric g is positive. Adjusting the zero of the potential $V(\phi) \to V(\phi) - V_0$, V_0 a constant, corresponds to adding $\sqrt{g} V_0$ to the action, which is exactly the same as modifying or adding a cosmological constant. Thus the gravitational spacetime inside the bubble and outside the bubble will necessarily be quite different, with different values of the cosmological constant. This makes perfect sense with our understanding that gravitation is sensitive to and couples to the total energy in a system, including the vacuum energy density. Thus we have to specify the cosmological constant of our initial false vacuum, of which we are going to compute the decay. The cosmological constant being exceptionally small at the present time, we will consider two cases of potential interest. First we will consider the possibility that we are living in a false vacuum with zero cosmological constant and this false vacuum decays to a true vacuum of negative cosmological constant, *i.e.* $V(\phi_+) = 0$. Second, we will consider that a false vacuum with a finite, positive cosmological constant decays to the true vacuum without cosmological constant where we live, *i.e.* $V(\phi_-) = 0$.

6.6.1 Gravitational Bounce

We assume that the bounce in the presence of gravity will have maximal symmetry, $O(4)$ symmetry. The metric, remember that we are now in Euclidean spacetime, then must be of the form

$$ds^2 = d\xi^2 + \rho(\xi)^2 d\Omega^2,$$

(6.96)

where $d\Omega^2$ is the metric on the three-sphere S^3, and ξ is the Euclidean radial coordinate and corresponds to the proper radial distance along a radial trajectory. $\rho(\xi)$ is the radius of curvature of each concentric S^3 that foliate the space. $d\Omega^2$ can be expressed in a number of coordinates, for example the analogue of spherical polar coordinates in $\mathbf{R}^4$, or in a more sophisticated manner in terms of left invariant 1-forms on the group manifold of $SU(2)$ which is exactly S^3. But we will not need this part of the metric explicitly and hence we will not exhibit it, as we will assume everything is spherically symmetric and hence independent of the angular degrees of freedom.

We can then compute the Euclidean equations of motion. These for the scalar field are

$$\partial_\mu \left(\sqrt{g} g^{\mu\nu} \partial_\nu \phi \right) - \sqrt{g} V'(\phi) = 0.$$

(6.97)

Using the rather simple form for the metric and the assumption that our field ϕ does not depend on the angular coordinates, we find the equation simplifies to

$$\partial_\xi \left(\sqrt{g} g^{\xi\xi} \partial_\xi \phi \right) - \sqrt{g} V'(\phi) = 0. \tag{6.98}$$

Furthermore, $g^{\xi\xi} = 1$ and $\sqrt{g} = \rho^3(\xi)\sqrt{g_\Omega}$ where g_Ω is the determinant of the metric of the angular coordinates, which is just the metric on a unit three-sphere. g_Ω depends explicitly on the angular coordinates but it does not depend on ξ. Since the only derivative that appears in the equation of motion is with respect to ξ, g_Ω simply factors out of both terms and then can be cancelled. This gives

$$
\begin{aligned}
0 &= \partial_\xi \left(\rho^3(\xi) \partial_\xi \phi \right) - \rho^3(\xi) V'(\phi) \\
&= \rho^3(\xi) \partial_\xi^2 \phi + 3\rho^2(\xi) \partial_\xi \rho \partial_\xi \phi - \rho^3(\xi) V'(\phi).
\end{aligned}
\tag{6.99}
$$

Dividing through by ρ^3 yields

$$\partial_\xi^2 \phi + \frac{3\partial_\xi \rho}{\rho} \partial_\xi \phi = V'(\phi). \tag{6.100}$$

This field equation is augmented by the Einstein equation $G_{\mu\nu} = -8\pi G T_{\mu\nu}$. The sign in this equation is convention-dependent, corresponding to the definition of the curvature tensor, the signature of the metric and the definition of the Ricci tensor. We will use the sign convention in Coleman–De Luccia [33], which is not our favourite convention, but we will stick with it to be close to the original paper. The Einstein equation yields only one net equation,

$$G_{\xi\xi} = -8\pi G T_{\xi\xi}. \tag{6.101}$$

The other components, which are just the diagonal spatial components, are either trivial identities or equivalent to this equation. The energy momentum tensor of the scalar field is

$$T_{\mu\nu} = \partial_\mu \phi \partial_\nu \phi - g_{\mu\nu} \mathcal{L}_E. \tag{6.102}$$

To obtain the Einstein equation, Equation (6.101), one has to compute the Ricci curvature through the Christoffel symbols, which is straightforward but somewhat tedious. We will not spell out the details here; with the use of symbolic manipulation software, the calculation is actually trivial. We find that there is only one independent equation,

$$(\partial_\xi \rho)^2 = 1 + \frac{1}{3} 8\pi G \rho^2 \left(\frac{1}{2} (\partial_\xi \phi)^2 - V(\phi) \right). \tag{6.103}$$

It makes perfect sense that there are only two independent equations of motion, as there are only two independent fields, ρ and ϕ. The two equations of motion can be obtained from an effective one-dimensional Euclidean action

$$S_E[\phi, \rho] = 2\pi^2 \int d\xi \left(\rho^3 \left(\frac{1}{2} (\partial_\xi \phi)^2 + V(\phi) \right) + \frac{3}{8\pi G} \left(\rho^2 \partial_\xi^2 \rho + \rho (\partial_\xi \rho)^2 - \rho \right) \right). \tag{6.104}$$

The equation of motion for ϕ is straightforward, that for ρ appears only after self-consistently using the derivative of Equation (6.103) to eliminate the second-order derivative in its usual equation of motion.

We solve Equation (6.100) in the approximation that the first derivative term is negligible, and the assumption that the potential term can be approximated by a function $V(\phi) = V_0(\phi) + o(\epsilon)$ with the condition that $V'(\phi_\pm) = 0$ and $V_0(\phi_+) = V_0(\phi_-)$. This latter assumption is very reasonable if the actual potential is obtained from a small perturbation of a degenerate double-well potential. We do not assume that the double well is symmetric, however, just that the minima have the same value for the potential. Then Equation (6.100) becomes

$$\partial_\xi^2 \phi = V_0'(\phi), \tag{6.105}$$

which admits an immediate first integral

$$\frac{1}{2}\left(\partial_\xi \phi\right)^2 = V_0(\phi) + C, \tag{6.106}$$

where C is the integration constant. C is determined by the value of V_0 at ϕ_+, as we are looking for a solution that interpolates from ϕ_- at the initial value of ξ, which is normally taken to be zero, to ϕ_+ as $\xi \to \infty$. Thus

$$\frac{1}{2}\left(\partial_\xi \phi\right)^2 = V_0(\phi) - V_0(\phi_+). \tag{6.107}$$

This equation can be easily integrated as

$$\int_{(\phi_+ + \phi_-)/2}^{\phi} d\phi \sqrt{2\left(V_0 - V_0(\phi_+)\right)} = \int_{\bar{\xi}}^{\xi} d\xi = \xi - \bar{\xi}, \tag{6.108}$$

where $\bar{\xi}$ is the value at which the field is mid-way between ϕ_+ and ϕ_-, which can be taken as the position of the wall. In principle, then, we should solve for ϕ which is implicitly defined by this equation. This will not be done explicitly and, continuing implicitly, once we have ϕ, we can solve Equation (6.103) for ρ. To solve this first-order differential equation requires the specification of one integration constant, we choose that as

$$\bar{\rho} = \rho(\bar{\xi}), \tag{6.109}$$

which is the radius of curvature of the wall. We do not need to have ϕ or ρ explicitly, if all we want is the value of the action for the bounce. This will depend on $\bar{\rho}$; however, we can determine $\bar{\rho}$ by imposing that the action be stationary with respect to variations of $\bar{\rho}$.

We start with the Euclidean action, Equation (6.104), and integrate by parts on the two-derivative term to bring it all in terms of single derivatives. We will only be calculating the action relative to its value for the false vacuum, thus it is calculated in a limiting fashion as the difference of two terms which separately do not make sense and diverge in principle, but the difference is finite. Thus the surface term is irrelevant as we will do the same to the action without the bounce instanton with just the false vacuum. This gives

$$S_E = 4\pi^2 \int d\xi \left(\rho^3 \left(\frac{1}{2}\phi'^2 + V \right) - \frac{3}{8\pi G} (\rho\rho'^2 + \rho) \right) \qquad (6.110)$$

and then we eliminate ρ' with Equation (6.103). This gives the rather compact expression

$$S_E = 4\pi^2 \int d\xi \left(\rho^3 V - \frac{3\rho}{8\pi G} \right) = -\frac{12\pi^2}{8\pi G} \int d\xi\rho \left(1 - \frac{8\pi G}{3}\rho^2 V \right). \qquad (6.111)$$

Now we use the thin-wall approximation, *i.e.* we assume that the bounce instanton will be much like the same in the absence of gravity, and for $\epsilon \to 0$, it will be of the form of a thin-wall bubble. We will justify the thin-wall approximation after the analysis. Outside the bubble the bounce configuration is entirely in the false vacuum and we are comparing the bounce action to the action of exactly the false vacuum, thus the contribution to the action is zero

$$S_{E,\,\text{outside}} = 0. \qquad (6.112)$$

Within the wall, we can put $\rho = \bar\rho$, and $V \to V_0$ up to $o(\epsilon)$ terms, giving

$$S_{E,\,\text{wall}} = 4\pi^2\bar\rho^2 \int d\xi \left(V_0(\phi) - V_0(\phi_+) \right) = 2\pi^2\bar\rho^3 S_1, \qquad (6.113)$$

where S_1 was defined by Equation (6.27) in the absence of gravity. Finally, for the inside of the bubble, $\phi = \phi_\pm$ is a constant, for both cases when we are computing the action for the bounce or for the false vacuum, thus we have from Equation (6.103)

$$d\xi = d\rho \left(1 - \frac{8\pi G}{3}\rho^2 V(\phi_\pm) \right)^{-1/2}. \qquad (6.114)$$

Thus choosing ϕ_- for the bounce and ϕ_+ for the false vacuum we have

$$S_{E,\,\text{inside}} = -\frac{12\pi^2}{8\pi G} \int_0^{\bar\rho} \rho d\rho \left(\left(1 - \frac{8\pi G}{3}\rho^2 V(\phi_-) \right)^{1/2} - \left(1 - \frac{8\pi G}{3}\rho^2 V(\phi_+) \right)^{1/2} \right)$$

$$= \frac{12\pi^2}{(8\pi G)^2} \left(\frac{1}{V(\phi_-)} \left(\left(1 - \frac{8\pi G}{3}\bar\rho^2 V(\phi_-) \right)^{3/2} - 1 \right) \right.$$

$$\left. - \frac{1}{V(\phi_+)} \left(\left(1 - \frac{8\pi G}{3}\bar\rho^2 V(\phi_+) \right)^{3/2} - 1 \right) \right) \qquad (6.115)$$

where,

$$S_E = S_{E,\,\text{outside}} + S_{E,\,\text{wall}} + S_{E,\,\text{inside}}. \qquad (6.116)$$

This yields an unwieldy expression; however, for the cases which interest us, it is quite simple.

Firstly, for the case $\phi_+ = \epsilon, \phi_- = 0$, the case where we are living in a spacetime after the formation of a bubble, we have the simple expression (after taking the limit $V(\phi_-) \to 0$ in the action $S_{E,\text{inside}}$)

$$S_E = 2\pi^2 \bar{\rho}^3 S_1 + \frac{12\pi^2}{(8\pi G)^2} \left(-4\pi G \bar{\rho}^2 - \frac{1}{\epsilon} \left(\left(1 - \frac{8\pi G}{3} \bar{\rho}^2 \epsilon \right)^{3/2} - 1 \right) \right). \quad (6.117)$$

Then setting the derivative with respect to $\bar{\rho}$ to vanish, gives

$$\frac{dS_E}{d\bar{\rho}} = 0 = 6\pi^2 \bar{\rho}^2 S_1 + \frac{12\pi^2}{8\pi G} \bar{\rho} \left(-1 + \left(1 - \frac{8\pi G}{3} \bar{\rho}^2 \epsilon \right)^{1/2} \right), \quad (6.118)$$

which is easily solved as

$$\bar{\rho} = \frac{12 S_1}{4\epsilon + 24\pi G S_1^2} \equiv \frac{\bar{\rho}_0}{1 + (\bar{\rho}_0/2\Lambda)^2}, \quad (6.119)$$

where $\bar{\rho}_0 = 3S_1/\epsilon$, which is the bubble radius in the absence of gravity, and $\Lambda = \sqrt{3/(8\pi G\epsilon)}$, the radius at which the Schwarzschild radius of the energy from converting a false vacuum to a true vacuum is equal to the bubble radius as defined in Equation (6.94). Evaluating the action at the value of $\bar{\rho}$ yields

$$S_E = \frac{1}{\left(1 + (\bar{\rho}_0/2\Lambda)^2 \right)^2} \frac{27\pi^2 S_1^4}{2\epsilon^3} = \frac{S_E^0}{\left(1 + (\bar{\rho}_0/2\Lambda)^2 \right)^2}, \quad (6.120)$$

where S_E^0 is the action of the bounce in the absence of gravity. We can obtain this formula by brute force replacement for $\bar{\rho}$; however, we can minimize the algebra by noting the Euclidean action, as a function of $\bar{\rho}$, has the form

$$S_E = \alpha \bar{\rho}^3 - \beta \bar{\rho}^2 + \gamma - \delta \left(1 - \zeta \bar{\rho}^2 \right)^{3/2} \quad (6.121)$$

with $\alpha = 2\pi^2 S_1$, $\beta = 3\pi/4G$, $\gamma = 3/16G^2\epsilon$, $\delta = 8\pi G\epsilon/3$ and $\bar{\rho} = (3S_1/\epsilon)/(1 + (\bar{\rho}_0/2\Lambda)^2)$ and the above definitions of $\bar{\rho}_0$ and Λ. The action is stationary at $\bar{\rho}$ hence

$$3\alpha \bar{\rho}^2 - 2\beta \bar{\rho} - 3\delta \left(1 - \zeta \bar{\rho}^2 \right)^{1/2} (-\zeta \bar{\rho}) = 0. \quad (6.122)$$

Then factoring out by 3, multiplying by $\bar{\rho}$ and adding and subtracting terms we can reconstruct S_E

$$3 \left(\alpha \bar{\rho}^3 - \beta \bar{\rho}^2 + \gamma - \delta \left(1 - \zeta \bar{\rho}^2 \right)^{1/2} (1 - \zeta \bar{\rho}^2) + \delta \left(1 - \zeta \bar{\rho}^2 \right)^{1/2} - \gamma + \frac{\beta}{3} \bar{\rho}^2 \right) = 0 \quad (6.123)$$

so then we get

$$S_E = \gamma - \frac{\beta}{3} \bar{\rho}^2 - \delta \left(1 - \zeta \bar{\rho}^2 \right)^{1/2}. \quad (6.124)$$

From the derivative, Equation (6.122), we can easily find

$$\delta \left(1 - \zeta \bar{\rho}^2 \right)^{1/2} = \frac{2\beta}{3\zeta} - \frac{\alpha \bar{\rho}}{\zeta} \quad (6.125)$$

and then we have

$$S_E = \gamma - \frac{\beta}{3}\bar{\rho}^2 - \frac{2\beta}{3\zeta} + \frac{\alpha\bar{\rho}}{\zeta}, \tag{6.126}$$

which now is straightforward to evaluate, yielding Equation (6.120).

For the second case, $V(\phi_+) = 0$, $V(\phi_-) = -\epsilon$, where we are now living in a false vacuum that may decay at any moment, we obtain with similar algebra

$$\bar{\rho} = \frac{\bar{\rho}_0}{1 - (\bar{\rho}/2\Lambda)^2} \tag{6.127}$$

while

$$S_E = \frac{S_E^0}{\left(1 - (\bar{\rho}/2\Lambda)^2\right)^2}. \tag{6.128}$$

For the thin-wall approximation to be valid, we required that the radius of the bubble was much larger than the length scale over which ϕ changed significantly. The friction term, $(3/\rho)(d\phi/d\rho)$, was neglected in Equation (6.12) as the factor $(3/\rho) \sim (3/\bar{\rho}) \approx 0$. Now in the presence of gravitation we have a different friction term, $(3\partial_\xi\rho)/\rho$, which is given by Equation (6.103)

$$\frac{1}{\rho^2}\left(\frac{d^2\rho}{d\xi^2}\right)^2 = \frac{1}{\rho^2} + \frac{8\pi G}{3}\left(\frac{1}{2}\left(\frac{d\phi}{d\xi}\right)^2 - V\right). \tag{6.129}$$

The first term is the same as without gravity and small if $\bar{\rho}$ is large. The second term vanishes on one side of the wall, is constant and of $o(\epsilon)$ on the other, and over the wall it interpolates between these two values. From Equation (6.107) it is to lowest order a constant, $-V_0(\phi_+)$, which in our two cases is of $o(\epsilon)$ plus corrections which are also of $o(\epsilon)$. Hence we lose nothing by replacing it with ϵ. This turns the second term into $1/\Lambda^2$. Hence the two terms which control the size of $\frac{1}{\rho^2}\left(\frac{d^2\rho}{d\xi^2}\right)^2$ are negligible, justifying self-consistently the thin-wall approximation, if $\bar{\rho}$ and Λ are large compared to the variation of ϕ. The variation of ϕ is from ϕ_+ to ϕ_-, over the thickness of the wall. This thickness is determined by the masses and coupling constants that are in V_0 which are not taken to be remarkable, *i.e.* neither very large nor very small. Thus the wall thickness will be independent of ϵ and hence the variation of ϕ is of $o(1)$. Thus self-consistently, for small ϵ, we can impose that $\bar{\rho}$ and Λ are large compared to the variation of ϕ, and the thin-wall approximation is justified. It is important to note that this puts no constraint on $\bar{\rho}_0/\Lambda$ which governs the difference in the solutions Equations (6.119), (6.120), (6.127), (6.128) with gravitation and those without, Equations (6.29), (6.30), for $\bar{\rho}$ and S_E above. Thus $\bar{\rho}_0/\Lambda$ can be taken as large as we want. Although this may not be phenomenologically relevant, it is interesting to consider the possibility.

In the first case with $\phi_+ = \epsilon, \phi_- = 0$ we see that the effect of gravitation is to increase the probability of vacuum decay, as the denominator in Equation (6.120) is greater than 1 and hence reduces S_E. Gravitation also diminishes the bubble radius. For the second case, $V(\phi_+) = 0$, $V(\phi_-) = -\epsilon$, the effects of gravitation are in the opposite direction, making it harder for the vacuum to decay as the denominator in Equation (6.128) is less than 1 and can even vanish, hence increasing S_E to arbitrarily large values. In this case, the bubble radius is increased by gravity, in the limiting case, pushing it to infinite radius at a finite value of $\bar{\rho}_0/\Lambda = S_1\sqrt{24\pi G/\epsilon}$. Thus for fixed S_1 and ϵ but for increasing G, we reach a point when the bubble has infinite radius and its action is infinite, completely suppressing vacuum decay. Thus gravitation totally suppresses vacuum decay for $\bar{\rho}_0 = 2\Lambda$, which means

$$\epsilon = 6\pi G S_1^2. \tag{6.130}$$

An explanation of the quenching of vacuum decay is because of energy conservation. If we calculate the energy of a bubble of radius $\bar{\rho}$ first in the absence of gravitation, we have the volume term and the surface term

$$E = -\frac{4\pi}{3}\epsilon\bar{\rho}^2 + 4\pi S_1\bar{\rho}^2. \tag{6.131}$$

In this (second) case of interest, $V(\phi_+) = 0$, $V(\phi_-) = -\epsilon$, thus we are living in a false vacuum of zero-energy density and the true vacuum has negative energy density. Then using the expression $\bar{\rho}_0 = 3S_1/\epsilon$ we have

$$E = \frac{4\pi}{3}\epsilon\bar{\rho}^2(\bar{\rho}_0 - \bar{\rho}), \tag{6.132}$$

thus we see that the energy vanishes for the bubble, which is expected as the energy before the bubble materialized was zero. Then the effects of gravitation can be taken into account, imposing energy conservation. If gravitation increases the total energy of the bubble, then the bubble must grow in size to compensate and if the gravitation decreases the energy it must shrink. In the case at hand, evidently the bubble must grow.

The gravitational contribution to the energy has two terms. First, the ordinary Newtonian potential energy, which is computed by integrating the gravitational field squared over all space

$$E_{\text{Newton}} = -\frac{\epsilon\pi\bar{\rho}_0^5}{15\Lambda^2}. \tag{6.133}$$

This follows from the straightforward calculation of the Newtonian energy of the gravitational field inside a sphere with negative mass density $-\epsilon$. That energy is

$$E_{\text{Newton}} = \frac{1}{2}\int d^3x(-\epsilon)\Phi(\vec{x}), \tag{6.134}$$

where the gravitational potential satisfies

$$\nabla^2\Phi(\vec{x}) = 4\pi G(-\epsilon). \tag{6.135}$$

Then the energy is given by

$$E_{\text{Newton}} = \frac{1}{2}\int d^3x\frac{\nabla^2\Phi}{4\pi G}\Phi = -\frac{1}{8\pi G}\int d^3x\,|\vec{g}|^2, \tag{6.136}$$

where $\vec{g} = -\vec{\nabla}\Phi$ is the gravitational field. Applying Gauss' law to

$$\vec{\nabla} \cdot \vec{g} = -4\pi G(-\epsilon) \tag{6.137}$$

yields

$$\vec{g} = \frac{4\pi G}{3}\epsilon \vec{r} \tag{6.138}$$

for the interior of the bubble. The gravitational field vanishes in the exterior. The integral Equation (6.136) quickly yields the result, Equation (6.133). The second contribution comes because the existence of the energy distorts the geometry correcting the volume of the bubble and hence correcting the volume term in the energy. From Equation (6.114) we can write the volume element of the bubble

$$4\pi\rho^2 d\xi = 4\pi\rho^2 d\rho \left(1 - \frac{1}{2}\frac{\rho^2}{\Lambda^2}\right) + o(G^2). \tag{6.139}$$

Then integrating over the bubble, the energy density $-\epsilon$ yields a change

$$E_{\text{geom}} = \frac{2\pi\epsilon\bar{\rho}_0^5}{5\Lambda^2} \tag{6.140}$$

giving a total change

$$E_{\text{grav}} = \frac{\pi\epsilon\bar{\rho}_0^5}{3\Lambda^2}. \tag{6.141}$$

Thus the change in energy is positive, which means that, with gravitation, the radius of the bubble must increase. It appears that for finite values of the couplings and parameters, when $\bar{\rho}_0 = 2\Lambda$, the bubble size becomes infinite. Increasing the gravitational coupling then gives no solution, *i.e.* the false vacuum becomes stable.

Once the bubble has materialized through quantum tunnelling, we can describe its subsequent evolution essentially classically. For Minkowski space-like separated points with respect to the centre of the bubble, all we have to do is analytically continue the solution back to Minkowski time. Thus for flat space we had $\rho^2 = \vec{x} \cdot \vec{x} + \tau^2 \to \vec{x} \cdot \vec{x} - t^2$. However, we must continue both the solution and the metric back to Minkowski time. Thus an $O(4)$-invariant Euclidean manifold becomes a $O(3,1)$-invariant Minkowskian manifold. The metric starts as

$$ds^2 = -d\xi^2 - (\rho(\xi))^2 d\Omega^2, \tag{6.142}$$

the negative definite metric being chosen as we wish to continue to a metric of signature $(+,-,-,-)$ where $d\Omega^2$ becomes the metric on a unit hyperboloid with space-like normal once continued to Minkowski spacetime. For this region $\phi = \phi(\rho)$ the solution that we have implicitly assumed to exist (although we have not been required to find it explicitly) and for a thin wall, the bubble wall is always at $\rho = \bar{\rho}$ and lies in this region. If we are outside a materializing bubble, then this is all we have to know about the manifold. It is possible to describe further the evolution of the bubble for the two cases that we have considered;

however, we will not continue the discussion further, it no longer requires the methods of instantons. We recommend the reader to consult the original article of Coleman and De Luccia [33].

6.7 Induced Vacuum Decay

We continue our study of the decay of the false vacuum precipitated by the existence of topological defects in that vacuum [79, 85]; we restrict our attention to the example of the decay of a "false cosmic string". Such a topological soliton corresponds to a topologically stable, non-trivial configuration inside a spacetime that is in the false vacuum. We will not worry about gravitational corrections. Topological solitons exist when the vacuum is degenerate and, generically, we have spontaneous symmetry breaking.

6.7.1 Cosmic String Decay

Cosmic strings occur in a spontaneously broken $U(1)$ gauge theory, a generalized Abelian Higgs model [61]. This model contains a complex scalar field interacting with an Abelian gauge field, hence scalar electrodynamics. However, we consider the inverse from the usual case, the potential for the complex scalar field ϕ, has a local minimum at a non-zero value $\phi^2 = a^2$, where the symmetry is broken, while the true minimum occurs at vanishing scalar field, $\phi = 0$. The scalar field potential is considered an effective potential, we do not worry about renormalizability. We assume the energy density splitting between the false vacuum and the true vacuum is very small. The spontaneously broken vacuum is the false vacuum.

In a scenario where from a high-temperature, unbroken symmetry phase the theory passes through an intermediate phase of spontaneous symmetry breaking, it is generic that there will be topological defects trapped in the symmetry-broken vacuum. Furthermore, the system could be trapped in the spontaneously broken phase, even though, as the temperature cools, the true vacuum returns to the unbroken symmetry phase. For the complex scalar field, its phase $e^{i\theta}$, can wrap the origin an integer number of times so that $\Delta\theta = 2n\pi$, as we go around a given line in three-dimensional space. The line can be infinite or form a closed loop. Corresponding to the given line there must exist a line of zeros of the scalar field, where the scalar field vanishes and corresponds to the true vacuum. The corresponding minimum energy configuration (when the roles of the false vacuum and true vacuum are reversed) is called a cosmic string, alternatively a Nielsen–Olesen string [96] or a vortex string [3]. In the scenario that we have described, the true vacuum lies at the regions of vanishing scalar field, thus the interior of the cosmic string is in the true vacuum while the exterior is in the false vacuum. It is already interesting that such classically stable configurations actually exist. Such strings must be unstable to quantum mechanical tunnelling decay. In this section we show how to calculate the amplitude for this decay, in the thin-wall limit.

In [85], the decay of vortices in the strictly two-spatial-dimensional context was considered. There, the vortex was classically stable at a given radius R_0. Through quantum tunnelling, the vortex could tunnel to a larger vortex of radius R_1, which was no longer classically stable. Dynamically the interior of the vortex was at the true vacuum, thus energetically lower by the energy density splitting multiplied by the area of the vortex. The vacuum energy behaves as $\sim -\epsilon R^2$, while the magnetic field energy behaves like $\sim 1/R^2$ and the energy in the wall behaves like $\sim R$. Thus the energy functional has the form

$$E = \alpha/R^2 + \beta R - \epsilon R^2. \qquad (6.143)$$

For sufficiently small ϵ, this energy functional is dominated by the first two terms. It is infinitely high for a small radius due to the magnetic energy, and will diminish to a local minimum when the linear wall energy begins to become important. This occurs at a radius R_0, well before the quadratic area energy, due to the energy splitting between the false vacuum and the true vacuum becoming important, when ϵ is sufficiently small. Clearly, though, for large enough radius of the thin-wall string configuration, the energy splitting will be the most important term, and a thin-walled vortex configuration of sufficiently large radius will be unstable to expanding to infinite radius. However, a vortex of radius R_0 will be classically stable and only susceptible to decay via quantum tunnelling. The amplitude for such tunnelling, in the semi-classical approximation in the strictly two-dimensional context, has been calculated in [85].

Here we consider the model in a 3+1-dimensional setting. The vortex can be continued along the third additional dimension as a string, called a cosmic string. The interior of the string contains a large magnetic flux distributed over a region of the true vacuum. It is separated by a thin wall from the outside, where the scalar field is in the false vacuum. The analysis of the decay of two-dimensional vortices cannot directly apply to the decay of the cosmic string, as the cosmic string must maintain continuity along its length. Thus the radius of the string at a given position cannot spontaneously make the quantum tunnelling transition to the larger iso-energetic radius, called R_1, as it is continuously connected to the rest of the string. The whole string could, in principle, spontaneously tunnel to the fat string along its whole length, but the probability of such a transition is strictly zero for an infinite string, and correspondingly small for a closed string loop. Here we will describe the tunnelling transition to a state that corresponds to a spontaneously formed bulge in the putatively unstable thin cosmic string.

6.7.2 Energetics and Dynamics of the Thin, False String

6.7.2.1 Set-up We consider the Abelian Higgs model (spontaneously broken scalar electrodynamics) with a modified scalar potential as in [85] but now generalized to 3+1 dimensions. The Lagrangian density of the model has the form

$$\mathcal{L} = -\frac{1}{4}F_{\mu\nu}F^{\mu\nu} + (D_\mu\phi)^*(D^\mu\phi) - V(\phi^*\phi), \qquad (6.144)$$

where $F_{\mu\nu} = \partial_\mu A_\nu - \partial_\nu A_\mu$ and $D_\mu\phi = (\partial_\mu + eA_\mu)\phi$. The potential is a sixth-order polynomial in ϕ [79, 111], written

$$V(\phi^*\phi) = \lambda(|\phi|^2 - \epsilon v^2)(|\phi|^2 - v^2)^2. \qquad (6.145)$$

Note that the Lagrangian is no longer renormalizable in 3+1 dimensions; however, the understanding is that it is an effective theory obtained from a well-defined renormalizable fundamental Lagrangian. The fields ϕ and A_μ, the vacuum expectation value v have mass dimension 1, the charge e is dimensionless and λ has mass dimension 2 since it is the coupling constant of the sixth-order scalar potential. The potential energy density of the false vacuum $|\phi| = v$ vanishes, while that of the true vacuum has $V(0) = -\lambda v^6\epsilon$. We rescale as

$$\phi \to v\phi \quad A_\mu \to vA_\mu \quad e \to \lambda^{1/2}ve \quad x \to x/(v^2\lambda^{1/2}) \qquad (6.146)$$

so that all fields, constants and the spacetime coordinates become dimensionless, then the Lagrangian density is still given by Equation (6.144) where now the potential is

$$V(\phi^*\phi) = (|\phi|^2 - \epsilon)(|\phi|^2 - 1)^2. \qquad (6.147)$$

and there is an overall factor of $1/(\lambda v^2)$ in the action.

Initially, the cosmic string will be independent of z, the coordinate along its length, and will correspond to a tube of radius R with a trapped magnetic flux in the true vacuum inside, separated by a thin wall from the false vacuum outside. R will vary in Euclidean time τ and in z to yield an instanton solution. Thus we promote R to a field $R \to R(z,\tau)$. Hence we will look for axially symmetric solutions for ϕ and A_μ in cylindrical coordinates (r, θ, z, τ). We use the following ansatz for a vortex of winding number n:

$$\phi(r,\theta,z,\tau) = f(r, R(z,\tau))e^{in\theta}, \qquad A_i(r,\theta,z,\tau) = -\frac{n}{e}\frac{\varepsilon^{ij}r_j}{r^2}a(r, R(z,\tau)), \quad (6.148)$$

where ε^{ij} is the two-dimensional Levi–Civita symbol. This ansatz is somewhat simplistic; it is clear that if the radius of the cosmic string swells out at some range of z, the magnetic flux will dilute and hence through the (Euclidean) Maxwell's equations some "electric" fields will be generated. In three-dimensional, source-free Euclidean electrodynamics, there is no distinct electric field, the Maxwell equations simply say that the three-dimensional magnetic field is a divergence-free and rotation-free vector field that satisfies superconductor boundary conditions at the location of the wall. It is clear that the correct form of the electromagnetic fields will not simply be a diluted magnetic field that always points along the length of the cosmic string as with our ansatz; however, the correction will not give a major contribution and we will neglect it. Indeed, the induced fields will always be smaller by a power of $1/c^2$ when the usual units are used.

The Euclidean action functional for the cosmic string then has the form

$$
S_E[A_\mu, \phi] = \frac{1}{\lambda v^2} \int d^4 x \left[\sum_i \left(\frac{1}{2} F_{0i} F_{0i} + \frac{1}{2} F_{i3} F_{i3} \right) + \frac{1}{2} F_{03} F_{03} + \sum_{ij} \frac{1}{4} F_{ij} F_{ij} \right.
$$

$$
\left. + (\partial_\tau \phi)^* (\partial_\tau \phi) + (\partial_z \phi)^* (\partial_z \phi) + \sum_i D_i(\phi)^* (D_i \phi) + V(\phi^* \phi) \right] \quad (6.149)
$$

where i, j take values just over the two transverse directions and we have already incorporated that $A_0 = A_3 = 0$.

Substituting Equations (6.147) and (6.148) into Equation (6.149), we obtain

$$
S_E = \frac{2\pi}{\lambda v^2} \int dz d\tau \int_0^\infty dr\, r \left[\frac{n^2 \dot{a}^2}{2e^2 r^2} + \frac{n^2 a'^2}{2e^2 r^2} + \frac{n^2 (\partial_r a)^2}{2e^2 r^2} + \dot{f}^2 + f'^2 + (\partial_r f)^2 \right.
$$

$$
\left. + \frac{n^2}{r^2} (1-a)^2 f^2 + (f^2 - \epsilon)(f^2 - 1)^2 \right], \quad (6.150)
$$

where the dot and prime denote differentiation with respect to τ and z, respectively. Then $\dot{a} = \left(\frac{\partial a(r,R)}{\partial R} \right) \dot{R}$ and $a' = \left(\frac{\partial a(r,R)}{\partial R} \right) R'$, and likewise for f, hence the action becomes

$$
S_E = \frac{2\pi}{\lambda v^2} \int dz d\tau \int_0^\infty dr\, r \left[\frac{n^2 \left(\left(\frac{\partial a(r,R)}{\partial R} \right) \dot{R} \right)^2}{2e^2 r^2} + \frac{n^2 \left(\left(\frac{\partial a(r,R)}{\partial R} \right) R' \right)^2}{2e^2 r^2} + \frac{n^2 (\partial_r a)^2}{2e^2 r^2} \right.
$$

$$
+ \left(\frac{\partial f(r,R)}{\partial R} \dot{R} \right)^2 + \left(\frac{\partial f(r,R)}{\partial R} R' \right)^2
$$

$$
\left. + (\partial_r f)^2 + \frac{n^2}{r^2} (1-a)^2 f^2 + (f^2 - \epsilon)(f^2 - 1)^2 \right]
$$

$$
= \frac{2\pi}{\lambda v^2} \int dz \int_0^\infty dr\, r \left[\left(\frac{n^2}{2e^2 r^2} \left(\frac{\partial a(r,R)}{\partial R} \right)^2 + \left(\frac{\partial f(r,R)}{\partial R} \right)^2 \right) (\dot{R}^2 + R'^2) \right.
$$

$$
\left. + \frac{n^2 (\partial_r a)^2}{2e^2 r^2} + (\partial_r f)^2 + \frac{n^2}{r^2} (1-a)^2 f^2 + (f^2 - \epsilon)(f^2 - 1)^2 \right]. \quad (6.151)
$$

We note the two- (Euclidean) dimensional, rotationally invariant form $(\dot{R}^2 + R'^2)$ which appears in the kinetic term. This allows us to make the $O(2)$ symmetric ansatz for the instanton, and the easy continuation of the solution to Minkowski time, to a relativistically invariant $O(1,1)$ solution, once the tunnelling transition has been completed.

In the thin-wall limit, the Euclidean action can be evaluated essentially analytically, up to corrections which are smaller by at least one power of $1/R$. The method of evaluation is identical to that in [85] and we shall not give the details here; we get

$$S_E = \frac{1}{\lambda v^2} \int d^2 x \frac{1}{2} M(R(z,\tau))(\dot{R}^2 + R'^2) + E(R(z,\tau)) - E(R_0), \qquad (6.152)$$

where

$$M(R) = \left[\frac{2\pi n^2}{e^2 R^2} + \pi R \right] \qquad (6.153)$$

$$E(R) = \frac{n^2 \Phi^2}{2\pi R^2} + \pi R - \epsilon \pi R^2. \qquad (6.154)$$

Φ is the total magnetic flux and R_0 is the classically stable thin tube string radius.

6.7.3 Instantons and the Bulge

6.7.3.1 Tunnelling Instanton We look for an instanton solution that is $O(2)$ symmetric. The appropriate ansatz is

$$R(z,\tau) = R(\sqrt{z^2 + \tau^2}) = R(\rho) \qquad (6.155)$$

with the imposed boundary condition that $R(\infty) = R_0$. It is useful to understand what this ansatz means. We expect that the solution will be localized in Euclidean two space, say around the origin. Far from the origin, the solution will be $R = R_0$. Thus if we go to $\tau = -\infty$, the string will be in its dormant, thin state, all at $R = R_0$. As Euclidean time progresses, at some Euclidean time $\tau = -R_1$ a small bulge, an increase in the radius, will start to form at $z = 0$. This bulge will then increase dramatically, until at $\tau = 0$ it will be distributed over a region of the original string of length $2R_1$, the factor of 2 because the radius of the $O(2)$ symmetric bubble is R_1 in both directions. Then the bubble will "bounce" back and shrink and the string will return to its original radius. An alternative description is in terms of the creation of a soliton–anti-soliton pair. The instanton solution will describe the transition from a string of radius R_0 at $\tau = -\infty$, to a point in $\tau = -R_1$ at $z = 0$ when a soliton–anti-soliton pair starts to be created. The configuration then develops a bulge which forms when the pair separates to a radius which again has to be R_1 because of $O(2)$ invariance and which is the bounce point of the instanton along the z-axis at $\tau = 0$. Finally the subsequent Euclidean time evolution continues in a manner which is just the (Euclidean) time reversal of evolution leading up to the bounce point configuration, until a simple cosmic string of radius R_0 is re-established for $\tau \geq R_1$ and all z, *i.e.* for $\rho \geq R_1$. The action functional is given by

$$S_E = \frac{2\pi}{\lambda v^2} \int d\rho\, \rho \left[\frac{1}{2} M(R(\rho)) \left(\frac{\partial R(\rho)}{\partial \rho} \right)^2 + E(R(\rho)) - E(R_0) \right]. \qquad (6.156)$$

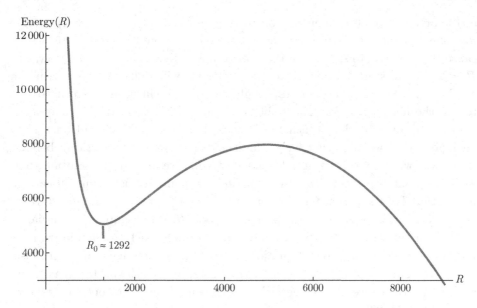

Figure 6.14. The energy as a function of R, for $n = 100, e = .005$ and $\epsilon = .001$

The instanton equation of motion is

$$\frac{d}{d\rho}\left(\rho M(R)\frac{dR}{d\rho}\right) - \frac{1}{2}\rho M'(R)\left(\frac{dR}{d\rho}\right)^2 - \rho E'(R) = 0 \qquad (6.157)$$

with the boundary condition that $R(\infty) = R_0$, and we look for a solution that has $R \approx R_1$ near $\rho = 0$, where R_1 is the large radius for which the string is approximately iso-energetic with the string of radius R_0. The solution necessarily "bounces" at $\tau = 0$ since $\partial R(\rho)/\partial\tau|_{\tau=0} = R'(\rho)(\tau/\rho)|_{\tau=0} = 0$. (The potential singularity at $\rho = 0$ is not there since a smooth configuration requires $R'(\rho)|_{\rho=0} = 0$.)

The equation of motion is better cast as an essentially conservative, dynamical system with a "time"-dependent mass and the potential given by the inversion of the energy function as pictured in Figure 6.14, but in the presence of a "time"-dependent friction where ρ plays the role of time:

$$\frac{d}{d\rho}\left(M(R)\frac{dR}{d\rho}\right) - \frac{1}{2}M'(R)\left(\frac{dR}{d\rho}\right)^2 - E'(R) = -\frac{1}{\rho}\left(M(R)\frac{dR}{d\rho}\right). \qquad (6.158)$$

As the equation is "time"-dependent, there is no analytic trick to evaluating the bounce configuration and the corresponding action. The solution must be found numerically, which starts with a given $R \approx R_1$ at $\rho = 0$ and achieves $R = R_0$ for $\rho > \rho_0$. We can be confident of the existence of a solution by showing the existence of an initial condition that gives an overshoot and another initial condition that gives an undershoot, as pioneered by Coleman [32, 23]. If we start at the origin at $\rho = 0$ high enough on the far right side of the (inverted) energy functional pictured in Figure 6.14, the equation of motion, Equation (6.158), will cause the radius R to slide down the potential and then roll up the hill towards $R = R_0$. If we start too far up to the right, we will roll over the maximum at $R = R_0$, while if we do not start high enough we will never make it to the top of the hill at $R = R_0$. The right-hand side of Equation (6.158) acts as a "time"-dependent friction, which becomes negligible as $\rho \to \infty$, and once it is negligible, the motion is effectively conservative. It is not unrealistic to believe that there will be a correct initial point that will give exactly the solution that we desire, that as $\rho \to \infty$, $R(\rho) \to R_0$. We find the solution exists using numerical integration. For the parameter choice $n = 100, e = .005$ and $\epsilon = .001$, if we start at $R \approx 11,506.4096$, we generate the profile function $R(\rho)$ in Figure 6.15. Actually, numerically integrating to $\rho \approx 80,000$ the function falls back to the minimum of the inverted energy functional Equation (6.14). On the other hand, if we increase the starting point by .0001, the numerical solution overshoots the maximum at $R = R_0$. Hence we have numerically implemented the overshoot/undershoot criterion of [32, 23].

The cosmic string emerges with a bulge described by the function numerically evaluated and represented in Figure 6.15 which corresponds to $R(z, \tau = 0)$. A three-dimensional depiction of the bounce point is given in Figure 6.16. One should imagine the radius $R(z)$ along the cosmic string to be R_0 to the left, then bulging out to the the large radius as described by the mirror image of the function in Figure 6.15 and then returning to R_0 according to the function in Figure 6.15. This radius function has argument $\rho = \sqrt{z^2 + \tau^2}$. Due to the Lorentz invariance of the original action, the subsequent Minkowski time evolution is given by $R(\rho) \to R(\sqrt{z^2 - t^2})$, which is only valid for $z^2 - t^2 \geq 0$. Fixed $\rho^2 = z^2 - t^2$ describes a space-like hyperbola that asymptotes to the light cone. The value of the function $R(\rho)$ therefore remains constant along this hyperbola. This means that the point at which the string has attained the large radius moves away from $z \approx 0$ to $z \to \infty$ at essentially the speed of light. The other side moves towards $z \to -\infty$. Thus the soliton–anti-soliton pair separates quickly, moving at essentially the speed of light, leaving behind a fat cosmic string, which is subsequently classically unstable to expand and fill all space.

The rate at which the classical fat string expands depends on the actual value of ϵ. Once the string radius is large enough, its boundary wall is completely analogous to a domain wall that separates a true vacuum from a false vacuum. The true vacuum exerts a constant pressure on the wall, and it accelerates into the region of false vacuum. Obviously, if there is nothing to retard its expansion, it will accelerate to move at a velocity that eventually approaches the speed of light. The only effects retarding the velocity increase are the inertia and possible radiation. Radiation should be negligible as there are no massless fields in the exterior and there are no accelerating charges. The acceleration,

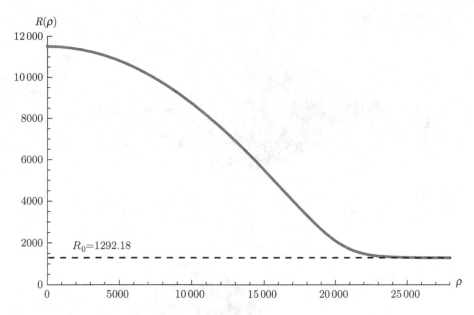

Figure 6.15. The radius as a function of ρ

a, is proportional to pressure divided by the mass per unit area. The pressure is simply the energy density difference, $p = \epsilon$. The mass per unit area can be obtained from Equation (6.153). Here the contribution to the mass per unit length from the wall is simply πR. Thus the mass per unit area, μ, is obtained from $\pi R \times L = \mu 2\pi R \times L$ for a given length L, which gives $\mu = 1/2$. Then we have

$$a \approx \epsilon/\mu = 2\epsilon. \tag{6.159}$$

Thus it is clear that this acceleration can be arbitrarily small, for small ϵ, and it is possible to imagine that once the tunnelling transition has occurred the fat cosmic string will exist and be identifiable for a long time.

6.7.4 Tunnelling Amplitude

It is difficult to say too much about the tunnelling amplitude or the decay rate per unit volume analytically in the parameters of the model. The numerical solution we have obtained for some rather uninspired choices of the parameters gives rise to the profile of the instanton given in Figure 6.15. This numerical solution could then be inserted into the Euclidean action to determine its numerical value, call it $S_0(\epsilon)$. It seems difficult to extract any analytical dependence on ϵ; however, it is reasonable to expect that as $\epsilon \to 0$ the tunnelling barrier, as can be seen in Figure 6.14, will get bigger and bigger and hence the tunnelling amplitude will vanish. On the other hand, there should exist a limiting value, call it ϵ_c, where the tunnelling barrier disappears at the so-called dissociation point [126, 81, 80],

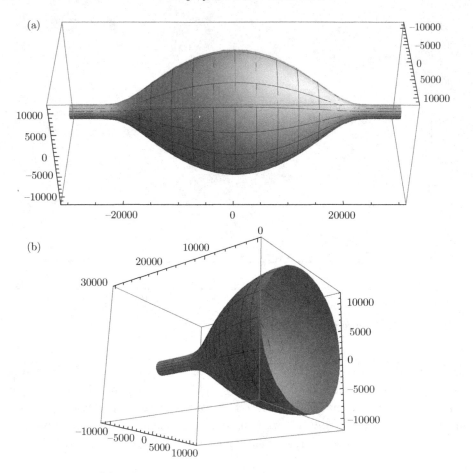

Figure 6.16. (a) Cosmic string profile at the bounce point. (b) Cut away of the cosmic string profile at bounce point

such that as $\epsilon \to \epsilon_c$, the action of the instanton will vanish, analogous to what was found in [85]. In general, the decay rate per unit length of the cosmic string will be of the form

$$\Gamma = A^{\text{c.s.}} \left(\frac{S_0(\epsilon)}{2\pi} \right) e^{-S_0(\epsilon)}, \qquad (6.160)$$

where $A^{\text{c.s.}}$ is the determinantal factor excluding the zero modes and $\left(\frac{S_0(\epsilon)}{2\pi} \right)$ is the correction obtained after taking into account the two zero modes of the bulge instanton. These correspond to invariance under Euclidean time translation and spatial translation along the cosmic string [32, 23]. In general, there will be a length L of cosmic string per volume L^3. For a second-order phase transition to the meta-stable vacuum, L is the correlation length at the temperature of the transition which satisfies $L^{-1} \approx \lambda v^2 T_c$ [70, 69, 130, 129]. For first-order transitions, it is not clear what the density of cosmic strings will be. We will keep L as a parameter, but we expect that it is microscopic. Then in a large volume Ω, we will have a total length NL of cosmic string, where $N = \Omega/L^3$. Thus the decay rate for the volume Ω will be

$$\Gamma \times (NL) = \Gamma\left(\frac{\Omega}{L^2}\right) = A^{\text{c.s.}}\left(\frac{S_0(\epsilon)}{2\pi}\right) e^{-S_0(\epsilon)} \frac{\Omega}{L^2} \tag{6.161}$$

or the decay rate per unit volume will be

$$\frac{\Gamma \times (NL)}{\Omega} = \frac{\Gamma}{L^2} = \frac{A^{\text{c.s.}}\left(\frac{S_0(\epsilon)}{2\pi}\right) e^{-S_0(\epsilon)}}{L^2}. \tag{6.162}$$

A comparable calculation with point-like defects [85] would give a decay rate per unit volume of the form

$$\frac{\Gamma^{\text{point like}}}{L^3} = \frac{A^{\text{point like}}\left(\frac{S_0^{\text{point like}}(\epsilon)}{2\pi}\right)^{3/2} e^{-S_0^{\text{point like}}(\epsilon)}}{L^3} \tag{6.163}$$

and the corresponding decay rate from vacuum bubbles (without topological defects) [32, 23] would be

$$\Gamma^{\text{vac. bubble}} = A^{\text{vac. bubble}}\left(\frac{S_0^{\text{vac. bubble}}(\epsilon)}{2\pi}\right)^2 e^{-S_0^{\text{vac. bubble}}(\epsilon)}. \tag{6.164}$$

Since the length scale L is expected to be microscopic, we would then find that the number of defects in a macroscopic volume (*i.e.* universe) could be incredibly large, suggesting that the decay rate from topological defects would dominate over the decay rate obtained from simple vacuum bubbles [32, 23]. Of course the details depend on the actual values of the Euclidean action and the determinantal factor that is obtained in each case.

There are many instances where the vacuum can be meta-stable. The symmetry-broken vacuum can be meta-stable. Such solutions for the vacuum can be important for cosmology, and for the case of supersymmetry breaking see [1, 47] and the many references therein. In string cosmology, the inflationary scenario that has been obtained in [67] also gives rise to a vacuum that is meta-stable, and it must necessarily be long-lived to have cosmological relevance.

In a condensed matter context, symmetry-breaking ground states are also of great importance. For example, there are two types of superconductors [7]. The cosmic string is called a vortex-line solution in this context, and it is relevant to type II superconductors. The vortex line contains an unbroken symmetry region that carries a net magnetic flux, surrounded by a region of broken symmetry. If the temperature is raised, the true vacuum becomes the unbroken vacuum, and it is possible that the system exists in a superheated state where the false vacuum is meta-stable [41]. This technique has actually been used to construct detectors for particle physics [11, 105]. Our analysis might even describe the decay of vortex lines in superfluid liquid Helium III [86]. The decay of all of these meta-stable states could be described through the tunnelling transition mediated by instantons in the manner that we have computed. For appropriate limiting values of the parameters, for example when $\epsilon \to \epsilon_c$, the suppression of tunnelling is absent, and the existence of vortex lines or cosmic strings could cause the decay of the meta-stable vacuum without bound.

7

Large Orders in Perturbation Theory

7.1 Generalities

We can use instanton methods to obtain the size of the terms in large orders of perturbation theory. We will first consider particle quantum mechanics [21] and then generalize to quantum field theory. The general idea concerns actions of the form

$$S(\phi) = \int d^d x \left(\frac{1}{2} \partial_\mu \phi \partial_\mu \phi + \frac{g}{(2N)!} \phi^{2N} \right) \tag{7.1}$$

with $N = \frac{d}{d-2}$. With

$$\mathcal{I}(g) = \int \mathcal{D}\phi e^{-\frac{S(\phi)}{\hbar}} = \sum_k \mathcal{I}_k g^k \tag{7.2}$$

we have from Cauchy's formula

$$\mathcal{I}_k = \frac{1}{2\pi i} \oint dg \left(\frac{\mathcal{I}(g)}{g^{k+1}} \right), \tag{7.3}$$

where the integral is over a contour containing the origin. For large k we want to perform this integral by Gaussian approximation about a critical point in ϕ and g. The critical point must satisfy the equations of motion, $\bar{h} = 1$,

$$0 = \frac{\partial}{\partial g} \frac{e^{-\frac{S(\phi,g)}{\hbar}}}{g^k} \Rightarrow \frac{k}{g} = -\frac{1}{(2N)!} \int d^d x \phi^{2N} \tag{7.4}$$

and the usual equation for ϕ

$$\partial_\mu \partial^\mu \phi = \frac{g}{(2N-1)!} \phi^{(2N-1)}. \tag{7.5}$$

Changing the scale

$$\phi \to (-g)^{-\frac{1}{2N-2}} \psi \tag{7.6}$$

gives

$$\frac{k}{g} = -\frac{1}{(2N)!}(-g)^{-\frac{N}{N-1}}\int d^d x \psi^{2N}, \qquad (7.7)$$

which implies

$$(2N)!\frac{k}{\int d^d x \psi^{2N}} = (-g)^{-\frac{N}{N-1}+1} = (-g)^{-\frac{1}{N-1}}, \qquad (7.8)$$

which in turn means

$$-g \sim \frac{1}{k^{N-1}}. \qquad (7.9)$$

The other equation is simply

$$(-g)^{-\frac{1}{2N-2}}\partial_\mu \partial^\mu \psi = g\frac{\psi^{2N-1}}{(2N-1)!}(-g)^{-\frac{2N-1}{2N-2}}, \qquad (7.10)$$

which should have a solution with

$$\int d^d x \psi^{2N} < \infty. \qquad (7.11)$$

We find such a critical point in various examples, and then perform the integrals by Gaussian approximation.

7.2 Particle Mechanics

In particle quantum mechanics we consider the Hamiltonian

$$H = \frac{1}{2}\sum_{i=1}^{n}p_i^2 + \frac{1}{2}\sum_{i=1}^{n}x_i^2 + g\left(\sum_{i=1}^{n}x_i^2\right)^N, \qquad (7.12)$$

which describes n anharmonic oscillators which interact with each other. Then

$$\lim_{\beta\to\infty} -\frac{1}{\beta}\ln\left(\frac{\mathrm{tr}\left(e^{-\beta H}\right)}{\mathrm{tr}\left(e^{-\beta H_0}\right)}\right) = \lim_{\beta\to\infty} -\frac{1}{\beta}\ln\left(\frac{e^{-\beta\mathcal{E}}+\cdots}{e^{-\beta\mathcal{E}_0}+\cdots}\right) = \mathcal{E} - \mathcal{E}_0, \qquad (7.13)$$

where $\mathcal{E}$ and $\mathcal{E}_0$ are the ground-state energy of the system and the corresponding free system. The ratio of the traces can be expressed as a path integral

$$\frac{\mathrm{tr}\left(e^{-\beta H}\right)}{\mathrm{tr}\left(e^{-\beta H_0}\right)} = \mathcal{N}\int_{periodic}\mathcal{D}\vec{x}(\tau)e^{-\int_0^\beta d\tau\left(\frac{1}{2}\dot{\vec{x}}^2+\frac{1}{2}\vec{x}^2+g|\vec{x}|^N\right)}. \qquad (7.14)$$

$\mathcal{N}$ is chosen so that the ratio is equal to 1 for $g = 0$. Periodic $\vec{x}(\tau)$ converts the path integral into a trace. The term of order k is extracted via the Cauchy theorem. This integral corresponds to the integral in Equation (7.2).

For particle quantum mechanics, however, we can actually perform the g integration exactly,

$$\frac{1}{2\pi i}\oint dg\frac{e^{-g\int_0^\beta d\tau|\vec{x}|^{2N}}}{g^{k+1}} = (-1)^k\frac{\left(\int_0^\beta d\tau|\vec{x}|^{2N}\right)^k}{k!} \qquad (7.15)$$

thus we find

$$\left.\frac{\text{tr}\left(e^{-\beta H}\right)}{\text{tr}\left(e^{-\beta H_0}\right)}\right|_k = \frac{(-1)^k}{k!}N\int_{periodic}\mathcal{D}x(\tau)e^{-\left(\int_0^\beta d\tau\left(\frac{1}{2}\dot{\vec{x}}^2+\frac{1}{2}\vec{x}^2\right)-k\ln\left(\int_0^\beta d\tau|\vec{x}|^{2N}\right)\right)}. \quad (7.16)$$

To perform the path integral, we look for a critical point in $\vec{x}$, to the equation of motion

$$\ddot{\vec{x}} = \vec{x} - \frac{2Nk\vec{x}|\vec{x}|^{2(N-1)}}{\int_0^\beta d\tau|\vec{x}|^{2N}}. \quad (7.17)$$

Changing the scale by $\vec{x}\to\sqrt{\frac{2Nk}{\int_0^\beta d\tau|\vec{x}|^{2N}}}\vec{x}$ yields the equation

$$\ddot{\vec{x}} = \vec{x} - \vec{x}|\vec{x}|^{2(N-1)}. \quad (7.18)$$

The solution is easily found,

$$\vec{x} = \vec{u}x_0\left(\tau-\tau_0\right) \quad (7.19)$$

with $|\vec{u}|^2=1$ and the function x_0 given by,

$$(x_0\left(\tau\right))^{2(N-1)} = \frac{N}{\cosh^2\left((N-1)\tau\right)}, \quad (7.20)$$

where $x_0(\tau)$ satisfies

$$\ddot{x}_0 = x_0 - x_0(x_0)^{2(N-1)}. \quad (7.21)$$

This is most easily verified by observing

$$\frac{d}{d\tau}x_0^{2(N-1)} = \frac{-2N(N-1)\sinh(N-1)\tau}{\cosh^3(N-1)\tau} = -2x_0^{2(N-1)}(N-1)\tanh(N-1)\tau. \quad (7.22)$$

But

$$\frac{d}{d\tau}x_0^{2(N-1)} = 2(N-1)x_0^{2(N-1)-1}\dot{x}_0 \quad (7.23)$$

and therefore

$$\frac{d}{d\tau}x_0 = -x_0\tanh(N-1)\tau. \quad (7.24)$$

Finally

$$\begin{aligned}\ddot{x}_0 &= -\dot{x}_0\tanh(N-1)\tau + x_0(N-1)\text{sech}^2(N-1)\tau\\ &= x_0\tanh^2(N-1)\tau + x_0(N-1)\text{sech}^2(N-1)\tau\\ &= x_0\left(1-\text{sech}^2(N-1)\tau-(N-1)\text{sech}^2(N-1)\tau\right)\\ &= x_0\left(1-\frac{N}{\cosh^2(N-1)\tau}\right)\\ &= x_0 - x_0(x_0)^{2(N-1)}\end{aligned} \quad (7.25)$$

as required. Periodicity is satisfied if we begin at $\tau=\frac{\beta}{2}$ and end at $\tau=\frac{\beta}{2}$. As $\beta\to\infty$ the action is calculable,

$$S\left[\vec{u}x_0\left(\tau-\tau_0\right)\right] = Nk - Nk\ln\left(2Nk\right) + k(N-1)\ln(J), \quad (7.26)$$

where

$$J = \int_{-\infty}^{\infty} d\tau \, x_0^{2N}(\tau) = \frac{N^{\frac{N}{N-1}} 2^{\frac{N+1}{N-1}} \left(\Gamma \left(\frac{N}{N-1} \right) \right)^2}{N-1} \frac{}{\Gamma \left(\frac{2N}{N-1} \right)}. \tag{7.27}$$

It remains to calculate the determinant corresponding to the Gaussian fluctuations about this critical point. The operator coming from the second variation of the action is

$$\frac{\delta^2}{\delta x_\alpha(\tau_1) \, \delta x_\beta(\tau_2)} S(\vec{x}) \Big|_{\vec{x} = \vec{u} x_0(\tau)} = \left(\left(-\frac{d^2}{d\tau_1^2} + 1 - \frac{N}{\cosh^2((N-1)\tau_1)} \right) \delta_{\alpha\beta} \right.$$

$$\left. - \frac{2(N-1) N u_\alpha u_\beta}{\cosh^2((N-1)\tau_1)} \right) \delta(\tau_1 - \tau_2)$$

$$+ \frac{2N}{J} u_\alpha u_\beta x_0^{2N-1}(\tau_1) \, x_0^{2N-1}(\tau_2)$$

$$= M_L u_\alpha u_\beta + M_T \left(\delta_{\alpha\beta} - u_\alpha u_\beta \right) \tag{7.28}$$

with

$$M_L = \left(-\frac{d^2}{d\tau_1^2} + 1 - \frac{(2N-1)N}{\cosh^2((N-1)\tau_1)} \right) \delta(\tau_1 - \tau_2) + \frac{2N}{J} x_0^{2N-1}(\tau_1) \, x_0^{2N-1}(\tau_2) \tag{7.29}$$

and

$$M_T = \left(-\frac{d^2}{d\tau_1^2} + 1 - \frac{N}{\cosh^2((N-1)\tau_1)} \right) \delta(\tau_1 - \tau_2). \tag{7.30}$$

For the transverse operator M_T, the corresponding "quantum mechanical" Hamiltonian is

$$H = p^2 - \frac{\lambda(\lambda+1)}{\cosh^2(x)}, \tag{7.31}$$

which is exactly solvable. The eigenfunctions are the Jacobi functions. The ratio of the determinants is given by

$$\frac{\det(H-z)}{\det(H_0-z)} = \frac{\Gamma\left(1+\sqrt{-z}\right)\Gamma\left(\sqrt{-z}\right)}{\Gamma\left(1+\lambda+\sqrt{-z}\right)\Gamma\left(\sqrt{-z}-\lambda\right)}, \tag{7.32}$$

which is calculated using the Affleck Coleman method [31, 114, 36], where Γ is the usual gamma function. For the case at hand, $\lambda = \frac{1}{N-1}$, $z = -\frac{1}{(N-1)^2}$ which gives the transverse operator up to a factor of $1/(N-1)^2$. We must separate out the zero modes. These arise because of the invariance of the original Hamiltonian under global rotations of $\vec{x}$ (equivalently of $\vec{u}$). Rotations about a direction orthogonal to $\vec{u}$ should all be in the transverse operator. Thus the zero mode for M_T is simply $x_0(\tau)$.

$$\delta_{rot.} \left(\vec{u} x_0(\tau) \right) = (\delta \vec{u}) x_0(\tau), \tag{7.33}$$

where $(\delta \vec{u})$ counts the number of independent rotations. To find $\frac{\det'(H-z)}{\det(H_0-z)}$, we must divide the ratio by the smallest eigenvalue for λ not equal to its critical value, and then take the limit. Now

$$H - z = -\frac{d^2}{dx^2} - \frac{\lambda(\lambda+1)}{\cosh^2(x)} - z; \tag{7.34}$$

however, if we scale $x \to (N-1)t$, with $\lambda = \frac{1}{N-1}$, $z = -\frac{1}{(N-1)^2}$, we get

$$H - z = \frac{1}{(N-1)^2}\left(-\frac{d^2}{dt^2} - \frac{N}{\cosh^2((N-1)t)} + 1\right). \tag{7.35}$$

Then with $z = -\frac{1}{(N-1)^2} + \epsilon$ the zero mode becomes an eigenmode with eigenvalue ϵ. Each eigenvalue of $H - z$ is $\frac{1}{(N-1)^2}$ times the eigenvalue of $-\frac{d^2}{dt^2} - \frac{N}{\cosh^2((N-1)t)} + 1$. Thus we must divide by $\epsilon(N-1)^2$ to get $\det'$. Hence

$$\lim_{z \to -\frac{1}{(N-1)^2}} \frac{2\pi}{\left(-\frac{1}{(N-1)^2} - z\right)(N-1)^2} \frac{\det(H-z)}{\det(H_0-z)} = \frac{2\pi(N+1)}{2(N-1)} \frac{\Gamma^2\left(\frac{N}{N-1}\right)}{\Gamma\left(\frac{2N}{N-1}\right)} \tag{7.36}$$

The 2π comes from the definition of the measure in the Gaussian integral. There are $n-1$-independent transverse directions, for each one we have the same $\det'$, to the power $-\frac{1}{2}$, giving the total power $-\frac{n-1}{2}$. The Jacobian factor coming from changing the integration variable from the zero mode "Gaussian fluctuation" to the integration over the position gives a factor of $\sqrt{\int_{-\infty}^{\infty} d\tau \dot{x}_0^2(\tau)} = (k(N+1))^{\frac{1}{2}}$. Thus the total contribution of the transverse modes is

$$\frac{(2\pi)^{\frac{n}{2}}}{\Gamma\left(\frac{n}{2}\right)} (k(N+1))^{\frac{n-1}{2}} \left[\frac{\pi(N+1)}{(N-1)} \frac{\Gamma^2\left(\frac{N}{N-1}\right)}{\Gamma\left(\frac{2N}{N-1}\right)}\right]^{-\frac{n-1}{2}}. \tag{7.37}$$

The first factor is the volume from integrating over the directions of $\vec{u}$. The longitudinal operator can also be treated in a similar fashion. With

$$\begin{aligned} M_L &= \bar{M}_L + |u\rangle\langle u| \\ \bar{M}_L &= -\frac{d^2}{dt^2} + 1 - \frac{(2N-1)N}{\cosh^2((N-1)t)} \end{aligned} \tag{7.38}$$

where $|u\rangle\langle u|$ projects on the mode $x_0^{2N-1}(\tau)$. There is one zero mode coming from time translation invariance, $\frac{dx_0(\tau)}{d\tau}$. It is orthogonal to $x_0^{2N-1}(\tau)$

$$\int_{-\infty}^{\infty} d\tau x_0^{2N-1}(\tau) \frac{dx_0(\tau)}{d\tau} = \int_{-\infty}^{\infty} d\tau \frac{d}{d\tau} \frac{x_0^{2N}(\tau)}{2N} = \left.\frac{x_0^{2N}(\tau)}{2N}\right|_{-\infty}^{\infty} = 0 \tag{7.39}$$

Then with $|v\rangle$ denoting eigenstates orthogonal to $|u\rangle$

$$\det\left(\bar{M}_L + |u\rangle\langle u|\right) = \det\left(\bar{M}_L\right)\det\left(1 + \bar{M}_L^{-1}|u\rangle\langle u|\right)$$

$$= \det\left(\bar{M}_L\right)\det\left(1 + \left(|u\rangle\langle u| + \sum_v |v\rangle\langle v|\right)\bar{M}_L^{-1}|u\rangle\langle u|\right)$$

$$= \det\left(\bar{M}_L\right)\det\left(1 + \langle u|\bar{M}_L^{-1}|u\rangle|u\rangle\langle u| + \sum_v \langle v|\bar{M}_L^{-1}|u\rangle|v\rangle\langle u|\right)$$

$$= \det\left(\bar{M}_L\right)\left(1 + \langle u|\bar{M}_L^{-1}|u\rangle\right) \tag{7.40}$$

where the final equality follows because the second determinant is of a matrix that is upper triangular. From the equation of motion

$$\frac{\bar{M}_L}{2(1-N)}x_0 = x_0^{2N-1}. \tag{7.41}$$

Thus

$$\langle u|\bar{M}_L^{-1}|u\rangle = \frac{\int_{-\infty}^{\infty} d\tau x_0^{2N-1}\bar{M}_L^{-1}x_0^{2N-1}}{\int_{-\infty}^{\infty} d\tau \left(x_0^{2N-1}\right)^2}$$

$$= \frac{1}{2(1-N)}\frac{\int_{-\infty}^{\infty} d\tau x_0^{2N-1}x_0}{\int_{-\infty}^{\infty} d\tau \left(x_0^{2N-1}\right)^2}. \tag{7.42}$$

The integrals can be done exactly, giving

$$\det\left(M_L\right) = \frac{-1}{(N-1)}\det\left(\bar{M}_L\right) \tag{7.43}$$

$\bar{M}_L$ has a negative mode, which cancels the minus sign. This is not an instability, $\bar{M}_L$ is just an auxilliary operator. Now we have an operator of the same form as the transverse part before

$$H - z = -\frac{d^2}{dx^2} + \frac{\lambda(\lambda+1)}{\cosh^2(x)} - z \tag{7.44}$$

with $\lambda = \frac{N}{N-1}$, and $z = \frac{-1}{(N-1)^2}$. Then the det$'$ is,

$$\lim_{z \to \frac{-1}{(N-1)^2}} \frac{\det'(H-z)}{\det(H_0-z)}\bigg|_{\lambda=\frac{N}{N-1}} = -2\pi\frac{1}{2}\frac{\Gamma^2\left(\frac{N}{N-1}\right)}{\Gamma\left(\frac{2N}{N-1}\right)} \tag{7.45}$$

The Jacobian from the usual change of variables in the integration is

$$\left(\int_{-\infty}^{\infty} d\tau \left(\frac{dx_0}{d\tau}\right)^2\right)^{\frac{1}{2}} = (k(N-1))^{\frac{1}{2}}, \tag{7.46}$$

which then gives the total contribution

$$\beta\left(\frac{1}{N-1}2\pi\frac{1}{2}\frac{\Gamma^2\left(\frac{N}{N-1}\right)}{\Gamma\left(\frac{2N}{N-1}\right)}\right)^{-\frac{1}{2}}(k(N-1))^{\frac{1}{2}}, \tag{7.47}$$

where the β comes from the integration over the position of the instanton. Finally, putting all the pieces together, we get the correction to the kth order in perturbation to the energy splitting

$$\mathcal{E} - \mathcal{E}_0 = (-1)^{k+1} g^k \left(\frac{2}{\pi}\right)^{\frac{1}{2}} \frac{((N-1))^{\frac{n+1}{2}} k^{\frac{n-1}{2}}}{\Gamma\left(\frac{n}{2}\right) 2^k} \left(\frac{\Gamma\left(\frac{2N}{N-1}\right)}{\Gamma^2\left(\frac{N}{N-1}\right)}\right)^{k(N-1)+\frac{n}{2}}$$

$$\times e^{k(N-1)\ln\left(\frac{k(N-1)}{e}\right)} (1 + o(g)). \tag{7.48}$$

7.3 Generalization to Field Theory

This result can be generalized to the case of quantum field theory; we leave this for the reader. We should make one point, though. Generally, we do not believe that the functional integral is an analytic function in an annulus around the point $g = 0$ in the g-plane. Indeed, for g negative the Hamiltonian is not self-adjoint for sufficiently large N. We expect that in reality $\mathcal{I}(g)$ is defined in the complex plane by analytic continuation, and this analytic function has a branch cut along the negative real axis which terminates at $g = 0$. We must use the once subtracted dispersion relation

$$\mathcal{I}(g) = -\frac{1}{2\pi i} \int_0^R d\lambda \frac{1}{\lambda + g} \left(\mathcal{I}(-\lambda + i\epsilon) - \mathcal{I}(-\lambda - i\epsilon)\right) + \frac{1}{2\pi i} \int_{|g'|=R} \frac{\mathcal{I}(g)}{(g' - g)}, \tag{7.49}$$

which corresponds to the contour in Figure 7.1. If the second integral vanishes as $R \to \infty$ we get

$$\mathcal{I}(g) = -\frac{1}{2\pi i} \int_0^\infty d\lambda \frac{1}{\lambda + g} \left(\text{discontinuity}\left(\mathcal{I}(-\lambda)\right)\right). \tag{7.50}$$

For

$$\mathcal{I}(g) = \sum_k \mathcal{I}_k g^k \tag{7.51}$$

we have

$$\mathcal{I}_k = -(-1)^k \frac{1}{2\pi i} \int_0^\infty d\lambda \frac{1}{\lambda^{k+1}} \left(\text{discontinuity}\left(\mathcal{I}(-\lambda)\right)\right). \tag{7.52}$$

The factor $1/\lambda^{k+1}$ becomes more and more singular at the end point $\lambda = 0$; thus, if we know how the discontinuity of $\mathcal{I}(-\lambda)$ behaves for small λ, we can find the behaviour of $\mathcal{I}_k$ for large k. For an expected asymptotic behaviour

$$\text{discontinuity}\left(\mathcal{I}(-\lambda)\right) \sim 2iBe^{-\frac{S_c}{\lambda}} \lambda^{-\alpha} \sum_l a_l \lambda^l \tag{7.53}$$

implies directly

$$\mathcal{I}_k \sim -\frac{1}{\pi}(-1)^k B \sum_l a_l \Gamma(k + \alpha - l) S_c^{-(k+\alpha-l)}. \tag{7.54}$$

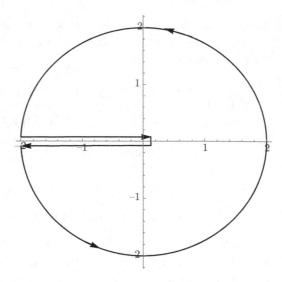

Figure 7.1. Integration contour in the complex g plane for field theory

The discontinuity of discontinuity $(\mathcal{I}(-\lambda))$ can actually be computed using the semi-classical methods that we have been learning about. Collins and Soper [35] show that it has an expansion of exactly the form given in Equation (7.53). Thus the formal calculations that we have done, not worrying about the cut in the complex g plane, produce the same results with much less difficulty.

7.4 Instantons and Quantum Spin Tunnelling

We continue this chapter with an application to quantum spin tunnelling. This calculation starts out as an independent tunnelling calculation that, in principle, has nothing to do with large orders in perturbation theory. However, it turns out that the tunnelling calculations are all attainable through large orders in perturbation theory. We will have to understand what spin-coherent states are and the corresponding path integral.

7.5 Spin-Coherent States and the Path Integral for Spin Systems

For a spin system, instead of the orthogonal position $|x\rangle$ and momentum $|p\rangle$ basis we define a basis of spin-coherent states [106, 100, 75, 87]. Let $|s,s\rangle$ be the highest weight vector in a particular representation of the rotation group. This state is taken to be an eigenstate of the operators $\hat{S}_z$ and $\hat{S}$, two mutually commuting operators:

$$\hat{S}_z |s,s\rangle = s|s,s\rangle \quad \hat{S}^2 |s,s\rangle = s(s+1)|s,s\rangle. \tag{7.55}$$

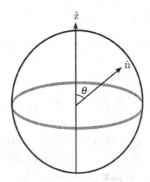

Figure 7.2. The directions of the unit vectors $\hat{z}$ and $\hat{n}$ on a two-sphere

The spin operators $\hat{S}_i$, $i = x, y, z$ satisfy the Lie algebra of $SU(2)$,

$$[\hat{S}_i, \hat{S}_j] = i\epsilon_{ijk}\hat{S}_k \tag{7.56}$$

where ϵ_{ijk} is the totally antisymmetric tensor symbol and summation over repeated indices is implied in Equation (7.56).

The coherent state is defined as [100, 87, 75, 127, 49, 48]

$$|\hat{n}\rangle = e^{i\theta\hat{m}\cdot\hat{S}}|s,s\rangle = \sum_{m=-s}^{s} \mathcal{M}^s(\hat{n})_{ms}|s,m\rangle, \tag{7.57}$$

where $|\hat{n}\rangle$ is an element of the $2s+1$-dimensional Hilbert (representation) space for the spin states, $\hat{n} = (\cos\phi\sin\theta, \sin\phi\sin\theta, \cos\theta)$ is a unit vector, $i.e.$ $\hat{n}^2 = 1$ and $\hat{m} = (\hat{n} \times \hat{z})/|\hat{n} \times \hat{z}|$ is a unit vector orthogonal to $\hat{n}$ and $\hat{z}$. $\hat{z}$ is the quantization axis pointing from the origin to the north pole of a unit sphere and $\hat{n} \cdot \hat{z} = \cos\theta$ as shown in Figure 7.2. Rotating the unit vector $\hat{n}$ counterclockwise about the $\hat{m}$ direction by the angle θ brings it exactly to the unit vector $\hat{z}$. $|\hat{n}\rangle$ corresponds to a rotation of an eigenstate of $\hat{S}_z$, i.e $|s,s\rangle$, to an eigenstate with a quantization axis along $\hat{n}$ on a two-dimensional sphere $\mathcal{S}^2 = SU(2)/U(1)$. It turns out that the matrices $\mathcal{M}^s(\hat{n})$ satisfy a non-trivial relation

$$\mathcal{M}^s(\hat{n}_1)\mathcal{M}^s(\hat{n}_2) = \mathcal{M}^s(\hat{n}_3)e^{i\mathcal{G}(\hat{n}_1,\hat{n}_2,\hat{n}_3)\hat{S}_z} \tag{7.58}$$

where $\mathcal{G}(\hat{n}_1, \hat{n}_2, \hat{n}_3)$ is the area of a spherical triangle with vertices $\hat{n}_1, \hat{n}_2, \hat{n}_3$. Note that Equation (7.58) is not a group multiplication, thus the matrices $\mathcal{M}^s(\hat{n})$ do not form a group representation and $\mathcal{G}(\hat{n}_1, \hat{n}_2, \hat{n}_3)$ is called a co-cycle, which represents the obstruction that the matrices $\mathcal{M}^s(\hat{n})$ exhibit to forming a true representation of the rotation group.

Unlike normal position and momentum eigenstates, the inner product of two coherent states is not orthogonal:

$$\langle\hat{n}|\hat{n}'\rangle = e^{is\mathcal{G}(\hat{n},\hat{n}',\hat{z})}[\frac{1}{2}(1+\hat{n}\cdot\hat{n}')]^s \tag{7.59}$$

It has the following property:

$$\hat{n} \cdot \hat{S} |\hat{n}\rangle = s |\hat{n}\rangle \Rightarrow \langle \hat{n}| \hat{S} |\hat{n}\rangle = s\hat{n} \tag{7.60}$$

while the resolution of identity is given by

$$\hat{I} = \frac{2s+1}{4\pi} \int d^3 \hat{n} \delta(\hat{n}^2 - 1) |\hat{n}\rangle \langle \hat{n}|, \tag{7.61}$$

where $\hat{I}$ is a $(2s+1) \times (2s+1)$ identity matrix, and the delta function ensures that $\hat{n}^2 = 1$. Using the expression in Equations (7.59) and (7.61) we can express the imaginary time transition amplitude between $|\hat{n}_i\rangle$ and $|\hat{n}_f\rangle$ as a path integral, which for the spin system is given by [48, 127]

$$\langle \hat{n}_f | e^{-\beta \hat{H}(\hat{S})} |\hat{n}_i\rangle = \int \mathcal{D}\hat{n} \, e^{-S_E[\hat{n}]}, \tag{7.62}$$

where

$$S_E[\hat{n}] = is S_{WZ} + \int d\tau U(\hat{n}(\tau)), \quad U(\hat{n}(\tau)) = \langle \hat{n}| \hat{H} |\hat{n}\rangle \tag{7.63}$$

and S_{WZ} arises because of the additional phase $e^{is\mathcal{G}(\hat{n},\hat{n}',\hat{z})}$ in Equation (7.59). We have set $\hbar = 1$ in the path integral.

The Wess–Zumino (WZ) action, S_{WZ} in the coordinate independent formalism, is given by[1] [97, 122, 120, 49, 48]

$$S_{WZ} = \int_{\frac{1}{2}\mathcal{S}^2} d\tau d\xi \, \hat{n}(\tau,\xi) \cdot [\partial_\tau \hat{n}(\tau,\xi) \times \partial_\xi \hat{n}(\tau,\xi)], \tag{7.64}$$

where $\hat{n}(\tau)$ has been extended over a topological half-sphere $\frac{1}{2}\mathcal{S}^2$ in the variables τ, ξ. We call this the coordinate independent expression since no system of coordinates is specified for the unit vector $\hat{n}$. In the topological half-sphere we define $\hat{n}$ with the boundary conditions

$$\hat{n}(\tau,0) = \hat{n}(\tau), \; \hat{n}(\tau,1) = \hat{z} \tag{7.65}$$

so that the original configuration lies at the equator and the point $\xi = 1$ is topologically compactified by the boundary condition. This can be easily obtained by imagining that the original closed loop $\hat{n}(\tau)$ at $\xi = 0$ is simply pushed up along the meridians to $\hat{n}(\tau) = \hat{z}$ at $\xi = 1$. The WZ term originates from the non-orthogonality of spin-coherent states in Equation (7.59). Geometrically, it defines the area of the closed loop on the spin space, defined by the nominally periodic, original configuration $\hat{n}(\tau)$. It is crucial to note that there is an ambiguity of modulo 4π, since different ways of pushing the original configuration up to the north pole can give different values for the area enclosed by the closed loop as we can imagine that the closed loop englobes the whole two sphere any integer number of times, but this ambiguity has no physical significance since

---

[1] An alternative way of deriving this equation can be found in [16].

$e^{iN4\pi s} = 1$ for integer and half-odd integer s. The action, Equation (7.63), is valid for a semi-classical spin system whose phase space is $\mathcal{S}^2$. It is the starting point for studying macroscopic quantum spin tunnelling between the minima of the energy $U(\hat{n})$.

7.6 Coordinate-Independent Formalism

In the coordinate independent formalism, the spin is represented by a unit vector $\hat{n}(\tau)$ but no parametrization of the unit vector is assumed. It is best to exemplify the coordinate independent analysis through an explicit system.

We will study the simplest biaxial single-molecule magnet whose spin Hamiltonian is given by

$$H = -K_z \hat{S}_z^2 + K_y \hat{S}_y^2, \quad K_z \gg K_y > 0. \tag{7.66}$$

The above Hamiltonian possesses an easy-axis in the z-direction and a hard-axis in the y-direction. When $K_y = 0$, the spin is localized along the z-axis, which is usually parameterized by the variable $\theta = 0, \pi$ and possesses two degenerate minima localized at the north and the south poles of the two-sphere of phase space. Addition of small $K_y \neq 0$ introduces dynamics into the system and causes tunnelling. The real tunnelling variable is expected to be θ in the easy-axis direction.

The Hamiltonian defined above has been studied in the presence of a magnetic field by many authors [28, 27, 53]. However, the quantum-phase interference for this model has not been computed, due to the subtlety involved in computing the action for the instanton. Since the relation $\hat{S}^2 = \hat{S}_x^2 + \hat{S}_y^2 + \hat{S}_z^2 = s(s+1)$ holds for any spin system, it is evident that any biaxial single-molecule magnet is related to any other either by rescaling the anisotropy constant or by rotation of axes. For instance, the Hamiltonian studied by Enz and Schilling [43]:

$$H = -A\hat{S}_x^2 + B\hat{S}_z^2, \quad (h = 0), \tag{7.67}$$

possesses an easy x-axis and a hard z-axis. This model in the conventional spherical parametrization in terms of the phase space variables, $\hat{n} = (\sin\theta\cos\phi, \sin\theta\sin\phi, \cos\theta)$ is exactly our Hamiltonian Equation (2.15) in the unconventional spherical parametrization $\hat{n} = (\sin\theta\sin\phi, \cos\theta, \sin\theta\cos\phi)$ with $K_z = A$ and $K_y = B$.

7.6.1 Coordinate-Dependent Analysis

To demonstrate the technique for investigating the quantum-phase interference in the z-easy-axis model, we will first keep to the conventional, coordinate-dependent spherical parametrization, $\hat{n} = (\sin\theta\cos\phi, \sin\theta\sin\phi, \cos\theta)$. It was shown [52] that perturbation theory in the K_y term for integer spin leads to an energy-splitting proportional to $(K_y)^s$ while for half-odd integer spin, the splitting vanishes in accordance with Kramers' theorem. We will recover this result using the spin-coherent state path integral, and we will explicitly demonstrate in all detail that the result can be obtained without recourse to a coordinate-dependent parametrization.

The transition amplitude in the spin-coherent state path integral, in the coordinate-dependent formalism, is given by [75]

$$\langle \theta_f, \phi_f | e^{-\beta H} | \theta_i, \phi_i \rangle = \int \mathcal{D}[\cos\theta]\, \mathcal{D}[\phi]\, e^{-S_E/\hbar} \tag{7.68}$$

where the Euclidean action is

$$S_E = \int d\tau \left[is\dot{\phi}(1 - \cos\theta) + E(\theta, \phi) \right] \tag{7.69}$$

where the first term is the WZ term in the coordinate dependent formalism and the classical anisotropy energy Equation (2.15) is

$$E(\theta, \phi) = (K_z + K_y \sin^2\phi)\sin^2\theta. \tag{7.70}$$

We note that the WZ term, being first-order in time derivatives, remains imaginary upon analytic continuation to Euclidean time. This has important ramifications for the putative instanton solutions: they too must have non-trivial imaginary parts. The classical degenerate ground states correspond to $\phi = 0$, $\theta = 0, \pi$, that is the spin is pointing in the north or south pole of the two-sphere. The classical equations of motion obtained by varying the action with respect to θ and ϕ, respectively, are

$$is\dot{\phi}\sin\theta = -\frac{\partial E(\theta, \phi)}{\partial\theta} \tag{7.71}$$

$$is\dot{\theta}\sin\theta = \frac{\partial E(\theta, \phi)}{\partial\phi}. \tag{7.72}$$

It is evident from these two equations, because of the explicit i, that one variable has to be imaginary for the equations to be consistent. The only appropriate choice is to take real θ and imaginary ϕ, since the real tunnelling coordinate (z-easy-axis) is θ. This comes out naturally from the conservation of energy, which follows by multiplying Equation (7.72) by $\dot{\phi}$ and Equation (7.71) by $\dot{\theta}$ and subtracting the two:

$$\frac{dE(\theta, \phi)}{d\tau} = 0 \quad \text{i.e,} \quad E(\theta, \phi) = \text{const.} = 0, \tag{7.73}$$

the normalization coming from the value at $\theta = 0$. Thus,

$$E(\theta, \phi) = (K_z + K_y \sin^2\phi)\sin^2\theta = 0. \tag{7.74}$$

Since $\sin^2\theta \neq 0$, as it must vary from 0 to π, it follows that,

$$\sin\phi = \pm i\sqrt{\frac{K_z}{K_y}}, \tag{7.75}$$

therefore ϕ is imaginary and constant. Let $\phi = \phi_R + i\phi_I$, then $\sin\phi = \sin\phi_R\cosh\phi_I + i\cos\phi_R\sinh\phi_I$. We must take $\phi_R = n\pi$ so that $\sin\phi_R = 0$ as the right-hand side of Equation (7.75) is imaginary. Hence

$$(-1)^n \sinh\phi_I = \pm\sqrt{\frac{K_z}{K_y}}, \tag{7.76}$$

as $\cos\phi_R = (-1)^n$. There are four solutions of this equation: $n = 0$, $\phi = i\phi_I$ and $n = 1$, $\phi = \pi - i\phi_I$ for the positive sign and $n = 0$, $\phi = -i\phi_I$ and $n = 1$, $\phi = \pi + i\phi_I$ for the negative sign, where ϕ_I is the same in both cases. Taking into account that $K_z \gg K_y > 0$, we have $\phi_I = \text{arcsinh}\left(\sqrt{\frac{K_z}{K_y}}\right) \approx \frac{1}{2}\ln\left(\frac{4K_z}{K_y}\right)$.

The classical equation of motion (7.72) simplifies to

$$is\frac{\dot{\theta}}{\sin\theta} = K_y\sin 2\phi = iK_y\sinh 2\phi_I \tag{7.77}$$

The solution is easily found as

$$\theta(\tau) = 2\arctan[\exp(\omega(\tau - \tau_0))], \tag{7.78}$$

where $\omega = \frac{K_y}{s}\sinh 2\phi_I$. This corresponds to the tunnelling of the state $|\uparrow\rangle$ from $\theta(\tau) = 0$ at $\tau = -\infty$ to the state $|\downarrow\rangle$, $\theta(\tau) = \pi$ at $\tau = \infty$. The two solutions $\phi = i\phi_I$ and $\phi = \pi + i\phi_I$ in the upper-half plane of complex ϕ correspond to the instanton, $(\dot{\theta} > 0)$ while the solutions $\phi = -i\phi_I$ and $\phi = \pi - i\phi_I$ in the lower-half plane of complex ϕ correspond to the anti-instanton, $(\dot{\theta} < 0)$.

Since the energy, $E(\theta, \phi)$, in the action Equation (7.69) is conserved and therefore always remains zero along this trajectory, the action for this path is determined only by the WZ term which is given by

$$S_E = S_{WZ} = is\int_{-\infty}^{\infty} d\tau\,\dot{\phi}(1 - \cos\theta) = is\int_{\phi_i}^{\phi_f} d\phi(1 - \cos\theta). \tag{7.79}$$

Naively, one can use the fact that ϕ is constant and hence $\dot{\phi} = 0$, which seems to give $S_{WZ} = 0$; however, care must be taken when computing the action. A non-zero Euclidean action is found by realizing, as in [99], that we must take into account the fact that ϕ must be translated from $\phi = 0$ to $\phi = n\pi + i\phi_I$ before the instanton can occur and then back to $\phi = 0$ after the instanton has occurred. Since the action is linear in time derivative of ϕ, the actual path taken does not matter, only the boundary values matter. In the present problem, we have two solutions for ϕ, *i.e.* $\phi = i\phi_I$ and $\phi = \pi + i\phi_I$ corresponding to two instanton paths, call them *I* and *II*. The full action is then

$$S_E^I = is\int_0^{\pi + i\phi_I} d\phi(1 - \cos\theta)|_{\theta=0} + is\int_{\pi + i\phi_I}^0 d\phi(1 - \cos\theta)|_{\theta=\pi}$$

$$= -2\pi is + 2s\phi_I \tag{7.80}$$

and

$$S_E^{II} = is \int_0^{i\phi_I} d\phi (1 - \cos\theta)|_{\theta=0} \tag{7.81}$$

$$+ is \int_{i\phi_I}^0 d\phi (1 - \cos\theta)|_{\theta=\pi} = 2s\phi_I,$$

where it is clear that the total derivative term $d\phi$ contributes nothing as the two contributions cancel in the round trip, while the $d\phi \cos\theta$ gives all the answer, since $\cos\theta = 1$ initially, before the instanton has occurred, while $\cos\theta = -1$ after. The action for the corresponding anti-instantons is identical. The amplitude for the transition from $\theta = 0$ to $\theta = \pi$, as usual, is calculated by summing over a sequence of one instanton followed by an anti-instanton with an odd total number of instantons and anti-instantons [31], but we must add the two exponentials of the actions S_E^I and S_E^{II} for both instanton and anti-instanton. We note

$$e^{S_E^I} + e^{S_E^{II}} = e^{-2s\phi_I} \left(1 + e^{2\pi is}\right) = e^{-2s\phi_I} \left(1 + \cos 2\pi s\right), \tag{7.82}$$

where the last factor vanishes for half-odd integer spin. Then we get that the expression for the amplitude is given by

$$\langle \pi | e^{-\beta \hat{H}} | 0 \rangle = \mathcal{N} \sinh\left(\kappa \beta (1 + \cos(2\pi s)) e^{-2s\phi_I}\right) \tag{7.83}$$

where κ is the properly normalized square root of the determinant of the operator governing the second-order fluctuations without the zero mode, which we have not computed, and $\mathcal{N}$ is the usual normalization factor. The energy splitting can be read off from this expression

$$\Delta E = \kappa (1 + \cos(2\pi s)) e^{-2s\phi_I}. \tag{7.84}$$

For half-odd integer spin the splitting vanishes, while for integer spin we have

$$\Delta E = 4\kappa \left(\frac{K_y}{4K_z}\right)^s \tag{7.85}$$

which agrees with the result found by perturbation theory [52].

7.6.2 Coordinate-Independent Analysis

Now we wish to see that the spherical-polar coordinate-dependent parametrization of the unit vector $\hat{n}$ is not at all necessary. Then the action for the Hamiltonian in Equation (7.66) can be written as

$$S_E = \int d\tau \mathcal{L}_E = \int d\tau \left[-K_z (\hat{n} \cdot \hat{z})^2 + K_y (\hat{n} \cdot \hat{y})^2 \right]$$

$$+ is \int d\tau d\xi \left[\hat{n} \cdot (\partial_\tau \hat{n} \times \partial_\xi \hat{n}) \right]. \tag{7.86}$$

The first term is the anisotropy energy while the second term is the WZ term written in a coordinate-independent form. The WZ term is integrated over a two manifold whose boundary is physical, Euclidean time τ. Thus the configuration in τ is extended into a second dimension with coordinate ξ. The equations of motion arise from variation with respect to $\hat{n}$. However, $\hat{n}$ is a unit vector, hence its variation is not arbitrary, indeed, $\hat{n} \cdot \delta\hat{n} = 0$. Thus, to obtain the equations of motion, we vary $\hat{n}$ as if it is not constrained, but then we must project on to the transverse part of the variation:

$$\delta_{\hat{n}} S_E = 0 \;\Rightarrow\; \int d\tau (\delta_{\hat{n}} \mathcal{L}_E) \cdot \delta\hat{n} = 0 \;\Rightarrow\; \hat{n} \times (\delta_{\hat{n}} \mathcal{L}_E) = 0. \qquad (7.87)$$

Taking the cross-product of the resulting equation one more time with $\hat{n}$ does no harm, and this process yields the equations of motion

$$is\partial_\tau \hat{n} - 2K_z(\hat{n} \cdot \hat{z})(\hat{n} \times \hat{z}) + 2K_y(\hat{n} \cdot \hat{y})(\hat{n} \times \hat{y}) = 0. \qquad (7.88)$$

Taking the cross-product of this equation with $\partial_\tau \hat{n}$, the first term vanishes as the vectors are parallel, yielding

$$-2K_z(\hat{n} \cdot \hat{z})\partial_\tau \hat{n} \times (\hat{n} \times \hat{z}) + 2K_y(\hat{n} \cdot \hat{y})\partial_\tau \hat{n} \times (\hat{n} \times \hat{y}) = 0. \qquad (7.89)$$

Simplifying the triple vector product, using $\partial_\tau \hat{n} \cdot \hat{n} = 0$, and then taking the scalar product of the subsequent equation with $\hat{n}$ gives

$$\partial_\tau \left(-K_z(\hat{n} \cdot \hat{z})^2 + K_y(\hat{n} \cdot \hat{y})^2 \right) = 0, \qquad (7.90)$$

which is the conservation of energy. The initial value of $\hat{n} = \hat{z}$ says that the energy must equal

$$\left(-K_z(\hat{n} \cdot \hat{z})^2 + K_y(\hat{n} \cdot \hat{y})^2 \right) = -K_z. \qquad (7.91)$$

From this equation and because $\hat{n}$ is a unit vector we find

$$\hat{n} \cdot \hat{y} = \pm\sqrt{\frac{K_z}{K_y}((\hat{n} \cdot \hat{z})^2 - 1)} = \pm i\sqrt{\frac{K_z}{K_y}(1 - (\hat{n} \cdot \hat{z})^2)}$$

$$\hat{n} \cdot \hat{x} = \pm\sqrt{\frac{K_y + K_z}{K_y}(1 - (\hat{n} \cdot \hat{z})^2)}, \qquad (7.92)$$

where the $\pm$ signs are not correlated. Then

$$\frac{\hat{n} \cdot \hat{y}}{\hat{n} \cdot \hat{x}} = \pm i\sqrt{\frac{K_z}{K_y + K_z}} = \tan\phi; \qquad (7.93)$$

hence, we recover the result immediately that ϕ is a complex constant, just as before. Taking the scalar product of Equation (7.88) with $\hat{z}$ yields

$$is\partial_\tau(\hat{n} \cdot \hat{z}) + 2K_y(\hat{n} \cdot \hat{y})(\hat{n} \cdot \hat{x}) = 0 \qquad (7.94)$$

and replacing from Equation (7.92) gives

$$is\partial_\tau(\hat{n}\cdot\hat{z}) \pm 2i\sqrt{K_z(K_y+K_z)}(1-(\hat{n}\cdot\hat{z})^2)=0 \qquad (7.95)$$

Notice that the i's neatly cancel leaving a trivial, real differential equation for $\hat{n}\cdot\hat{z}$, which we can write as

$$\frac{\partial_\tau(\hat{n}\cdot\hat{z})}{1-(\hat{n}\cdot\hat{z})}+\frac{\partial_\tau(\hat{n}\cdot\hat{z})}{1+(\hat{n}\cdot\hat{z})}=\pm\frac{4}{s}\sqrt{K_z(K_y+K_z)}. \qquad (7.96)$$

This integrates as

$$\ln\frac{1+(\hat{n}\cdot\hat{z})}{1-(\hat{n}\cdot\hat{z})}=\pm\frac{4}{s}\sqrt{K_z(K_y+K_z)}(\tau-\tau_0). \qquad (7.97)$$

Exponentiating and solving for $\hat{n}\cdot\hat{z}$ gives

$$\hat{n}\cdot\hat{z}=\pm\tanh\left(\frac{2}{s}\sqrt{K_z(K_y+K_z)}(\tau-\tau_0)\right), \qquad (7.98)$$

which is exactly the same as the solution found for θ in Equation (7.78). The instanton (upper sign) interpolates from $n_z=1$ to $n_z=-1$ as $\tau\to\pm\infty$.

Thus it is important to know that the equations of motion can be solved without recourse to a specific choice for the coordinates. We will now evaluate the tunnelling amplitude and the quantum interference directly in terms of the coordinate-independent variables. Since the energy remains constant along the instanton trajectory, the action is determined entirely from the WZ term

$$S_{WZ}=is\int d\tau\int_0^1 d\xi\,[\hat{n}\cdot(\partial_\tau\hat{n}\times\partial_\xi\hat{n})]. \qquad (7.99)$$

The integration over ξ can be done explicitly by writing the unit vector as

$$\hat{n}(\tau,\xi)=f(\tau,\xi)n_z(\tau)\hat{z}+g(\tau,\xi)[n_x(\tau)\hat{x}+n_y(\tau)\hat{y}] \qquad (7.100)$$

with the boundary conditions $\hat{n}(\tau,\xi=0)=\hat{n}(\tau)$ and $\hat{n}(\tau,\xi=1)=\hat{z}$, where we write n_z for $\hat{n}(\tau)\cdot\hat{z}$, etc. Using the expression in Equations (7.100) and the condition that $\hat{n}\cdot\hat{n}=1$, one obtains

$$g^2=\frac{1-f^2n_z^2}{1-n_z^2} \qquad (7.101)$$

These functions obey the boundary conditions

$$f(\tau,\xi=0)=1, f(\tau,\xi=1)=\frac{1}{n_z(\tau)},$$
$$g(\tau,\xi=0)=1, g(\tau,\xi=1)=0 \qquad (7.102)$$

The integrand of Equation (7.99) can now be written in terms of the functions defined in Equation (7.100). After a straightforward, but rather tedious, calculation we obtain

$$\hat{n} \cdot (\partial_\tau \hat{n} \times \partial_\xi \hat{n}) = n_z (g^2 f' - fgg')(n_x \dot{n}_y - n_y \dot{n}_x)$$

$$= \frac{n_z f'}{1 - n_z^2}(n_x \dot{n}_y - n_y \dot{n}_x), \tag{7.103}$$

where $f' \equiv \partial_\xi f$, $\dot{n}_{x,y} \equiv \partial_\tau n_{x,y}$. The second equality follows from Equation (7.101). Replacing Equation (7.103) into the WZ term, the ξ integration in Equation (7.99) can be done explicitly which yields

$$S_{WZ} = is \int d\tau \frac{(n_x \dot{n}_y - n_y \dot{n}_x)}{1 + n_z}. \tag{7.104}$$

This expression defines the WZ term in the coordinate-independent form as a function of time alone. We can always make recourse to any specific coordinates, taking the z-easy-axis system, with the spherical parameterization we recover the usual form of the WZ term in condensed matter physics, *i.e.* Equation (7.79). Multiplying the top and the bottom of the integrand in Equation (7.104) by $(1 - n_z)$, the resulting integrand simplifies to

$$S_{WZ} = is \int \frac{d(n_y/n_x)}{1 + (n_y/n_x)^2}(1 - n_z)$$

$$= is \int d[\arctan(n_y/n_x)](1 - n_z)$$

$$= is \int d\phi(1 - n_z), \tag{7.105}$$

which is rather analogous to the coordinate-dependent expression in Equation (7.79).

It was already noted from Equation (7.93) that ϕ has to be imaginary. To recover the quantum-phase interference in the coordinate-independent formalism, ϕ must be translated from the initial point, say $\phi = 0$, to the final point, $\phi = n\pi + i\phi_I$, $n = 0,1$ before and after the instanton occurs [99]. The two contributions to the action from these paths are given by

$$S_{WZ}^I = is \int_0^{\pi + i\phi_I} d\phi(1 - n_z)|_{n_z = 1} + is \int_{\pi + i\phi_I}^0 d\phi(1 - n_z)|_{n_z = -1}$$

$$= -2\pi i s + 2s\phi_I \tag{7.106}$$

and

$$S_{WZ}^{II} = is \int_0^{i\phi_I} d\phi(1 - n_z)|_{n_z = 1} \tag{7.107}$$

$$+ is \int_{i\phi_I}^0 d\phi(1 - n_z)|_{n_z = -1} = 2s\phi_I$$

which are exactly the expressions as before. Then the previous evaluation quantum interference goes through unchanged.

7.7 Instantons in the Spin Exchange Model

We will study a second example where instantons give rise to quantum tunnelling in spin systems and breaks the degeneracy, the case of two large, coupled, quantum spins in the presence of a large, simple, easy-axis anisotropy, interacting with each other through a standard spin–spin exchange coupling, corresponding to the Hamiltonian

$$H = -K(S_{1z}^2 + S_{2z}^2) + \lambda \vec{S}_1 \cdot \vec{S}_2. \tag{7.108}$$

We will take $K > 0$ and the case of equal spins $\vec{S}_1 = \vec{S}_2 = \vec{S}$. $\lambda > 0$ gives an anti-ferromagnetic coupling while $\lambda < 0$ sign corresponds to ferromagnetic coupling. The first term gives rise to the anisotropy, favouring an easy-axis, the z-axis, the first term's contribution to the energy is obviously minimized if the spin is pointing along the z-direction and is as large as possible. The second term is called the Heisenberg exchange energy interaction. The spins $\vec{S}_i$ could correspond to quantum spins of macroscopic multi-atomic molecules [113, 116, 90], or the quantum spins of macroscopic ferromagnetic grains [28, 27], or the average spin of each of the two staggered Neél sub-lattices in a quantum anti-ferromagnet [116, 91, 92].

A Néel lattice is simply a spin system where adjacent spins are maximal and point in opposite directions. It is the epitome of anti-ferromagnetic order. We will be exclusively looking at one dimension, thus what are called spin chains. As the spins on a lattice are distinguishable, one choice starting at a given spin of up, down, up, down, $\cdots$ is a different configuration from down, up, down, up, $\cdots$, starting from the same spin. This twofold degeneracy is akin to the two-fold degeneracy of a ferromagnetic system, where all spins could point up or all spins could point down. The direction of the up and down is determined by the anisotropy, which picks out a favoured direction for the spins. In this section, we will only consider two spins, but in the next section we will generalize our results to a spin chain.

The non-interacting system of our Hamiltonian is defined by $\lambda = 0$, here the spin eigenstates of S_{iz}, notationally $|s, s_{1z}\rangle \otimes |s, s_{2z}\rangle \equiv |s_{1z}, s_{2z}\rangle$, are obviously exact eigenstates. The ground state is fourfold degenerate, corresponding to the states $|s, s\rangle$, $|-s, -s\rangle$, $|s, -s\rangle$ and $|-s, s\rangle$, which we will write as $|\uparrow, \uparrow\rangle, |\downarrow, \downarrow\rangle, |\uparrow, \downarrow\rangle, |\downarrow, \uparrow\rangle$, each with energy $E = -2Ks^2$. The first excited state, which is eightfold degenerate, is split from the ground state by energy $\Delta E = K(2s - 1)$.

In the weak coupling limit, $\lambda/K \to 0$, an interesting question to ask is what is the ground state and the first few excited states of the system for large spin $\vec{S}$. For spin $1/2$, the exact eigenstates are trivially found; for spin 1, the problem is a 9×9 matrix, which again can be diagonalized, but for the general case we must diagonalize a $(2s + 1)^2 \times (2s + 1)^2$ matrix, although that is rather sparse. For weak coupling the anisotropic potential continues to align or anti-align the spins along the z-axis in the ground state.

As the non-interacting ground state is fourfold degenerate, in first-order degenerate perturbation theory, we should diagonalize the exchange interaction in the degenerate subspace. However, it turns out to be already diagonal in that subspace. The full Hamiltonian can be alternatively written as

$$H = -K(S_{1z}^2 + S_{2z}^2) + \lambda\left(S_{1z}S_{2z} + \frac{1}{2}(S_1^+ S_2^- + S_1^- S_2^+)\right), \qquad (7.109)$$

where $S_i^\pm = S_{ix} \pm iS_{iy}$ for $i = 1,2$. $S_i^\pm$ act as raising and lowering operators for S_{iz}, and hence they must annihilate the states $|\uparrow,\uparrow\rangle, |\downarrow,\downarrow\rangle$. Thus the two states $|\uparrow,\uparrow\rangle, |\downarrow,\downarrow\rangle$ are actually exact eigenstates of the full Hamiltonian with exact energy eigenvalue $(-2K + \lambda)s^2$. These two states do not mix with the two states $|\uparrow,\downarrow\rangle, |\downarrow,\uparrow\rangle$ as the eigenvalue of $S_{1z} + S_{2z}$, which is conserved, is $+2s$, $-2s$ for the two ferromagnetic states and 0 for the two anti-ferromagnetic states. The perturbation, apart from the diagonal term $\lambda S_{1z}S_{2z}$, acting on the two states $|\uparrow,\downarrow\rangle, |\downarrow,\uparrow\rangle$ takes them out of the degenerate subspace, thus this part does not give any correction to the energy. The action of the diagonal term on either of these states is equal to $-\lambda s^2$. Thus the perturbation corresponds to the identity matrix within the degenerate subspace of the two states $|\uparrow,\uparrow\rangle, |\downarrow,\downarrow\rangle$, with eigenvalue $-\lambda s^2$ for the two anti-ferromagnetic states. This yields, in first-order degenerate perturbation theory, the perturbed energy eigenvalue of $(-2K - \lambda)s^2$ for the two states $|\uparrow,\downarrow\rangle, |\downarrow,\uparrow\rangle$. Thus the following picture emerges of the first four levels in first-order degenerate perturbation theory. For the $\lambda < 0$ (ferromagnetic coupling), the states $|\uparrow,\uparrow\rangle, |\downarrow,\downarrow\rangle$ are the exact, degenerate ground states of the theory, with energy eigenvalue $(-2K + \lambda)s^2 = (-2K - |\lambda|)s^2$. The first excited states are also degenerate, but only within first-order degenerate perturbation theory. They are given by $|\uparrow,\downarrow\rangle, |\downarrow,\uparrow\rangle$, with energy eigenvalue $(-2K - \lambda)s^2 = (-2K + |\lambda|)s^2$. For the $\lambda > 0$ (anti-ferromagnetic coupling), the roles are exactly reversed. The states $|\uparrow,\downarrow\rangle, |\downarrow,\uparrow\rangle$ give the degenerate ground state with energy $(-2K - \lambda)s^2$ in first-order degenerate perturbation, while the states $|\uparrow,\uparrow\rangle, |\downarrow,\downarrow\rangle$ give the exact, first (doubly degenerate) excited level with energy $(-2K + \lambda)s^2$. Thus the Hamiltonian in first-order degenerate perturbation theory is simply diagonal

$$\langle H \rangle = \begin{pmatrix} -2K + \lambda & 0 & 0 & 0 \\ 0 & -2K + \lambda & 0 & 0 \\ 0 & 0 & -2K - \lambda & 0 \\ 0 & 0 & 0 & -2K - \lambda \end{pmatrix} s^2 \qquad (7.110)$$

in the ordered basis $\{|\uparrow,\uparrow\rangle, |\downarrow,\downarrow\rangle, |\uparrow,\downarrow\rangle, |\downarrow,\uparrow\rangle\}$. The two ferromagnetic states are the exact degenerate ground states for $\lambda < 0$, while the two anti-ferromagnetic states are the approximate ground states for $\lambda > 0$.

However, we do not expect this result to stand in higher orders. We will show that, in fact, the states $|\pm\rangle = \frac{1}{\sqrt{2}}(|\uparrow,\downarrow\rangle \pm |\downarrow\uparrow\rangle)$ are the appropriate linear combinations implied by higher orders in degenerate perturbation theory, for the ground state in the anti-ferromagnetic case, and they are the second and third excited states in the ferromagnetic case. We will also show that the states $|\pm\rangle$ are no longer degenerate. The perturbing Hamiltonian links the state $|\pm s, \mp s\rangle$ only to the state $|\pm s \mp 1, \mp s \pm 1\rangle$. To reach the state $|\mp s, \pm s\rangle$ from the state $|\pm s, \mp s\rangle$ requires one to go to $2s$th order in perturbation, and s is assumed to be large. Indeed, we find our results via macroscopic quantum tunnelling using the spin-coherent state path integral. Using the path integral to determine large orders in perturbation theory has already been studied in field theory [35, 128].

Our two-spin system, in Minkowski time, is governed by an action $S = \int dt \mathcal{L}$ where,

$$\mathcal{L} = \int dx\, s\hat{n}_1 \cdot (\partial_x \hat{n}_1 \times \partial_t \hat{n}_1) - V_1(\hat{n}_1)$$

$$+ \int dx\, s\hat{n}_2 \cdot (\partial_x \hat{n}_2 \times \partial_t \hat{n}_2) - V_2(\hat{n}_2) - \lambda \hat{n}_1 \cdot \hat{n}_2, \qquad (7.111)$$

where now $\hat{n}_i = (\sin\theta_i \cos\phi_i, \sin\theta_i \sin\phi_i, \cos\theta_i), i = 1, 2$ are two different 3-vectors of unit norm, representing semi-classically the quantum spin [28, 27] and s is the value of each spin. We use the coordinate-dependent spherical-polar coordinate to describe the spins and the Lagrangian takes the form

$$\mathcal{L} = -s\dot{\phi}_1(1 - \cos\theta_1) - V_1(\theta_1, \phi_1)$$

$$-s\dot{\phi}_2(1 - \cos\theta_2) - V_2(\theta_2, \phi_2)$$

$$-\lambda\left(\sin\theta_1 \sin\theta_2 \cos(\phi_1 - \phi_2) + \cos\theta_1 \cos\theta_2\right). \qquad (7.112)$$

Our analysis is valid if we restrict our attention to any external potential with easy-axis, azimuthal symmetry, with a reflection symmetry (along the azimuthal axis), as in [68], $V_i(\theta_i, \phi_i) \equiv V(\theta_i) = V(\pi - \theta_i), i = 1, 2$. The potential is further assumed to have a minimum at the north pole and the south pole, at $\theta_i = 0$, and π. We will treat the special simple case of the potential given by

$$V(\hat{n}_i) \equiv V(\theta_i, \phi_i) = K\sin^2\theta_i. \qquad (7.113)$$

corresponding exactly to our Hamiltonian Equation (7.108). It was shown in [68], for uncoupled spins, that quantum tunnelling between the spin up and down states of each spin separately is actually absent because of conservation of the z-component of each spin. With the exchange interaction, only the total z-component is conserved, allowing transitions $|\uparrow, \downarrow\rangle \longleftrightarrow |\downarrow, \uparrow\rangle$. In general, tunnelling exists if there is an equipotential path that links the beginning and end points. We will see that such an equipotential path exists, but through complex values of the phase space variables.

We must find the critical points of the Euclidean action with $t \to -i\tau$, which gives

$$\mathcal{L}_E = is\dot{\phi}_1(1 - \cos\theta_1) + V(\theta_1) + is\dot{\phi}_2(1 - \cos\theta_2) + V(\theta_2)$$

$$+ \lambda\left(\sin\theta_1 \sin\theta_2 \cos(\phi_1 - \phi_2) + \cos\theta_1 \cos\theta_2\right). \qquad (7.114)$$

The solutions must start at $(\theta_1, \phi_1) = (0,0)$ and $(\theta_2, \phi_2) = (\pi, 0)$, say, and evolve to $(\theta_1, \phi_1) = (\pi, 0)$ and $(\theta_2, \phi_2) = (0,0)$. In Euclidean time, the WZ term has become imaginary and the equations of motion in general only have solutions for complexified field configurations. Varying with respect to ϕ_i gives equations that correspond to the conservation of angular momentum:

$$is\frac{d}{d\tau}\left(1 - \cos\theta_1\right) + \lambda \sin\theta_1 \sin\theta_2 \sin\left(\phi_1 - \phi_2\right) = 0 \qquad (7.115)$$

$$is\frac{d}{d\tau}\left(1 - \cos\theta_2\right) - \lambda \sin\theta_1 \sin\theta_2 \sin\left(\phi_1 - \phi_2\right) = 0 \qquad (7.116)$$

Varying with respect to θ_i gives the equations:

$$is\dot{\phi}_1 \sin\theta_1 + 2K \sin\theta_1 \cos\theta_1 + \lambda\left(\cos\theta_1 \sin\theta_2 \cos(\phi_1 - \phi_2) - \sin\theta_1 \cos\theta_2\right) = 0 \qquad (7.117)$$

$$is\dot{\phi}_2 \sin\theta_2 + 2K \sin\theta_2 \cos\theta_2 + \lambda\left(\cos\theta_2 \sin\theta_1 \cos(\phi_1 - \phi_2) - \sin\theta_2 \cos\theta_1\right) = 0. \qquad (7.118)$$

Adding Equations (7.115) and (7.116) we simply get

$$\frac{d}{d\tau}\left(\cos\theta_1 + \cos\theta_2\right) = 0. \qquad (7.119)$$

Hence $\cos\theta_1 + \cos\theta_2 = l = 0$, where the constant l is chosen to be zero using the initial condition $\theta_1 = 0, \theta_2 = \pi$ and therefore we can take $\theta_2 = \pi - \theta_1$. We can now eliminate θ_2 from the equations of motion, and writing $\theta = \theta_1$, $\phi = \phi_1 - \phi_2$ and $\Phi = \phi_1 + \phi_2$ we get the effective Lagrangian:

$$\mathcal{L} = is\dot{\Phi} - is\dot{\phi}\cos\theta + U(\theta, \phi), \qquad (7.120)$$

where $U(\theta, \phi) = 2K \sin^2\theta + \lambda\left(\sin^2\theta \cos\phi - \cos^2\theta\right) + \lambda$ is the effective potential energy. We have added a constant λ so that the potential is normalized to zero at $\theta = 0$. The first term in the Lagrangian is a total derivative and drops out. The equations of motion become:

$$is\dot{\phi}\sin\theta = -\frac{\partial U(\theta, \phi)}{\partial\theta} \qquad (7.121)$$

$$is\dot{\theta}\sin\theta = \frac{\partial U(\theta, \phi)}{\partial\phi}. \qquad (7.122)$$

These equations have no solutions on the space of real functions $\theta(\tau), \phi(\tau)$ due to the explicit i on the left-hand side. The analogue of conservation of energy follows immediately from these equations, multiplying (7.121) by $\dot{\theta}$ and (7.122) by $\dot{\phi}$ and subtracting, gives:

$$\frac{dU(\theta, \phi)}{d\tau} = 0, \quad i.e. \quad U(\theta, \phi) = const. = 0. \qquad (7.123)$$

The constant has been set to 0 again using the initial condition $\theta = 0$. Thus we have, specializing to our case, Equation (7.113)

$$U(\theta,\phi) = (2K + \lambda(\cos\phi+1))\sin^2\theta = 0 \qquad (7.124)$$

implying $(2K + \lambda(\cos\phi+1)) = 0$, since $\sin^2\theta \neq 0$, as is required for a non-trivial solution. Thus

$$\cos\phi = -\left(\frac{2K}{\lambda}+1\right) \qquad (7.125)$$

and we see that ϕ must be a constant. This is not valid in general, it is due to the specific choice of the external potential Equation (7.113). Since $K > |\lambda|$ we get $|\cos\phi| > 1$, which of course has no solution for real ϕ. We take $\phi = \phi_R + i\phi_I$ which gives $\cos\phi = \cos\phi_R\cosh\phi_I - i\sin\phi_R\sinh\phi_I$. As the right-hand side of Equation (7.125) is real, we must have either $\phi_I = 0$ or $\phi_R = n\pi$ or both. Clearly the $\phi_I = 0$ cannot yield a solution for Equation (7.125), hence we must take $\phi_R = n\pi$. As we must impose 2π periodicity on ϕ_R, only $n = 0$ or 1 exist. Then we get

$$\cos\phi = (-1)^n \cosh\phi_I = \begin{cases} -\left(\frac{2K}{\lambda}+1\right) & \text{if } \lambda > 0 \\ +\left(\frac{2K}{|\lambda|}-1\right) & \text{if } \lambda < 0 \end{cases}. \qquad (7.126)$$

Thus $n = 1$ for $\lambda > 0$ and $n = 0$ for $\lambda < 0$ allowing for the unified expression

$$\cosh\phi_I = \frac{2K+\lambda}{|\lambda|}. \qquad (7.127)$$

Equation (7.122) simplifies to

$$is\frac{\dot\theta}{\sin\theta} = -\lambda\sin\phi = -i\lambda(-1)^n\sinh\phi_I = i|\lambda|\sinh\phi_I \qquad (7.128)$$

as $\lambda(-1)^n = -|\lambda|$. Equation (7.127) has two solutions: positive ϕ_I corresponds to the instanton, $(\dot\theta > 0)$, and negative ϕ_I corresponds to the anti-instanton, $(\dot\theta < 0)$. The equation is trivially integrated with solution

$$\theta(\tau) = 2\arctan\left(e^{\omega(\tau-\tau_0)}\right), \qquad (7.129)$$

where $\omega = (|\lambda|/s)\sinh\phi_I$ and at $\tau = \tau_0$ we have $\theta(\tau) = \pi/2$, which has exactly the same form as the solution in the previous section, Equation (7.78). Thus $\theta(\tau)$ interpolates from 0 to π as τ interpolates from $-\infty$ to ∞ for an instanton and from π to 0 for an anti-instanton.

Using $\dot\phi = 0$ and Equation (7.123) that the effective energy is zero, we see that the action for this instanton trajectory, let us call it S_0, simply vanishes $S_0 = \int_{-\infty}^{\infty} d\tau\mathcal{L} = 0$. So where does the amplitude come from? As in the previous section, we have not taken into account the fact that ϕ must be translated from $\phi = 0$ (any initial point will do, as long as it is consistently used to compute the full amplitude) to $\phi = n\pi + i\phi_I$ before the instanton can occur and then back to

$\phi = 0$ after the instanton has occurred. Normally such a translation has no effect; either the contribution at the beginning cancels that at the end or, if the action is second-order in time derivative, moving adiabatically gives no contribution. But in the present case, for an instanton, before the instanton occurs, $\theta = 0$, while after it has occurred, $\theta = \pi$, and vice versa for an anti-instanton. As $\dot{\phi}$ is multiplied by $\cos\theta$ in the action, the two contributions actually add, and there is a net contribution to the action. Indeed, the full action for the combination of the instanton and the changes in ϕ is given by

$$\Delta S = \int_0^{n\pi + i\phi_I} -isd\phi \cos\theta|_{\theta=0} + S_0 + \int_{n\pi + i\phi_I}^0 -isd\phi \cos\theta|_{\theta=\pi}$$
$$= -is2n\pi + 2s\phi_I \tag{7.130}$$

we call it ΔS since it arises because of a change in ϕ, and where we have put $S_0 = 0$.

We will use this information to compute the following matrix element, using the spin-coherent states $|\theta,\phi\rangle$ and the two lowest energy eigenstates $|E_0\rangle$ and $|E_1\rangle$:

$$\langle\theta_f,\phi_f|e^{-\beta H}|\theta_i,\phi_i\rangle = e^{-\beta E_0}\langle\theta_f,\phi_f|E_0\rangle\langle E_0|\theta_i,\phi_i\rangle$$
$$+ e^{-\beta E_1}\langle\theta_f,\phi_f|E_1\rangle\langle E_1|\theta_i,\phi_i\rangle + \cdots \tag{7.131}$$

On the other hand, the matrix element is given by the spin-coherent state path integral

$$\langle\theta_f,\phi_f|e^{-\beta H}|\theta_i,\phi_i\rangle = \mathcal{N}\int_{\theta_i,\phi_i}^{\theta_f,\phi_f} \mathcal{D}\theta\mathcal{D}\phi\, e^{-S_E}. \tag{7.132}$$

The integration is done in the saddle point approximation. With $(\theta_i,\phi_i) = (0,0)$ corresponding to the state $|\uparrow,\downarrow\rangle$ and $(\theta_f,\phi_f) = (\pi,0)$ corresponding to the state $|\downarrow,\uparrow\rangle$, we get

$$\langle\downarrow,\uparrow|e^{-\beta H}|\uparrow,\downarrow\rangle = \mathcal{N}e^{-\Delta S}\kappa\beta(1 + \cdots), \tag{7.133}$$

where κ is the ratio of the square root of the determinant of the operator governing the second-order fluctuations about the instanton excluding the time translation zero mode, and that of the free determinant. It can, in principle, be calculated, but we will not do this here. The zero mode is taken into account by integrating over the position of the occurrence of the instanton giving rise to the factor of β. $\mathcal{N}$ is the overall normalization including the square root of the free determinant which is given by $Ne^{-E_0\beta}$, where E_0 is the unperturbed ground-state energy and N is a constant that depends on the form of the perturbative ground-state wave function. The result exponentiates, but since we must sum over all sequences of one instanton followed by any number of anti-instanton–instanton pairs, the total number of instantons and anti-instantons is odd, and we get

$$e^{-\Delta S}\kappa\beta \to \sinh\left(e^{-\Delta S}\kappa\beta\right). \tag{7.134}$$

Given $\Delta S = -is2n\pi + 2s\phi_I$ and solving Equation (7.127) for ϕ_I for $K \gg |\lambda|$

$$\phi_I = \text{arccosh}\left(\frac{2K+\lambda}{|\lambda|}\right) \approx \ln\left(\frac{4K}{|\lambda|}\right) \tag{7.135}$$

gives:

$$e^{-\Delta S} = \begin{cases} e^{is2\pi - 2s\phi_I} & \text{if } \lambda > 0 = \begin{cases} \left(\frac{|\lambda|}{4K}\right)^{2s} & \text{if } s \in \mathbf{Z} \\ -\left(\frac{|\lambda|}{4K}\right)^{2s} & \text{if } s \in \mathbf{Z}+1/2 \end{cases} \\ \left(\frac{|\lambda|}{4K}\right)^{2s} & \text{if } \lambda < 0 \end{cases} \tag{7.136}$$

Then we get

$$\langle\downarrow,\uparrow|e^{-\beta H}|\uparrow,\downarrow\rangle = \pm\left(\frac{1}{2}e^{\left(\frac{|\lambda|}{4K}\right)^{2s}\kappa\beta} - \frac{1}{2}e^{-\left(\frac{|\lambda|}{4K}\right)^{2s}\kappa\beta}\right) Ne^{-\beta E_0}, \tag{7.137}$$

where the $-$ sign only applies for the case of anti-ferromagnetic coupling with half odd integer spin, *i.e.* $\lambda > 0, s = \mathbf{Z}+1/2$. An essentially identical analysis yields, for the persistence amplitudes

$$\langle\downarrow,\uparrow|e^{-\beta H}|\downarrow,\uparrow\rangle = \langle\uparrow,\downarrow|e^{-\beta H}|\uparrow,\downarrow\rangle$$
$$= \left(\frac{1}{2}e^{\left(\frac{|\lambda|}{4K}\right)^{2s}\kappa\beta} + \frac{1}{2}e^{-\left(\frac{|\lambda|}{4K}\right)^{2s}\kappa\beta}\right) Ne^{-\beta E_0}. \tag{7.138}$$

These calculated matrix elements can now be compared with what is expected for the exact theory:

$$\langle\downarrow,\uparrow|e^{-\beta H}|\uparrow,\downarrow\rangle = e^{-\beta(E_0-\frac{1}{2}\Delta E)}\langle\downarrow,\uparrow|E_0\rangle\langle E_0|\uparrow,\downarrow\rangle$$
$$+e^{-\beta(E_0+\frac{1}{2}\Delta E)}\langle\downarrow,\uparrow|E_1\rangle\langle E_1|\uparrow,\downarrow\rangle \tag{7.139}$$

and

$$\langle\downarrow,\uparrow|e^{-\beta H}|\downarrow,\uparrow\rangle = e^{-\beta(E_0-\frac{1}{2}\Delta E)}\langle\downarrow,\uparrow|E_0\rangle\langle E_0|\downarrow,\uparrow\rangle$$
$$+e^{-\beta(E_0+\frac{1}{2}\Delta E)}\langle\downarrow,\uparrow|E_1\rangle\langle E_1|\downarrow,\uparrow\rangle \tag{7.140}$$

The energy splitting can be read off from this result

$$\Delta E = E_1 - E_2 = 2\left(\frac{|\lambda|}{4K}\right)^{2s}\kappa \tag{7.141}$$

for all cases; however, the wave functions are different. The low-energy eigenstates are given by

$$|E_0\rangle = \frac{1}{\sqrt{2}}(|\downarrow,\uparrow\rangle + |\uparrow,\downarrow\rangle) \qquad |E_1\rangle = \frac{1}{\sqrt{2}}(|\downarrow,\uparrow\rangle - |\uparrow,\downarrow\rangle) \tag{7.142}$$

for the case $\lambda > 0$ for $s \in \mathbf{Z}$, where they are the actual ground and first excited state as well as for the case $\lambda < 0$ (although here these energy eigenstates should be labelled $|E_3\rangle$ and $|E_4\rangle$ as the actual ground states are the ferromagnetic states $|\uparrow,\uparrow\rangle, |\downarrow,\downarrow\rangle$). For the fermionic spin, anti-ferromagnetic case with $\lambda > 0$ and $s \in \mathbf{Z} + 1/2$ we get the reversal of the states

$$|E_0\rangle = \frac{1}{\sqrt{2}}(|\downarrow,\uparrow\rangle - |\uparrow,\downarrow\rangle) \qquad |E_1\rangle = \frac{1}{\sqrt{2}}(|\downarrow,\uparrow\rangle + |\uparrow,\downarrow\rangle), \qquad (7.143)$$

but the energy splitting remains the same.

This understanding of the ground state in the anti-ferromagnetic case is the main result. This difference in the ground states for integer and half-odd integer spins is understood in terms of the Berry phase [88, 38] (computed by the change in the WZ term) for the evolution corresponding to the instanton. It can also be understood by looking at perturbation theory to order $2s$; the details cannot be given here. Briefly, the action of the perturbation Equation (7.109) will lower one spin and raise the other. This can be done $2s$ times when we achieve a complete flip of both spins. We find that the effective 2×2 Hamiltonian for the degenerate subspace is proportional to the identity plus off-diagonal terms that are symmetric. For the integer spin case the off-diagonal terms are negative and for the half-odd integer case they are positive. Diagonalizing this 2×2 matrix gives the solutions for the ground states, exactly as we have found.

7.8 The Haldane-like Spin Chain and Instantons

The study of quantum spin chains has been a very important physical problem in condensed matter and mathematical physics over the past 100 years. They play an exemplary role in the study of strongly correlated quantum systems. In both experimental and theoretical physics, models of quantum spin chains are one of the most fundamental systems endowed with interesting phenomenon. The classic work on spin chains was that of Bethe [14] and Hulthén [63] for the one-dimensional ($D = 1$), isotropic Heisenberg spin-$\frac{1}{2}$ anti-ferromagnetic chain. They computed the exact anti-ferromagnetic ground state and its energy for an infinite chain. Anderson [6] worked out the ground-state energies and the spectrum for $D = 1, 2, 3$ by means of spin wave theory. The inclusion of an anisotropy term introduces much interesting physics ranging from quantum computing [90] to optical physics [110]. The resulting Hamiltonian is what we will study in this section. It now possesses two coupling constants which can compete against each other to lower the energy

$$\hat{H} = -K \sum_{i=1}^{N} S_{i,z}^2 + \lambda \sum_{i=1}^{N} \vec{S}_i \cdot \vec{S}_{i+1} \qquad (7.144)$$

and we consider the chain with periodic boundary conditions and consider $\lambda > 0$ so that we are in the anti-ferromagnetic regime, which is the more interesting regime.

This model is the generalization to a spin chain of the two-spin model that we studied in the previous section. Here each nearest-neighbour pair corresponds to the two-spin system that we have just studied. Each spin has magnitude $|\vec{S}_i| = s$ and we will consider the large s limit. The two limiting cases are weak anisotropy $\lambda \gg K$ and weak exchange coupling $\lambda \ll K$, where λ is the Heisenberg exchange interaction coupling constant and K is the anisotropy coupling constant. The limit of weak anisotropy was studied in a celebrated paper by Haldane [59] in a closely related model, hence we call our model a Haldane-like spin chain. Haldane demonstrated that in the large spin limit, $s \gg 1$, the system can be mapped to a non-linear sigma model in field theory with distinguishing effects between integer and half-odd integer spins. The full rotational symmetry is broken explicitly into rotational symmetry about the z-axis with the total z-component $S_{i,z} = \sum_i S_{i,z}$ conserved. The Hamiltonian also possesses a discrete reflection symmetry about the z-axis $S_{i,z} \to -S_{i,z}$. We will also study the model in the large spin limit, but we will take the limit of strong anisotropy, $K \gg \lambda$, the opposite limit to Haldane.

With $\lambda = 0$, the ground state is 2^N-fold degenerate, corresponding to each spin in the state $S_z = \pm s$. Then $s_z^2 = s^2$ and the energy is $-Ks^2 2N$, which is minimal. For an even number of sites, the model is called bi-partite and the two fully anti-aligned Neél states are good starting points for investigating the ground state. For an odd number of sites, the Neél states are frustrated; they must contain at least one defect, which are called domain wall solitons [115, 39, 93, 20, 95]. There is a high level of degeneracy as the soliton can be placed anywhere along the cyclic, periodic chain and this degenerate system is the starting point for investigating the ground state for the case of an odd number of sites. Frustrated systems are of great importance in condensed matter physics as they lead to exotic phases of matter such as spin liquid [9], spin glasses [15] and topological orders [73]. Solitons will also occur on the periodic chain with even number of sites, but they must occur in soliton–anti-soliton pairs.

Many physical magnetic systems such as $CsNiF_3$ and Co^{++} have been modelled with Hamiltonians of the form of Equation (7.144). Models of this form have been of research interest over the years since the work of Haldane [59]. To mention but a few, quite recently, the ground-state phase diagrams of the spin-2 XXZ anisotropic Heisenberg chain has been carefully investigated by the infinite system density-matrix-renormalization group (iDMRG) algorithm [74] and other numerical methods [58]. For the spin-1 XXZ anisotropic Heisenberg chain, the numerical exact diagonalization has been extensively investigated for finite size systems [25]. For an arbitrary spin, the phase diagrams and correlation exponents of an XXZ anisotropic Heisenberg chain has also been studied by representing the spins as a product of $2s$ spin $\frac{1}{2}$ operators [108]. This research has been focussed on ground-state phase diagrams and the existence of Haldane phase (conjecture). In this section, we will study the spin chain with Hamiltonian given by the simple form given in Equation (7.144) with periodic boundary condition $\vec{S}_{N+1} = \vec{S}_1$, and we consider $K \gg \lambda > 0$, *i.e.* strong easy-axis anisotropy and perturbative Heisenberg anti-ferromagnetic coupling. In the first subsection, we will study macroscopic quantum tunnelling of the Hamiltonian-defined Equation (7.144) for the case of an even spin chain. This analysis is based on spin-coherent state path integral formalism, which is appropriate for large spin systems. In the second subsection, which we include for completeness, we will deal with the case of an odd spin chain. Here, spin-coherent state path integral formalism fails to give a definitive result. Thus, our analysis is based on perturbation theory.

7.8.1 Even Number of Sites and Spin-Coherent State Path Integral

Let us consider our model, Equation (7.144), for N even. The ground state of the free theory (K term) is 2^N-fold degenerate corresponding to each spin in the highest (lowest) weight states $m = \pm s$. In the degenerate subspace, there are two fully aligned states $|\uparrow,\uparrow,\uparrow,\uparrow,\cdots,\uparrow,\uparrow\rangle$ and $|\downarrow,\downarrow,\downarrow,\downarrow,\cdots,\downarrow,\downarrow\rangle$ and two fully anti-aligned Néel states $|p\rangle = |\uparrow,\downarrow,\uparrow,\downarrow,\cdots,\uparrow,\downarrow\rangle$ and $|-p\rangle = |\downarrow,\uparrow,\downarrow,\uparrow,\cdots,\downarrow,\uparrow\rangle$, where the arrow denotes the highest (lowest) weight states, *i.e.* $m = s \equiv\uparrow(m = -s \equiv\downarrow)$ for each individual spin and the remaining degenerate states are produced by flipping individual spins relative to these extremal states. These two Néel states $|\pm p\rangle$ have the lowest energy at first-order in perturbation theory; however, they are not exact eigenstates of the quantum Hamiltonian in Equation (7.144), thus we expect ground-state quantum tunnelling coherence between them. Such tunnelling is usually mediated by an instanton trajectory, and the exponential of the instanton action (multiplied by a prefactor) yields the energy splitting. We will obtain this instanton trajectory via the spin-coherent state path integral formalism [5, 76, 75, 99], which is the appropriate formalism for large spin systems. In this formalism, the spin operators become unit vectors parameterized by spherical coordinates. The corresponding Euclidean Lagrangian in this formalism is given by

$$L_E = is \sum_{i=1}^{N} \dot{\phi}_i (1 - \cos\theta_i) + K \sum_{i=1}^{N} \sin^2\theta_i$$

$$+ \lambda \sum_{i=1}^{N} \left(\sin\theta_i \sin\theta_{i+1} \cos(\phi_i - \phi_{i+1}) + \cos\theta_i \cos\theta_{i+1} \right), \quad (7.145)$$

where the periodicity condition $i = N + 1 = 1$ is imposed. The first term is the usual WZ [120] term which arises from the non-orthogonality of spin-coherent states while the other two terms correspond to the anisotropy energy and the Heisenberg exchange energy. Quantum amplitudes are obtained via the path integral and the saddle point approximation. Solutions of the Euclidean classical

equations of motion give information about quantum tunnelling amplitudes. The
Euclidean classical equation of motion for ϕ_i is

$$is\frac{d(1-\cos\theta_i)}{d\tau} = \sin\theta_{i-1}\sin\theta_i\sin(\phi_{i-1}-\phi_i)$$

$$-\sin\theta_i\sin\theta_{i+1}\sin(\phi_i-\phi_{i+1}) \tag{7.146}$$

while the equation of motion for θ_i is

$$0 is\dot{\phi}_i\sin\theta_i + 2K\sin\theta_i\cos\theta_i$$

$$\lambda\left(\cos\theta_i(\sin\theta_{i+1}\cos(\phi_i-\phi_{i+1})) + \sin\theta_{i-1}\cos(\phi_{i-1}-\phi_i)\right)$$

$$= \lambda\left(\sin\theta_i(\cos\theta_{i+1}+\cos\theta_{i-1})\right) = 0. \tag{7.147}$$

Summing both sides of Equation (7.146) gives

$$is\sum_i\frac{d(1-\cos\theta_i)}{d\tau} = 0 \Rightarrow \sum_i\cos\theta_i = l = 0, \tag{7.148}$$

which corresponds to the conservation of the z-component of the total spin
$\sum_i S_i^z$, as the full Hamiltonian, Equation (7.144), is invariant under rotations
about the z-axis.

We will solve these equations using simplifying, physically motivated ansatze.
A particular solution of Equation (7.148) is $\theta_{2k-1}\equiv\theta$, and $\theta_{2k}=\pi-\theta$,
$k=1,2\cdots,N/2$. Making the further simplifying ansatz $\phi_i-\phi_{i+1}=(-1)^{i+1}\phi$
effectively reducing the system to a single spin problem, we get the effective
Lagrangian (adding an irrelevant constant)

$$L_E^{eff} = is\sum_{k=1}^{N}\dot{\phi}_k - is\cos\theta\sum_{k=1}^{N/2}(\dot{\phi}_{2k-1}-\dot{\phi}_{2k})$$

$$+ \sum_{i=1}^{N}\left[K+\lambda[1+\cos(\phi_i-\phi_{i+1})]\right]\sin^2\theta \tag{7.149}$$

$$= isN\dot{\Phi} - \frac{isN}{2}\dot{\phi}\cos\theta + U_{eff}, \tag{7.150}$$

where $U_{eff} = N[K+\lambda(1+\cos\phi)]\sin^2\theta$. The spin chain problem has reduced to
essentially the same problem we studied in the previous section with just two
spins. The instanton that we will find must go from $\theta=0$ to $\theta=\pi$. Conservation
of energy implies $\partial_\tau U_{eff} = 0$, which then must vanish, $U_{eff}=0$, since it is so at
$\theta=0$. This implies

$$\cos\phi = -\left(\frac{K}{\lambda}+1\right) \ll -1 \tag{7.151}$$

since $\sin\theta(\tau)\neq 0$ along the whole trajectory. Thus ϕ is a complex constant which
can be written as $\phi=\pi+i\phi_I$ identical to that of the two-spin case [99]. The
classical equation of motion for ϕ gives

$$is\dot{\theta} = -2\lambda\sin\theta\sin\phi = i2\lambda\sin\theta\sinh\phi_I \tag{7.152}$$

which integrates as

$$\theta(\tau) = 2\arctan\left(e^{\omega(\tau-\tau_0)}\right), \tag{7.153}$$

where $\omega = (2\lambda/s)\sinh\phi_I$. The instanton is independent of the number of spins and only depends on the initial and the final points. As found in [99], the instanton contributes to the action only through the WZ term, as $U_{eff} = 0$ all along the trajectory. The action is given by [99]

$$S_c = S_0 - \frac{isN}{2}\int_0^{\pi+i\phi_I} d\phi\cos\theta|_{\theta=0} - \frac{isN}{2}\int_{\pi+i\phi_I}^0 d\phi\cos\theta|_{\theta=\pi}$$

$$= 0 - isN\pi + Ns\phi_I = -isN\pi + Ns\phi_I. \tag{7.154}$$

The two Néel states reorganize into the symmetric and anti-symmetric linear superpositions, $|+\rangle$ and the $|-\rangle$ as in [99]. The energy splitting is then

$$\Delta E = 2\mathcal{D}e^{-S_c} = 2\mathcal{D}\left(\frac{\lambda}{2K}\right)^{Ns}\cos(sN\pi) \tag{7.155}$$

where $\mathcal{D}$ is a determinantal pre-factor which contains no λ dependence. The factor of λ^{Ns} signifies that this energy splitting arises from $2s\left(\frac{N}{2}\right)$ order in degenerate perturbation theory in the interaction term. The energy splitting, Equation (7.155), is the general formula for any even spin chain N. For $N = 2$, we recover the results obtained previously [29, 30, 71, 72, 99]. The factor sN can be even or odd, depending on the value of the spin. For half-odd integer spin $(2l+1)/2$ and for $N = 2(2k+1)$, the argument of the cosine in Equation (7.155) is $sN\pi = (2l+1)(2k+1)\pi$ and hence we find ΔE is negative, which means that $|-\rangle$ is the ground state and $|+\rangle$ is the first excited state. In all other cases, for any value of the spin s and $N = 2(2k)$ the argument of the cosine is $sN\pi = (2s)(2k)\pi$, which is an even integer multiple of π and hence we find ΔE is positive and then $|+\rangle$ is the ground state, $|-\rangle$ is the first excited state.

7.8.2 Odd Spin Chain, Frustration and Solitons

We include the analysis of the spin chain with an odd number of sites for the sake of completeness. This system can, in principle, be analysed using the spin-coherent state path integral. However, the tunnelling transitions are quite different, and no explicit, analytic expressions for the instantons that are required are known. In this situation, we revert back to the calculation using perturbative methods, which is actually quite interesting.

When we consider a periodic chain with an odd number of sites, a soliton-like defect arises due to the spin frustration. The fully anti-aligned Néel-like state cannot complete periodically, as it requires an even total number of spins. Thus there has to be at least one pair of spins that is aligned. This can come in the form up–up or down–down while all other pairs of neighbouring spins are in the up–down or down–up combination. As the total z-component of the spin is conserved, these states lie in orthogonal super-selection sectors and never transform into each other. The position of the soliton is arbitrary thus each sector is N-fold degenerate. In the first case the total z-component of the spin is s while in the second case it is $-s$. We will, without loss of generality, consider the s sector. These degenerate states are denoted by $|k\rangle$, $k = 1, \cdots, N$, where

$$|k\rangle = |\uparrow,\downarrow,\uparrow,\downarrow,\uparrow,\cdots, \underbrace{\uparrow,\uparrow,}_{k,k+1th\,place} ,\cdots,\uparrow,\downarrow\rangle \qquad (7.156)$$

in obvious notation. These states are not exact eigenstates of the quantum spin Hamiltonian in Equation (7.144), thus we also expect ground-state quantum tunnelling amongst these states, just as in the case of a particle in a periodic potential, which would lift the degeneracy and reorganize the soliton states into a band. The explicit form of the required instanton, which from the spin-coherent state path integral would give rise to the appropriate tunnelling, is not known. However, from Equation (7.155) we can see for the case of even spins that energy splitting actually arises at the $2s\left(\frac{N}{2}\right)$ order in (degenerate) perturbation theory. The path integral and instanton method only gives the result which must also be available at this high order in degenerate perturbation theory. This indicates that the appropriate formalism for the odd quantum spin chain would simply be (degenerate) perturbation theory at high order.

It is convenient to write the Hamiltonian as

$$\hat{H} = \hat{H}_0 + \hat{V} \qquad (7.157)$$

where $\hat{H}_0$ represents the K (free) term and $\hat{V}$ represents the λ (perturbative) term. The states in Equation (7.156) all have the same energy $E_s = -KNs^2$ from $\hat{H}_0$ and in first-order degenerate perturbation theory $E_s = -KNs^2 - \lambda(N-1)s^2 + \lambda s^2 = (-K-\lambda)Ns^2 + 2\lambda s^2$ and are split from the first excited level, which requires the introduction of a soliton/anti-soliton pair, by an energy of 4λ. As we take the limit $K \gg \lambda$, we assume that the action of lowering or raising the value of $\hat{S}_z$ incurs an energy cost proportion to K which is much more energy than creating a soliton/anti-soliton pair, which has an energy cost proportional to λ. Although the soliton/anti-soliton states are the next states in the spectrum, they cannot be attained perturbatively, except at order $2s$ in perturbation theory. In each order of perturbation theory less than $2s$, the degenerate multiplet of states mixes with the states of much higher energy, but due to invariance under translation, the corrections brought to each state are identical and their degeneracy cannot be split. However, at order $2s$, the degenerate multiplet is mapped to itself. Although the state of an additional soliton/anti-soliton pair is also reached at this order, since it is not degenerate in energy with the original multiplet of N states, its correction will be perturbatively small.

Reaching the degenerate multiplet at order $2s$ causes the multiplet to split in energy and the states to reorganize into a band. Indeed, $\hat{V}^{2s}$ contains the term

$(S_{k+1}^- S_{k+2}^+)^{2s}$ and $(S_{k-1}^+ S_k^-)^{2s}$. These operators represent quantum fluctuations close to the position of the soliton, which when acting on the ket $|k\rangle$, flips the anti-aligned pair of spins at positions $k+1, k+2$ and at $k-1, k$, respectively. It is easy to see that flipping this pair of spins has the effect of translating the soliton $|k\rangle \to |k+2\rangle$ and $|k\rangle \to |k-2\rangle$, respectively. All other terms in $\hat{V}^{2s}$ are quantum fluctuations away from the position of the soliton. They map to states out of the degenerate subspace, either inserting a soliton/anti-soliton pair or changing the value of S^z to non-extremal values, and hence do not contribute to breaking the degeneracy.

To compute the splitting and the corresponding eigenstates, we follow [30]. We have to diagonalize the $N \times N$ matrix with components $b_{\mu,\nu}$ given by

$$b_{\mu,\nu} = \langle \mu | \hat{V} \mathcal{A}^{2s-1} | \nu \rangle, \quad \mu, \nu = 1, 2, \cdots, N \tag{7.158}$$

where $\mathcal{A}^{2s-1} = \left(\frac{\mathcal{Q}}{E_s - \hat{H}_0} \hat{V} \right)^{2s-1}$, and $\mathcal{Q} = 1 - \sum |\mu\rangle \langle \mu|$. These matrices are a generalization of the 2×2 matrix in [30]. The calculation of the components is straightforward, for example, looking at $b_{\mu,1}$ we find

$$b_{\mu,1} = \left(\frac{\lambda}{2} \right)^{2s} \langle \mu | S_2^- S_3^+ \left(\frac{\mathcal{Q}}{E_s - \hat{H}_0} S_2^- S_3^+ \right)^{2s-1} |1\rangle$$
$$+ \left(\frac{\lambda}{2} \right)^{2s} \langle \mu | S_N^+ S_1^- \left(\frac{\mathcal{Q}}{E_s - \hat{H}_0} S_N^+ S_1^- \right)^{2s-1} |1\rangle. \tag{7.159}$$

Applying the operators $2s$ times on the right-hand side we obtain

$$b_{\mu,1} = \mathcal{C}[\langle \mu|3\rangle + \langle \mu|N-1\rangle], \tag{7.160}$$

where $\mathcal{C}$ is given by

$$\mathcal{C} = \pm \left(\frac{\lambda}{2} \right)^{2s} \prod_{m=1}^{2s} m(2s - m + 1) \prod_{m=1}^{2s-1} \frac{1}{Km(2s-m)}$$
$$= \pm K \left(\frac{\lambda}{2K} \right)^{2s} \left[\frac{(2s)!}{(2s-1)!} \right]^2 = \pm 4Ks^2 \left(\frac{\lambda}{2K} \right)^{2s}. \tag{7.161}$$

The first product in Equation (7.161) comes from the two square roots that accompany the action of the raising and lowering operators, and the second product is a consequence of the energy denominators. The plus or minus sign arises because we have $2s - 1$ products of negative energy denominators in Equation (7.159), so if s is integer, $2s - 1$ is odd and we get a minus sign, while for half-odd integer s, $2s - 1$ is even and we get a plus sign. Similarly, we can show that $b_{\mu,\nu} = \mathcal{C}[\langle \mu|\nu+2\rangle + \langle \mu|\nu-2\rangle]$, defined periodically of course. Thus we find that the matrix, $[b_{\mu,\nu}]$, that we must diagonalize is a circulant matrix [37]

$$[b_{\mu,\nu}] = C \begin{pmatrix} 0 & 0 & 1 & 0 & \cdots & 1 & 0 \\ 0 & 0 & 0 & 1 & \cdots & 0 & 1 \\ 1 & 0 & 0 & 0 & 1 & \cdots & 0 \\ \vdots & 1 & 0 & & \ddots & \cdots & \ddots \\ 1 & \cdots & \ddots & \cdots & 0 & 0 & 0 \\ 0 & 1 & \cdots & 1\cdots & 0 & 0 & 0 \end{pmatrix}. \tag{7.162}$$

In this matrix each row element is moved one step to the right, periodically, relative to the preceding row. The eigenvalues and eigenvectors are well known. The jth eigenvalue is given by

$$\varepsilon_j = b_{1,1} + b_{1,2}\omega_j + b_{1,3}\omega_j^2 + \cdots + b_{1,N}\omega_j^{N-1}, \tag{7.163}$$

where $\omega_j = e^{i\frac{2\pi j}{N}}$ is the jth, N^{th} root of unity with corresponding eigenvector $|\frac{2\pi j}{N}\rangle = (1, \omega_j, \omega_j^2, \cdots, \omega_j^{N-1})^T$, for $j = 0, 1, 2, \cdots, N-1$. For our matrix, Equation (7.162), the only non-zero coefficients are $b_{1,3}$ and $b_{1,N-1}$, thus the one-soliton energy bands are

$$\varepsilon_j = C(\omega_j^2 + \omega_j^{N-2}) = C(\omega_j^2 + \omega_j^{-2})$$
$$= 2C \cos\left(\frac{4\pi j}{N}\right). \tag{7.164}$$

Introducing the Brillouin zone momentum $q = j\pi/N$, the energy bands Equation (7.164) can be written as

$$\varepsilon_q = 2C \cos(4q) \tag{7.165}$$

which is gapless but is doubly degenerate as the cosine passes through two periods in the Brillouin zone. The exact spectrum is symmetric about the value $N/2$. With $[x]$ the greatest integer not greater than x, the states for $j = [N/2] - k$ and $j = [N/2] + k + 1$ for $k = 0, 1, 2, \cdots, [N/2] - 1$ are degenerate as $\cos\left(\frac{4\pi([N/2]-k)}{N}\right) = \cos\left(\frac{4\pi([N/2]+k+1)}{N}\right)$ since $[N/2] = N/2 - 1/2$. However, the state with $k = [N/2]$ is not paired, only $j = 0$ is allowed. When s is an integer, C is negative and the unpaired state $j = 0$ is the ground state which is then non-degenerate, but for s a half-odd integer, C is positive, and the ground states are the degenerate pair with $j = [N/2], [N/2] + 1$ in accordance with Kramers' theorem [78]. However, in the thermodynamic limit, $N \to \infty$, the spectrum simply becomes doubly degenerate for all values of the spin and gapless.

8

Quantum Electrodynamics in 1+1 Dimensions

8.1 The Abelian Higgs Model

Instantons imply drastic changes in the spectrum of theories with essentially Abelian gauge invariance in $1+1$ and $2+1$ dimensions. We say essentially Abelian, since we include in this class theories which are spontaneously broken to a residual $U(1)$ invariance. In $1+1$ dimensions we consider the theory defined by the Lagrangian density [61],

$$\mathcal{L} = (D_\mu \phi)^* (D^\mu \phi) - \frac{\lambda}{4} (\phi^* \phi)^2 - \frac{\mu^2}{2} \phi^* \phi - \frac{1}{4e^2} F_{\mu\nu} F^{\mu\nu}, \qquad (8.1)$$

where

$$D_\mu \phi = \partial_\mu \phi + i A_\mu \phi$$
$$F_{\mu\nu} = \partial_\mu A_\nu - \partial_\nu A_\mu. \qquad (8.2)$$

We take $D_\mu \phi = \partial_\mu \phi + i e A_\mu \phi$, but we have replaced $A_\mu \to \frac{1}{e} A_\mu$. The Lagrangian is invariant under a local gauge transformation which has a natural multiplication law corresponding to the group $U(1)$

$$\phi \to e^{i\Lambda(x,t)} \phi = g(x,t)\phi \quad g(x,t) \in U(1)$$
$$A_\mu \to e^{i\Lambda(x,t)} (A_\mu - i\partial_\mu) e^{-i\Lambda(x,t)} = A_\mu - \partial_\mu \Lambda. \qquad (8.3)$$

Then

$$D_\mu \phi \to \partial_\mu \left(e^{i\Lambda(x,t)} \phi \right) + i \left(A_\mu - \partial_\mu \Lambda(x,t) \right) e^{i\Lambda(x,t)} \phi$$

$$= e^{i\Lambda(x,t)} \partial_\mu \phi + e^{i\Lambda(x,t)} i\partial_\mu \Lambda(x,t)\phi + e^{i\Lambda(x,t)} i A_\mu \phi - e^{i\Lambda(x,t)} i\partial_\mu \Lambda(x,t)\phi$$

$$= e^{i\Lambda(x,t)} D_\mu \phi. \qquad (8.4)$$

We impose that $\lim_{|x|\to\infty} g(x,t) = 1$. This gives an effective topological compactification of the space since the gauge transformation at spatial infinity must be the same in all directions.

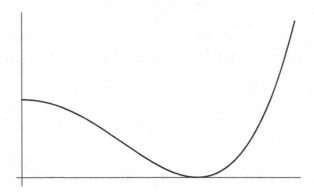

Figure 8.1. The symmetric breaking potential $U(|\phi|)$

There are two cases to consider depending on the sign of μ^2. For $\mu^2 > 0$ the quadratic part of the Lagrangian is

$$\mathcal{L} = (\partial_\mu \phi)^* (\partial^\mu \phi) - \frac{\mu^2}{2} \phi^* \phi - \frac{1}{4e^2} F_{\mu\nu} F^{\mu\nu} \qquad (8.5)$$

with corresponding equations of motion

$$\partial_\mu \partial^\mu \phi + \mu^2 \phi = 0$$
$$\partial_\mu F^{\mu\nu} = 0, \qquad (8.6)$$

which describe a free, massive, scalar particle and a massless vector field, the free, electromagnetic field. The conserved current corresponding to gauge invariance is

$$j^\mu = \phi^* \partial^\mu \phi - (\partial^\mu \phi^*)\,\phi. \qquad (8.7)$$

External charges that are well separated experience the usual Coulomb force. This is true in any dimension except in $1+1$ dimensions, the case that we are considering first. Here, the Coulomb force is independent of the separation and it costs an infinite amount of energy to separate two charges to infinity. We say that charges are confined. Furthermore, there is no photon. There is no transverse direction for the polarization states of the photon. The spectrum consists of bound states of particle–anti-particle pairs, which are stable. They cannot disintegrate since there is no photon.

For the other case with $\mu^2 < 0$, the potential (as depicted in Figure 8.1) is of the symmetry breaking type

$$U(|\phi|) = \frac{\lambda}{4} |\phi|^4 - \frac{|\mu^2|}{2} |\phi|^2 + C, \qquad (8.8)$$

where the C is adjusted so that the potential vanishes at the minimum. The minimum is at $|\phi|^2 = \frac{|\mu^2|}{\lambda}$. We fix the gauge so that $\Im m(\phi) = 0$, and we write

$$\phi = \frac{|\mu|}{\sqrt{\lambda}} + \eta \qquad (8.9)$$

with $\eta \in \mathbf{R}$. Then we get the Lagrangian density

$$\mathcal{L} = (\partial_\mu - iA_\mu)\left(\frac{|\mu|}{\sqrt{\lambda}} + \eta\right)(\partial^\mu + iA^\mu)\left(\frac{|\mu|}{\sqrt{\lambda}} + \eta\right)$$
$$- \frac{\lambda}{4}\left(\frac{|\mu|}{\sqrt{\lambda}} + \eta\right)^4 - \frac{\mu^2}{2}\left(\frac{|\mu|}{\sqrt{\lambda}} + \eta\right)^2 - \frac{1}{4e^2}F_{\mu\nu}F^{\mu\nu}, \qquad (8.10)$$

which yields the quadratic part

$$\mathcal{L}_0 = \partial_\mu \eta \partial^\mu \eta + \frac{\mu^2}{\lambda}A_\mu A^\mu + \mu^2 \eta^2 - \frac{1}{4e^2}F_{\mu\nu}F^{\mu\nu}. \qquad (8.11)$$

This now corresponds to a scalar particle with a mass of $\frac{\mu}{\sqrt{2}}$ and a vector particle of mass $\frac{\mu}{\sqrt{\lambda}}e$. Then the expectation is that the potential between particles should drop off exponentially with the usual Yukawa factor

$$e^{-\frac{r}{M}} \qquad (8.12)$$

with $M = \frac{|\mu|}{\sqrt{2}}$ or $M = \frac{|\mu|}{\sqrt{\lambda}}e$. We will find, surprisingly, that this is again not true in $1+1$ dimensions. Instantons change the force between the particles and actually imply confinement. The only difference between the cases $\mu^2 > 0$ and $\mu^2 < 0$ is that the force is exponentially smaller (in $\hbar$) for the case $\mu^2 < 0$; however, it is still independent of separation.

8.2 The Euclidean Theory and Finite Action

To see this result, we must analyse the Euclidean theory. Here the Lagrangian density is

$$\mathcal{L} = \frac{1}{4e^2}F_{\mu\nu}F^{\mu\nu} + (D_\mu\phi)^*(D^\mu\phi) + \frac{\lambda}{4}\left(\phi^*\phi - a^2\right)^2 \qquad (8.13)$$

adding a constant, where $a = \frac{|\mu|}{\sqrt{\lambda}}$. These are three positive terms. For a configuration of finite Euclidean action, each term must give a finite contribution when integrated over $\mathbf{R}^2$. This implies that $\phi^*\phi \to a^2$, $D_\mu\phi \to 0$ and $F_{\mu\nu} \to 0$ faster than $\frac{1}{r}$.

$$\phi^*\phi \to a^2 \Rightarrow \lim_{r\to\infty}\phi = g(\theta)a$$
$$F_{\mu\nu} \to 0 \Rightarrow iA_\mu \to \tilde{g}\partial_\mu(\tilde{g})^{-1} + o\left(\frac{1}{r^2}\right) = i\partial_\mu\Lambda$$
$$\text{for} \quad \tilde{g} = e^{i\Lambda}$$
$$D_\mu\phi \to 0 \Rightarrow \partial_\mu g(\theta)a + \tilde{g}\partial_\mu(\tilde{g})^{-1}g(\theta)a$$
$$= \left(-g(\theta)\partial_\mu g^{-1}(\theta) + \tilde{g}\partial_\mu(\tilde{g})^{-1}\right)g(\theta)a = 0. \qquad (8.14)$$

This is satisfied if $g(\theta) = \tilde{g}$. Thus the configurations with finite Euclidean action are characterized by $g(\theta)$. $g(\theta)$ defines a mapping of the circle at infinity parametrized by θ into the group $U(1)$, which is just the unit circle as in Figure 8.2.

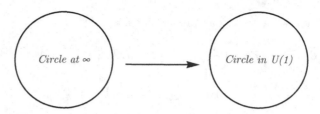

Figure 8.2. Mapping the spatial circle at ∞ to the circle in $U(1)$

8.2.1 Topological Homotopy Classes

The space of such maps separates into homotopically inequivalent classes. These classes are characterized by the winding number of the map. A map from one class cannot be continuously deformed to any other map from another class. It is intuitively obvious that there are an infinite number of classes each corresponding to a winding number. We can take

$$
\begin{aligned}
g^0(\theta) &= 1 \qquad n = 0 \\
g^1(\theta) &= e^{i\theta} \qquad n = 1 \\
&\quad . \\
&\quad . \\
&\quad . \\
g^\nu(\theta) &= e^{i\nu\theta} \qquad n = \nu.
\end{aligned}
\tag{8.15}
$$

Given $g(\theta)$ we can extract ν by the formula

$$
\nu = \frac{i}{2\pi} \int_0^{2\pi} d\theta\, g(\theta) \frac{d}{d\theta} g^{-1}(\theta).
\tag{8.16}
$$

If $g(\theta) = e^{i\nu\theta}$, $\frac{d}{d\theta} g^{-1}(\theta) = -i\nu g^{-1}(\theta)$ thus

$$
\nu = \frac{i}{2\pi} \int_0^{2\pi} g(\theta)\left(-i\nu\right)g^{-1}(\theta) = \frac{\nu}{2\pi} 2\pi = \nu.
\tag{8.17}
$$

If we make an arbitrary, infinitesimal change in $g(\theta)$,

$$
g(\theta) \to e^{i\epsilon\Lambda(\theta)} g(\theta) = g(\theta) + i\epsilon\Lambda(\theta)g(\theta)
\tag{8.18}
$$

with $\Lambda(\theta)$ of compact support in $[0, 2\pi)$,

$$
\begin{aligned}
\delta g(\theta) &= i\Lambda(\theta)g(\theta) \\
\delta\left(g(\theta)\frac{d}{d\theta}g^{-1}(\theta) \right) &= i\Lambda(\theta)g(\theta)\frac{d}{d\theta}g^{-1}(\theta) + g(\theta)\frac{d}{d\theta}\left(-i\Lambda(\theta)g^{-1}(\theta)\right) \\
&= i\Lambda(\theta)g(\theta)\frac{d}{d\theta}g^{-1}(\theta) - i\Lambda(\theta)g(\theta)\frac{d}{d\theta}g^{-1}(\theta) - i\frac{d}{d\theta}\Lambda(\theta) \\
&= -i\frac{d}{d\theta}\Lambda(\theta).
\end{aligned}
\tag{8.19}
$$

Thus

$$\delta\nu = \frac{i}{2\pi}\int_0^{2\pi}d\theta\,\delta\left(g(\theta)\frac{d}{d\theta}g^{-1}(\theta)\right) = \frac{i}{2\pi}(-i)\int_0^{2\pi}d\theta\frac{d}{d\theta}\Lambda(\theta)$$

$$= \frac{1}{2\pi}\left(\Lambda(2\pi) - \Lambda(0)\right) = 0. \tag{8.20}$$

Thus for each class, ν is an invariant under arbitrary continuous deformation. Furthermore, if $g(\theta) = g_{\nu_1}(\theta)g_{\nu_2}(\theta)$, then

$$\nu = \frac{i}{2\pi}\int_0^{2\pi}d\theta g_{\nu_1}(\theta)g_{\nu_2}(\theta)\frac{d}{d\theta}\left(g_{\nu_2}^{-1}(\theta)g_{\nu_1}^{-1}(\theta)\right)$$

$$= \frac{i}{2\pi}\int_0^{2\pi}d\theta g_{\nu_1}(\theta)\left(g_{\nu_2}(\theta)\frac{d}{d\theta}g_{\nu_2}^{-1}(\theta)\right)g_{\nu_1}^{-1}(\theta) + g_{\nu_1}(\theta)\frac{d}{d\theta}g_{\nu_1}^{-1}(\theta)$$

$$= \nu_1 + \nu_2. \tag{8.21}$$

Finally, using $iA_\mu = g\partial_\mu g^{-1} + o\left(\frac{1}{r^2}\right)$

$$\nu = \frac{i}{2\pi}\int_0^{2\pi}d\theta g(\theta)\frac{d}{d\theta}g^{-1}(\theta) = \frac{i}{2\pi}\int_0^{2\pi}d\theta r i\hat{r}_\mu\epsilon_{\mu\nu}A_\nu = -\frac{1}{2\pi}\oint_{r=\infty}dx_\mu A_\mu$$

$$= -\frac{1}{2\pi}\int d^2x\partial_\mu\epsilon_{\mu\nu}A_\nu = -\frac{1}{4\pi}\int d^2x\epsilon_{\mu\nu}F_{\mu\nu} = -\left(\frac{\Phi}{2\pi}\right), \tag{8.22}$$

giving that the flux is quantized in units of 2π. In each homotopy class, the configuration of minimum action must be stationary and hence satisfy the Euclidean equations of motion. Because the solutions with different ν cannot be obtained from each other by continuous deformation, there should be an infinite action barrier between each class.

8.2.2 Nielsen–Olesen Vortices

The solutions for each ν are known to exist and are called the Nielsen–Olesen vortices [96]. They are described by two radial functions, for $\nu = 1$

$$A_\mu = \epsilon_{\mu\nu}r_\nu\frac{\Phi(r)}{2\pi r^2}$$

$$\phi(r) = e^{i\theta}f(r). \tag{8.23}$$

This form implies the equations

$$-\frac{1}{r}\frac{d}{dr}\left(r\frac{d}{dr}f(r)\right) + \frac{1}{r^2}\left(1 - \frac{\Phi(r)}{2\pi}\right)^2 f(r) - \mu^2 f(r) + \lambda f^3(r) = 0 \tag{8.24}$$

and

$$-\frac{1}{e^2}\frac{d}{dr}\left(\frac{1}{r}\frac{d}{dr}\frac{\Phi(r)}{2\pi}\right) + \frac{f^2(r)}{r}\left(\frac{\Phi(r)}{2\pi} - 1\right) = 0. \tag{8.25}$$

A solution exists, as depicted in Figure 8.3, with the behaviour for the magnetic field $B(r)$

$$B(r) = \frac{1}{2\pi r}\frac{d}{dr}\frac{\Phi(r)}{2\pi} \to Ce^{-erf(r)}$$

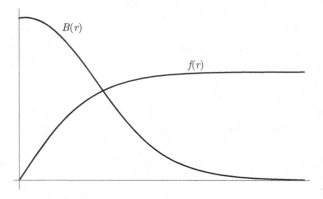

Figure 8.3. The form of the function $f(r)$ and the magnetic field $B(r)$

$$f(r) \to a$$
$$\Phi(r) \to \Phi, \qquad (8.26)$$

where $\Phi(r)$ has the interpretation of being equal to the magnetic flux inside the radius r while Φ is the total magnetic flux in the soliton, which is quantized in units of 2π. The magnetic field is concentrated around the origin and both fields approach the vacuum configuration exponentially fast with a non-trivial winding number.

This solution mediates tunnelling between inequivalent classical vacua, which correspond to classical configurations with zero energy. The energy is given by (for $A_0 = 0$)

$$\mathcal{E} = \int dx \frac{1}{2e^2} (\partial_0 A_1)^2 + (\partial_0 \phi)^* (\partial_0 \phi) + (D_1 \phi)^* (D_1 \phi) + \lambda (\phi^* \phi - a^2)^2. \quad (8.27)$$

The simplest zero-energy configuration is $\phi = a$, $A_\mu = 0$. There exists, however, the freedom to transform this solution by a local gauge transformation that depends only on space, which keeps the gauge condition $A_0 = 0$ invariant,

$$\phi \to g(x)a \qquad A_1 \to -ig(x)\partial_1 g^{-1}(x). \qquad (8.28)$$

We impose the additional condition, the $\lim_{x \to \infty} g(x) \to 1$, which is consistent with our desire to consider a theory with arbitrary local excitations but asymptotically no excitations. Then we get the effective compactification of the spatial hypersurface. Topologically it is now just a circle and $g(x)$ again maps the circle that is space onto the circle in $U(1)$. These maps are characterized by winding numbers. Thus the classical vacua

$$\phi = g_\nu(x)a$$
$$A_1 = -ig_\nu(x)\frac{d}{dx}g_\nu^{-1}(x) \qquad (8.29)$$

indexed by $\nu \in \mathbb{Z}$ are homotopically inequivalent. We cannot deform one into another while staying at $\mathcal{E} = 0$. However, the energy barrier between them is

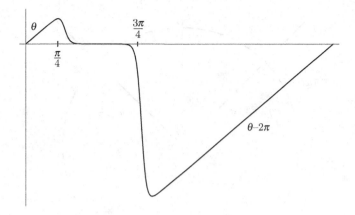

Figure 8.4. The form of the gauge transformation $\Lambda(\theta)$

not infinite. The Nielsen–Olesen vortices interpolate between these vacua. To see this we must transform the Nielsen–Olesen vortex into a form suitable for the description in terms of many vacua, *i.e.* to the gauge $A_0 = 0$.

We first perform a gauge transformation $g(\theta) = e^{-i\Lambda(\theta)}$, which has the limit at spacetime infinity given by

$$\Lambda(\theta) = \begin{cases} \theta & \text{for } \theta \in (0, \pi/4) \\ \pi/4 \to 0 & \text{for } \theta \in (\pi/4, \pi/4 + \epsilon) \\ 0 & \text{for } \theta \in (\pi/4 + \epsilon, 3\pi/4 - \epsilon) \\ 0 \to -5\pi/4 & \text{for } \theta \in (3\pi/4 - \epsilon, 3\pi/4) \\ (\theta - 2\pi) & \text{for } \theta \in (3\pi/4, 2\pi) \end{cases} \tag{8.30}$$

as drawn in Figure 8.4. The corresponding $g(\theta)$ is topologically trivial; we can simply deform the two saw-tooth humps to zero (non-trivial winding number requires a $\Lambda(\theta)$ discontinuous by $2n\pi$ between its value at $\theta = 0$ and $\theta = 2\pi$). Therefore, the gauge transformation can be continued everywhere inside the spacetime and define a gauge transformation at all points. This gauge transformation (note this is an inverse transformation, $\Lambda \to -\Lambda$) takes

$$A_\mu \to \tilde{A}_\mu = A_\mu + \partial_\mu \Lambda(\theta) \tag{8.31}$$

and it is easy to see that this vanishes asymptotically, except where $\Lambda(\theta) = 0$, *i.e.* for $\theta \in (\pi/4 + \epsilon, 3\pi/4 - \epsilon)$. Thus $\tilde{A}_\mu \to 0$ for $t \in [-\infty, T]$, T finite, exponentially fast as $|x| \to \infty$, as depicted in Figure 8.5.

Now we further perform the gauge transformation to put $\tilde{A}_0 = 0$ everywhere; this is easily implemented by the choice

$$\Lambda(x, t) = \int_{-\infty}^{t} dt' \tilde{A}_0(x, t')$$

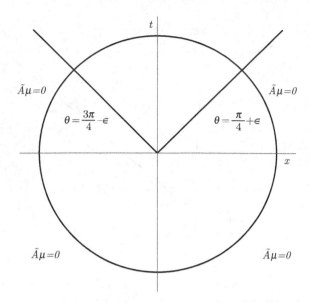

Figure 8.5. The regions of spacetime where the gauge field vanishes

$$A_1(x,t) \rightarrow \tilde{\tilde{A}}_1(x,t) = \tilde{A}_1(x,t) - \partial_x\Lambda(x,t) = \tilde{A}_1(x,t) - \partial_x\int_{-\infty}^{t} dt'\,\tilde{A}_0(x,t').$$

(8.32)

This makes $\tilde{\tilde{A}}_0 = 0$ everywhere, but maintains $\tilde{\tilde{A}}_1 \rightarrow 0$ for $t \in [-\infty, T]$, T finite, exponentially fast as $|x| \rightarrow \infty$, since both $\tilde{A}_1 \rightarrow 0$ and $\tilde{A}_0 \rightarrow 0$ exponentially fast at spatial infinity in the region $t \in [-\infty, T]$, due to the first gauge transformation. Thus as $t \rightarrow -\infty$, $\tilde{\tilde{A}}_1 = 0$, but as $t \rightarrow +\infty$ we have,

$$\tilde{\tilde{A}}_1 \rightarrow g(x)\partial_x g^{-1}(x)$$

(8.33)

with

$$g(x) = e^{i\int_{-\infty}^{\infty} dt'\,\tilde{A}_0(x,t')}g(\theta)e^{i\theta},$$

(8.34)

where $g(\theta) = e^{-i\Lambda(\theta)}$ is our first gauge transformation and $e^{i\int_{-\infty}^{\infty} dt'\,\tilde{A}_0(x,t')}$ is the second gauge transformation that put $\tilde{\tilde{A}}_0 = 0$. The final factor $e^{i\theta}$ is the asymptotic gauge transformation of the Nielsen–Olesen vortex. The first two factors are topologically trivial gauge transformations: in each case the exponent can be continuously switched to zero, thus the winding number of the gauge transformation $e^{i\theta}$, which is 1, is unchanged. However, the two trivial factors manage to bunch all of the non-trivial winding of $e^{i\theta}$ into the spatial line x at $t = \infty$.

Thus we have put the Nielsen–Olesen vortex in a gauge where it interpolates from the vacuum configuration $g(x) = 1$ at $t = -\infty$ to the non-trivially transformed vacuum configuration $g(x) = e^{i\int_{-\infty}^{\infty} dt'\,\tilde{A}_0(x,t')}g(\theta)e^{i\theta}$. The situation is exactly analogous to the problem of a periodic potential on a line. The classical vacua form a denumerable infinity of local minima indexed by the winding number n. There is a finite energy barrier between each one, and the Nielsen–Olesen vortex is the instanton that mediates the tunnelling between them.

8.3   Tunnelling Transitions

We can calculate the matrix element

$$\langle \nu = n | e^{\frac{-\hat{H}T}{\hbar}} | \nu = 0 \rangle = \mathcal{N} \int_{\nu[\phi_{in}]=0}^{\nu[\phi_f]=n} \mathcal{D}(A_1, \phi^*, \phi) e^{-\frac{S_\phi^E}{\hbar}} \qquad (8.35)$$

in the semi-classical approximation. The functional integral is simply identified with the integral over all finite action field configurations with $\nu = n$. The continuation from Euclidean space automatically projects on the vacuum in this sector. The critical point of the action contains the vortex with $\nu = n$; however, this is not the most important configuration. The most important configurations correspond to n_+ vortices with $\nu = 1$ and n_- vortices with $\nu = -1$, widely separated, such that $n_+ - n_- = n$. The action for such a configuration is very close to $(n_+ + n_-)S^E(\nu = 1)$. The entropy factor, counting the degeneracy of the configuration, is

$$\frac{(TL)^{n_+ + n_-}}{n_+! n_-!}. \qquad (8.36)$$

In comparison, for a single vortex with $\nu = n$, the action is presumably smaller, but the entropy factor is just TL, since there is only one object. Thus the dilute multi-instanton configurations are arbitrarily more important as $TL \to \infty$. Then

$$\langle \nu = n | e^{\frac{-\hat{H}T}{\hbar}} | \nu = 0 \rangle = \mathcal{N} \det_0^{-\frac{1}{2}} \sum_{n_+=0}^{n_+=\infty} \sum_{n_-=0}^{n_-=\infty} \frac{1}{n_+! n_-!} \left(TLe^{-\frac{S_0^E}{\hbar}} K \right)^{n_+ + n_-}$$

$$\times \delta_{n_+ - n_-, n}, \qquad (8.37)$$

where K^{-2} (so that it appears in the formula as just K) is given by the ratio of the determinant prime corresponding to the quadratic part of the Lagrangian in the presence of one vortex, divided by the determinant of the free quadratic part (written as $\det_0$), and the Jacobian factors from the usual change of variables that take into account zero modes. The prefactor is set equal to one by choosing the normalization $\mathcal{N}$. Now using the formula

$$\delta_{a,b} = \frac{1}{2\pi} \int_0^{2\pi} d\theta e^{i\theta(a-b)} \qquad (8.38)$$

we get

$$\langle \nu = n | e^{\frac{-\hat{H}T}{\hbar}} | \nu = 0 \rangle = \frac{1}{2\pi} \int_0^{2\pi} d\theta e^{i\theta n} \sum_{n_+, n_-=0}^{\infty} \frac{e^{in_+\theta} e^{-in_-\theta}}{n_+! n_-!} \left(TLe^{-\frac{S_\phi^E}{\hbar}} K \right)^{n_+ + n_-}$$

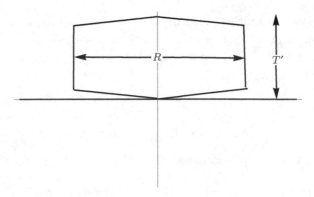

Figure 8.6. Creation of a pair of charges at the origin, separation by R, held for time T' and then annihilated

$$= \frac{1}{2\pi} \int_0^{2\pi} d\theta e^{i\theta n} e^{\left(TLe^{-\frac{S_0^E}{\hbar}} K\left(e^{i\theta} + e^{-i\theta}\right) \right)}$$

$$= \frac{1}{2\pi} \int_0^{2\pi} d\theta e^{i\theta n} e^{\left(2TLe^{-\frac{S_0^E}{\hbar}} K \cos\theta \right)}$$

$$= \int_0^{2\pi} d\theta e^{\frac{-\mathcal{E}(\theta)T}{\hbar}} \langle \nu = n| \theta\rangle\langle\theta |\nu = 0\rangle. \qquad (8.39)$$

Thus we find the infinite set of classical vacua rearrange themselves to form a band of states parametrized by θ with energy density

$$\frac{\mathcal{E}(\theta)}{L} = -2Ke^{-\frac{S_0^E}{\hbar}} \cos\theta \qquad (8.40)$$

and the matrix element

$$\langle \nu = n| \theta\rangle = \frac{e^{in\theta}}{\sqrt{2\pi}}. \qquad (8.41)$$

8.4 The Wilson Loop

This rearrangement of the vacua has important consequences for the force between charges. Consider the creation of an external charged particle and anti-particle pair. We create them at the origin, separate them by a large distance R, hold them at this separation for a long time T', and then we let them come together and annihilate, as depicted in Figure 8.6. A particle of charge q, in an electromagnetic field, experiences an additional change to its wave function by the factor

$$e^{-i\frac{q}{e}\int dx_\mu A_\mu}. \qquad (8.42)$$

Consider external charges governed by a dynamics with a Lagrangian

$$L = \frac{1}{2}\dot{x}_i^2 + q\dot{x}_i A_i - qA_0 - V(x_i). \tag{8.43}$$

The equation of motion is

$$\ddot{x}_i + q\dot{\vec{A}}_i(x_l) - q\dot{x}_j\partial_i\vec{A}_j(x_l) + q\partial_i A_0(x_l) + \partial_i V(x_l) = 0, \tag{8.44}$$

which can be rewritten

$$\ddot{x}_i - q\dot{x}_j\epsilon_{jik}B_k(x_l) = -\partial_i V(x_l) + q\vec{E}_i(x_l), \tag{8.45}$$

where $E_i(x_l) = \partial_t\vec{A}_i(x_l) + \partial_i A_0(x_l)$ is the electric field and $B_i(x_l) = \epsilon_{jik}\partial_j\vec{A}_k(x_l)$ is the magnetic field. Thus the action in the functional integral for the particle is augmented by the term

$$e^{-i\frac{S^0}{\hbar}} \to e^{-i\frac{S^0}{\hbar}} e^{-i\frac{q}{e}\int dt\left(\dot{x}_i\vec{A}_i(x_l) - qA_0(x_l)\right)}$$
$$= e^{-i\frac{S^0}{\hbar}} e^{-i\frac{q}{e}\int dx^\mu A_\mu}. \tag{8.46}$$

For an anti-particle the additional factor is, of course,

$$e^{i\frac{q}{e}\int dx^\mu A_\mu}. \tag{8.47}$$

Thus for our trajectory, the additional factor becomes a closed integral in the exponent,

$$e^{-i\frac{q}{e}\oint dx^\mu A_\mu}. \tag{8.48}$$

We perform the functional integral over the gauge and scalar fields treating our particles as external, with their dynamics controlled by $V(x_l)$. However, the wave functions of the particles will change by this additional factor, which we must take into account. When we integrate over A_μ, ϕ, ϕ^* we obtain the matrix element of the operator (in Euclidean space)

$$W = e^{-\frac{q}{e}\oint dx_\mu A_\mu}. \tag{8.49}$$

This is called the Wilson loop operator. The matrix element of the operator behaves approximately as

$$W \sim e^{-E(R)T'\left(\frac{q}{e}\right)}. \tag{8.50}$$

If $E(R) \sim CR$ for some constant C, the interaction between the charges is said to be confining, and the expectation value of the Wilson loop operator will behave like

$$\langle W \rangle \sim e^{-CA\left(\frac{q}{e}\right)}, \tag{8.51}$$

where A is the area of the loop. This is the celebrated criterion of area law behaviour of the Wilson loop for confining interactions. If, on the other hand, the $E(R) \sim D$ for some constant D, we get

$$\langle W \rangle \sim e^{-DP\left(\frac{q}{e}\right)}, \tag{8.52}$$

where P is the perimeter of the loop. Such behaviour of the Wilson loop does not imply confining interactions.

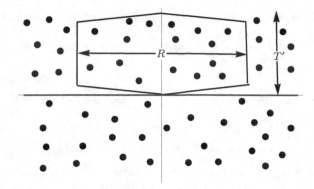

Figure 8.7. A dilute gas of instantons and anti-instantons surround the Wilson loop

8.4.1 Expectation Value of the Wilson Loop Operator

Thus we wish to evaluate

$$\langle \theta | W | \theta \rangle = \frac{\int \mathcal{D}(A_\mu, \phi, \phi^*) e^{-\frac{S^E}{\hbar}} e^{i\nu\theta} W}{\int \mathcal{D}(A_\mu, \phi, \phi^*) e^{-\frac{S^E}{\hbar}} e^{i\nu\theta}} \tag{8.53}$$

in the semi-classical approximation. For the numerator we divide the summation over the positions of the vortices (instantons) and the anti-vortices (anti-instantons) into those inside the loop and those outside the loop, as depicted in Figure 8.7. We drop the contribution from vortices situated on or near the boundary; these form a negligible part of the set of all configurations, if the size of the loop is much larger than the size of the vortices.

The integrand splits neatly into a part from outside and a part from inside the Wilson loop

$$S = S^{outside} + S^{inside}$$
$$\nu = \nu^{outside} + \nu^{inside} \tag{8.54}$$

however,

$$W = e^{2\pi i \frac{q}{e} \nu^{inside}}. \tag{8.55}$$

Inside the volume available is RT', while outside the volume available is $LT - RT'$, for each vortex. We sum independently over the vortices and the anti-vortices, inside and outside the loop, with no constraint on their numbers. The contribution inside has $\theta \to \theta + \frac{2\pi q}{e}$, thus we get

$$\langle \theta | W | \theta \rangle = e^{\left(2Ke^{-\frac{S_0^E}{\hbar}} \left((LT - RT') \cos\theta + RT' \cos\left(\theta + \frac{2\pi q}{e}\right) - LT \cos\theta \right) \right)}$$

$$= e^{\left(2Ke^{-\frac{S_0^E}{\hbar}} RT' \left(-\cos\theta + \cos\left(\theta + \frac{2\pi q}{e}\right) \right) \right)}. \tag{8.56}$$

Then comparing with Equation (8.50) we get

$$E(R) = 2R\left(\cos(\theta) - \cos\left(\theta + \frac{2\pi q}{e}\right)\right) K e^{-\frac{S_0^E}{\hbar}} \qquad (8.57)$$

and hence

$$E(R) \sim R \qquad (8.58)$$

implying confinement.

We can also calculate

$$\langle\theta|\frac{1}{2}\epsilon_{\mu\nu}F_{\mu\nu}|\theta\rangle = \frac{1}{2LT}\int d^2x\,\langle\theta|\epsilon_{\mu\nu}F_{\mu\nu}|\theta\rangle$$

$$= -\frac{4\pi}{2LT}\frac{\int\mathcal{D}(A_\mu,\phi,\phi^*)\nu e^{-\frac{S_0^E}{\hbar}}e^{i\nu\theta}}{\int\mathcal{D}(A_\mu,\phi,\phi^*)e^{-\frac{S_0^E}{\hbar}}e^{i\nu\theta}}$$

$$= \frac{4\pi}{2LT}i\frac{d}{d\theta}\ln\left(\int\mathcal{D}(A_\mu,\phi,\phi^*)e^{-\frac{S_0^E}{\hbar}}e^{i\nu\theta}\right)$$

$$= \frac{4\pi}{2LT}i\frac{d}{d\theta}\left(2Ke^{-\frac{S_0^E}{\hbar}}LT\cos(\theta)\right)$$

$$= -i4\pi Ke^{-\frac{S_0^E}{\hbar}}\sin(\theta). \qquad (8.59)$$

For small θ from Equation (8.40), removing a constant, we have

$$\frac{\mathcal{E}(\theta)}{L} = Ke^{-\frac{S_0^E}{\hbar}}\theta^2. \qquad (8.60)$$

Also,

$$\langle\theta|F_{12}|\theta\rangle = -i4\pi Ke^{-\frac{S_0^E}{\hbar}}\theta$$

$$E(R) = 2R\left(\theta^2 - \left(\theta + \frac{2\pi q}{e}\right)^2\right)Ke^{-\frac{S_0^E}{\hbar}}. \qquad (8.61)$$

This lends itself to the following interpretation. In the θ vacuum, there exists an electric field that is proportional to θ with a corresponding energy density proportional to θ^2. The external charges induce an electric field between them, proportional to their charges. The energy changes by the separation of the charges multiplied by the energy density, which in this case is clearly

$$\left(\theta + \frac{2\pi q}{e}\right)^2. \qquad (8.62)$$

There exist non-linear effects that convert these to periodic functions in $\frac{q}{e}$. This is because the theory contains particles of charge e. For $q > e$, a charged particle–anti-particle pair can be created, which then can migrate to the oppositely charged external charges, lowering their charge and hence the induced electric field. Thus q's are equivalent modulo e.

Our analysis, although encouraging, cannot work in higher dimensions. In $1+1$ dimensions, the flux of each instanton inside the loop must totally pass through the loop, independent of its position inside the loop. In $3+1$ dimensions the instantons are not like flux tubes, they are $O(4)$-symmetric objects. Instead of the Wilson loop, we would require some analogous "Wilson three-dimensional hypersurface" to reach the same conclusion. Confinement must, however, imply the area law for the usual Wilson loop, in any dimensions. Thus we do not expect instantons to be responsible for confinement in higher dimensions. We can, as we shall see in Chapter 9, circumvent this problem in $2+1$ dimensions by introducing a mild non-Abelian nature.

9

The Polyakov Proof of Confinement

In a totally surprising result, Polyakov demonstrated that instantons could provide the key to confinement in a particular model in 2+1 dimensions [103]. In this chapter, we will study in detail the Polyakov proof of confinement. We will see that it requires a mild non-Abelian aspect to the theory, but the confinement occurs essentially because of the existence of magnetic monopole solitons in the theory. Purely Abelian gauge theory also contains magnetic monopoles, but they are singular configurations of infinite energy, and hence of no import. The minor non-Abelian excursion simply allows for the existence of finite action (or energy) magnetic monopoles.

9.1 Georgi–Glashow model

We continue our study of quantum electrodynamics in 2+1 dimensions; however, now we shall consider a theory that is Abelian at low energy but non-Abelian at high energy. This occurs due to spontaneous symmetry breaking. We consider a non-Abelian gauge theory with gauge group $O(3) \sim SU(2)$ spontaneously broken to $U(1)$. The model is the 2+1-dimensional version of the Georgi–Glashow model [54]. The fields correspond to an iso-triplet of scalar fields interacting via non-Abelian gauge fields and self-interactions, the Lagrangian density is given by

$$\mathcal{L} = -\frac{1}{4e^2} F_{\mu\nu}^a F^{a\mu\nu} + |D_\mu \phi|^2 - \frac{1}{4}\lambda \left(|\phi|^2 - a^2 \right)^2, \tag{9.1}$$

where

$$F_{\mu\nu}^a = \partial_\mu A_\nu^a - \partial_\nu A_\mu^a + \epsilon^{abc} A_\mu^b A_\nu^c$$

$$\phi = \begin{pmatrix} \phi^1 \\ \phi^2 \\ \phi^3 \end{pmatrix}, \quad (D_\mu \phi)^a = \partial_\mu \phi^a + \epsilon^{abc} A_\mu^b \phi^c$$

$$|\phi|^2 = \phi^a \phi^a. \tag{9.2}$$

The theory is invariant under local redefinition of the fields by

$$\phi^a \to R^{ab}(x^\nu)\phi^b$$
$$A_\mu^a \to R^{ab}(x^\nu)A_\mu^b + \epsilon^{abc}R^{bd}(x^\nu)\partial_\mu R^{cd}(x^\nu), \qquad (9.3)$$

where $R^{ab}(x^\nu)$ is a smooth, $O(3)$-valued gauge transformation.

We may sometimes wish to use the matrix notation, hence we record the corresponding formulae here. The Higgs field is written as ϕ, which is a three-real entry column. The gauge field is a 3×3 real, anti-symmetric matrix A_μ for each spacetime index μ. There are exactly three independent anti-symmetric 3×3 matrices where a basis can be denoted as T^a with components numerically given by $T^a_{bc} = \epsilon^{abc}$ (here the placement of the index as upper or lower is of no import). Then $A_\mu = A_\mu^a T^a$. Then the gauge transformation is written as

$$\phi \to R(x^\nu)\phi$$
$$A_\mu \to R(x^\nu)A_\mu + R(x^\nu)\partial_\mu R^T(x^\nu), \qquad (9.4)$$

where $R(x^\nu)$ is a 3×3 orthogonal matrix (hence its inverse is given by its transpose).

We can easily see the perturbative, physical particle spectrum of the theory by making a choice of gauge

$$\phi^1 = \phi^2 = 0. \qquad (9.5)$$

To be honest, this is an incomplete gauge-fixing condition: it does not fix the gauge degree of freedom if ϕ is already in the three-direction and it does not fix the gauge transformations which leave ϕ_3 invariant. However, it is sufficient for us to extract the particle spectrum. Then, replacing $\phi^3 = a + \eta$ we have:

$$(D_\mu\phi)^1 = \partial_\mu\phi^1 + \epsilon^{1bc}A_\mu^b\phi^c = \epsilon^{123}A_\mu^2\phi^3 = A_\mu^2(a+\eta)$$
$$(D_\mu\phi)^2 = \partial_\mu\phi^2 + \epsilon^{2bc}A_\mu^b\phi^c = \epsilon^{213}A_\mu^1\phi^3 = -A_\mu^1(a+\eta)$$
$$(D_\mu\phi)^3 = \partial_\mu\phi^3 + \epsilon^{3bc}A_\mu^b\phi^c = \partial_\mu(a+\eta) = \partial_\mu\eta. \qquad (9.6)$$

Hence

$$|D_\mu\phi|^2 = \partial_\mu\eta\partial^\mu\eta + \left(A_\mu^1 A^{1\mu} + A_\mu^2 A^{2\mu}\right)\left(a^2 + 2a\eta + \eta^2\right), \qquad (9.7)$$

giving the Lagrangian density

$$\mathcal{L} = -\frac{1}{4e^2}F_{\mu\nu}^a F^{a\mu\nu} + \partial_\mu\eta\partial^\mu\eta + \partial_\mu\eta\partial^\mu\eta$$
$$+ \left(A_\mu^1 A^{1\mu} + A_\mu^2 A^{2\mu}\right)\left(a^2 + 2a\eta + \eta^2\right) - \frac{1}{4}\lambda\left(2a\eta + \eta^2\right)^2. \qquad (9.8)$$

This yields the quadratic part

$$\mathcal{L} = \frac{-1}{2e^2}\left(\partial_\mu A_\nu^a - \partial_\nu A_\mu^a\right)\left(\partial^\mu A^{a\nu}\right) + \partial_\nu\eta\partial^\mu\eta + \left(A_\mu^1 A^{1\mu} + A_\mu^2 A^{2\mu}\right)a^2 - \lambda a\eta^2. \quad (9.9)$$

The physical particle spectrum can now be read off from this equation; it corresponds to a massless vector field A_μ^3, two massive vector fields A_μ^1 and A_μ^2

of mass $M^2 = 4e^2a^2$, and a neutral massive scalar field η (neutral with respect to the gauge field A_μ^3) with mass $m^2 = \lambda a$. η is neutral since it does not couple to A_μ^3, while the massive vector fields A_μ^1 and A_μ^2 are charged as they do. The two fields ϕ_1 and ϕ_2 are, of course, absent. We might say this is due to our gauge choice; however, the fact that the corresponding physical excitations do not exist is independent of the gauge choice. What we are describing is the classic Higgs mechanism [61], where the putative massless Goldstone bosons associated with spontaneous symmetry-breaking are swallowed by the gauge bosons that correspond to the broken symmetry directions. These gauge bosons consequently become massive. Hence the Goldstone bosons are absent, but their degrees of freedom show up in the additional degrees of freedom of the massive vector bosons (as opposed to massless ones).

We will see in this chapter that, as in the case of the Abelian Higgs model in 1+1 dimensions in Chapter 8, the actual spectrum of the theory does not correspond to this naive spectrum. We will find that the theory in fact confines charged states due to the effects of instantons and that there are no massless states, especially there is no massless photon. The validity of the argument that the Wilson loop is able to subtend an appreciable amount of flux from the instantons, which was used in Chapter 8, becomes critical in $2+1$ dimensions. As the size of the Wilson loop becomes large, it can subtend an arbitrary amount of flux from nearby instantons, and hence the effect of instantons is significant. In 3+1 dimensions we will see that the argument fails.

9.2 Euclidean Theory

Analytically continuing our action to three-dimensional Euclidean space (although much of what we say is trivially generalized to d Euclidean dimensions) gives

$$S_E = \int d^3x \left(\frac{1}{4e^2} F_{\mu\nu}^a F_{\mu\nu}^a + \frac{1}{2} (D_\mu\phi)^a (D_\mu\phi)^a + \frac{1}{4} (\phi^a\phi^a - a^2)^2 \right), \qquad (9.10)$$

which is again composed of three positive semi-definite terms. We look for finite action configurations: these would correspond to instantons and should be relevant for tunnelling. Finite action requires that the fields behave in such a way that each term in the action goes to zero sufficiently fast at infinity, as each term is positive semi-definite. Sufficiently fast can include $\sim 1/r$ fall off of particular fields or their derivatives, the only condition is that the Euclidean action be finite, and hence each term vanishes sufficiently fast. This then implies that at infinity

$$\phi^a \to R_\phi^{ab}(\Omega)\phi_0^a \qquad \phi_0^a\phi_0^a = a^2 \qquad (9.11)$$
$$(D_\mu\phi)^a \to 0 \qquad (9.12)$$
$$F_{\mu\nu}^a \to 0, \qquad (9.13)$$

where Ω are the angular coordinates parametrizing the sphere at infinity. Equation (9.13) requires that the gauge fields approach a configuration that corresponds to a pure gauge transformation of the vacuum, sufficiently fast. We can write the gauge field in a matrix notation

$$A_\mu = A_\mu^a T^a, \tag{9.14}$$

where T^a are 3×3 matrices with components numerically given by $T_{bc}^a = \epsilon^{abc}$. Then Equation (9.13) implies, in this matrix notation, that the gauge field corresponds to a gauge transformation of zero,

$$A_\mu \to R_{A_\mu}(\Omega)\partial_\mu R_{A_\mu}^\dagger(\Omega). \tag{9.15}$$

Then automatically for the covariant derivative of the scalar field we get (suppressing the Ω dependence and its index a)

$$
\begin{aligned}
D_\mu\phi &\to (\partial_\mu + R_{A_\mu}\partial_\mu R_{A_\mu}^\dagger)R_\phi\phi_0 \\
&= R_\phi\left(R_\phi^\dagger\partial_\mu R_\phi + R_\phi^\dagger R_{A_\mu}(\partial_\mu R_{A_\mu}^\dagger)R_\phi\right)\phi_0 \\
&= R_\phi\left(R_\phi^\dagger R_{A_\mu}\partial_\mu(R_{A_\mu}^\dagger R_\phi)\right)\phi_0 = 0. \tag{9.16}
\end{aligned}
$$

This requires that $R_\phi^\dagger R_{A_\mu}\partial_\mu(R_{A_\mu}^\dagger R_\phi)$, which is a Lie algebra element, be in the direction that annihilates ϕ^0 or correspondingly $R_{A_\mu}^\dagger R_\phi$ leaves ϕ^0 invariant, that is $R_{A_\mu}^\dagger R_\phi = H$ where $H\phi^0 = \phi^0$. H may not be globally defined on the sphere at infinity; however, locally it is, and that is all we need. This defines the invariant subgroup or stabilizer of ϕ_0. But now we may redefine $R_\phi \to \tilde{R}_\phi = R_\phi H^{-1}$ as R_ϕ is only defined up to an element of the stabilizer of ϕ_0, as is obvious from Equation (9.11) (we will drop the tilde from now on). Thus we get $R_{A_\mu}^\dagger R_\phi = 1$ at least locally on the sphere at infinity. Although we started with different, independent gauge transformations, R_ϕ and R_{A_μ}, in Equations (9.11) and (9.15), respectively, we see that Equation (9.12) forces the gauge transformations to be the same. We will now call this gauge transformation $R(\Omega)$. We underline that $R(\Omega)$ may not be globally defined, and may actually be singular at some place on the sphere at infinity. In fact, for a non-trivial mapping it must be singular somewhere. However, its action on ϕ_0, which defines the values of the Higgs field at infinity, must be globally defined.

The condition of finite action is actually a little more subtle. Indeed, the gauge field must become pure gauge only as fast as $\sim 1/r$ for the $F_{\mu\nu}^a F_{\mu\nu}^a$ to give a finite contribution. Thus we should modify Equation (9.15) to

$$A_\mu \to R(\Omega)\partial_\mu R^\dagger(\Omega) + \tilde{A}_\mu, \tag{9.17}$$

where $\tilde{A}_\mu \sim o(1/r)$ (keeping in mind that the pure gauge terms also behave as $\sim 1/r$). However, such a modification could cause trouble in Equation (9.12), as the covariant derivative of the scalar field must vanish faster than $1/r^2$ for finite action. But this can again be solved if these additional possible terms in the gauge field are in the direction of the stabilizer of the Higgs field. Thus we can tolerate additional non-pure gauge terms in the gauge field as long as

$$\tilde{A}_\mu R\phi_0 = 0. \tag{9.18}$$

9.2.1 Topological Homotopy Classes

Thus finite action configurations are characterized by $R(\Omega)$ defined at $|\vec{x}| \to \infty$. This defines a map of the sphere at infinity S^{d-1} (generalizing temporarily to d dimensions) into the space of "vacuum" configurations, $\{\phi^a : \phi^a\phi^a = a^2\} \equiv \mathcal{M} = S^2$. The equivalence classes under homotopy of these maps form the homotopy groups

$$\Pi_{d-1}(\mathcal{M}). \tag{9.19}$$

There is a fascinating and complex set of corresponding homotopy groups [51]:

$$\Pi_{d-1}(\mathcal{M}) = \begin{cases} 0 & d=2 \\ \mathbb{Z} & d=3 \\ \mathbb{Z} & d=4 \\ \mathbb{Z}_2 & d=5 \\ \mathbb{Z}_2 & d=6 \\ \mathbb{Z}_{12} & d=7 \\ \mathbb{Z}_2 & d=8 \\ \mathbb{Z}_2 & d=9 \\ \mathbb{Z}_3 & d=10 \\ \mathbb{Z}_{15} & d=11 \\ \mathbb{Z}_2 & d=12 \\ \mathbb{Z}_2 \times \mathbb{Z}_2 & d=13 \\ \mathbb{Z}_{12} \times \mathbb{Z}_2 & d=14 \\ \mathbb{Z}_{84} \times \mathbb{Z}_2 \times \mathbb{Z}_2 & d=15 \\ \mathbb{Z}_2 \times \mathbb{Z}_2 & d=16 \\ \vdots \\ \vdots \end{cases} \tag{9.20}$$

Thus there exist topologically non-trivial configurations in each dimension and the possibility of non-trivial finite Euclidean action configurations. In $d = 3$, the corresponding instantons are actually the 't Hooft–Polyakov magnetic monopole solitons of the $3 + 1$-dimensional theory.

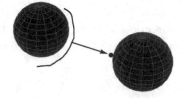

Figure 9.1. Mapping the whole S^2 at ∞ to a point

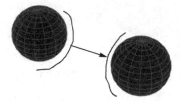

Figure 9.2. Mapping the S^2 at ∞ on to the vacuum manifold S^2

9.2.2 Magnetic Monopole Solutions

For $d = 3$, we have the maps

$$R(\Omega)\phi_0 : S^2 \to S^2, \tag{9.21}$$

where the first S^2 is defined by the set of all Ω's, *i.e.* the sphere at ∞, while the second S^2 is defined by the set of Higgs field values $\phi^2 = \phi^a\phi^a = a^2$. These fall into homotopically inequivalent classes, characterized by the winding number of the map, much like the previous case of maps of $S^1 \to S^1$ in the Abelian Higgs model. Pictured in Figures 9.1 and 9.2 are the trivial map to a point and the onto map, where each point in the first S^2 is mapped to the analogous point on the second S^2. We cannot continuously deform one configuration into another if they have different winding numbers, that is the definition of homotopy classes, and typically this implies that there exists an infinite action barrier between configurations in different classes. We will see that the topological winding number turns out to be associated with the magnetic charge of each sector. The minimum action configuration in each class must solve the equations of motion. The action must be stationary at the minimum action configuration since, if the first-order variation does not vanish, one can find a variation which lowers the action. The equations of motion are therefore satisfied. What is not necessary is that the minimum action configuration is non-trivial; it could, for example, collapse and shrink to a point or, conversely, spread out and dilute infinitely. We will show that it must be non-trivial.

The homotopy class with topological winding number $n = 1$ defines the standard instanton. We can prove that the action is bounded from below in each sector using a method first shown by Bogomolny [17]. We assume that the potential $V(\phi)$ is positive semi-definite. Defining the non-Abelian magnetic field as $B_i^a = \frac{1}{2}\epsilon_{ijk}F_{jk}^a$ we have

$$
\begin{aligned}
S_E &= \int d^3x \left(\frac{1}{2}\frac{B_i^a}{e}\frac{B_i^a}{e} + \frac{1}{2}(D_i\phi)^a(D_i\phi)^a + V(\phi) \right) \\
&\geq \int d^3x \frac{1}{2}\left(\frac{B_i^a}{e} \mp (D_i\phi)^a \right)^2 \pm \frac{B_i^a}{e}(D_i\phi)^a \\
&\geq \pm \int d^3x \frac{B_i^a}{e}(D_i\phi)^a \\
&= \pm\frac{1}{e}\int d^3x B_i^a \partial_i\phi^a + B_i^a \epsilon^{abc}A_i^b\phi^c \\
&= \pm\frac{1}{e}\int d^3x \partial_i(B_i^a\phi^a) - \left((\partial_i B_i^a)\phi^a - B_i^a\epsilon^{abc}A_i^b\phi^c\right) \\
&= \pm\frac{1}{e}\left(\oint dS^i(B_i^a\phi^a) - \int d^3x \left(\partial_i B_i^a + \epsilon^{abc}A_i^b B_i^c\right)\phi^a \right) \\
&\equiv \pm ga,
\end{aligned}
\tag{9.22}
$$

where in the second line we have simply completed the square and dropped the potential, in the third line we have dropped the positive semi-definite first term and in the penultimate equation the last term vanishes because of the Jacobi identity. The Jacobi identity is $\epsilon_{ijk}[D_i,[D_j,D_k]] = 0$ which is simply, trivially, algebraically valid (just spell out all of the terms and they cancel pairwise). This gives $D_i B_i^a = \partial_i B_i^a + \epsilon^{abc}A_i^b B_i^c = 0$ as $[D_j^a, D_k^b] = \epsilon_{jkl}\epsilon^{abc}B_l^c$ which is the non-Abelian analogue of Maxwell's equation $\nabla \cdot \vec{B} = 0$. Normally, in the purely Abelian theory, this equation denies the existence of magnetic monopoles. Here the magnetic monopoles do exist, since the non-Abelian divergence of the magnetic field contains inhomogeneous terms. The magnetic monopoles exist as instantons in the Euclideanized 2+1-dimensional theory or as actual static solitons in the 3+1-dimensional theory. g is the magnetic charge

$$
g = \frac{1}{ae}\oint dS^i B_i^a \phi^a
\tag{9.23}
$$

and a is the vacuum expectation value of the scalar field. Clearly, if g is positive we take the plus sign in Equation (9.22), and if g is negative we take the minus sign. This implies that the Euclidean action has a positive definite lower bound in each topological sector. We will show $g \neq 0$ except in the topologically trivial sector. Indeed, for the ansatz

$$
\begin{aligned}
\phi^a &= H(aer)\frac{x^a}{er^2} \\
A_i^a &= -\epsilon^{aij}\frac{x^j}{r^2}(1 - K(aer))
\end{aligned}
\tag{9.24}
$$

finite action requires

$$
\begin{aligned}
H(aer) &\to aer \quad, \quad r \to \infty \\
K(aer) &\to 0 \quad, \quad r \to \infty \\
H(aer) &< o(aer) \quad, \quad r \to 0 \\
K(aer) &< o(aer) \quad, \quad r \to 0.
\end{aligned} \tag{9.25}
$$

Thus for large r

$$
\phi^a \approx a \frac{x^a}{r} = (R\phi_0)^a
$$

$$
A_i^a \approx -\epsilon^{aij} \frac{x^j}{r^2} = R\partial_i R^\dagger + \tilde{A}_i^a \tag{9.26}
$$

giving

$$
F_{ij}^a \approx \epsilon_{ijk} \frac{x^k x^a}{r^4}. \tag{9.27}
$$

Defining the Abelian magnetic field as

$$
B_i = \frac{1}{2}\epsilon_{ijk} F_{jk}^a \frac{\phi^a}{a} \approx \frac{x^i}{r^3} \tag{9.28}
$$

we have

$$
g = \frac{1}{e} \oint dS_i B_i = \frac{4\pi}{e} \neq 0. \tag{9.29}
$$

This is in fact the Dirac quantization condition on magnetic charge, $gq = 2\pi$, for the minimal electric charge $q = e/2$. Not surprisingly, the theory knows that it can, in principle, have fields in the spinor representation of the iso-spin group $(SO(3))$ that do carry charge $e/2$.

For the Higgs field satisfying the conditions of the "Higgs" vacuum

$$
\begin{aligned}
\phi^a \phi^a &= a^2 \\
(D_\mu \phi)^a &= 0
\end{aligned} \tag{9.30}
$$

we can write the explicit solution, using the iso-vector notation $\vec{\phi}$ for the Higgs field

$$
A_\mu^a = \frac{1}{a^2} \left(\vec{\phi} \times \partial_\mu \vec{\phi} \right)^a + \frac{1}{a} \phi^a A_\mu
$$

$$
F_{\mu\nu}^a = \frac{1}{a} \phi^a F_{\mu\nu}, \tag{9.31}
$$

where

$$
F_{\mu\nu} = \frac{1}{a^3} \phi^a \left(\partial_\mu \vec{\phi} \times \partial_\nu \vec{\phi} \right)^a + \partial_\mu A_\nu - \partial_\nu A_\mu. \tag{9.32}
$$

A_μ generates only a source-free magnetic field, but

$$
B_i = \frac{1}{2}\epsilon_{ijk} F_{jk} \tag{9.33}
$$

can have non-zero magnetic charge due to the first term in Equation (9.32). The magnetic charge in any region is

$$g = \frac{1}{e} \oint_\Sigma \vec{B} \cdot d\vec{S} = \frac{1}{2ea^3} \oint_\Sigma dS_i \epsilon_{abc} \epsilon^{ijk} \phi^a \partial_j \phi^b \partial_k \phi^c. \tag{9.34}$$

We will show that this integral is actually a topological invariant and equal to the result $4\pi/e$ that we found for the configuration in Equation (9.29) above. It counts the winding number of the map from the surface Σ which is topologically S^2 into the S^2 defined by $\phi^a \phi^a = a^2$. Indeed, consider a variation $\delta\phi$ which is of compact support, then $\vec{\phi} \to \vec{\phi} + \delta\vec{\phi}$ but since $\left(\vec{\phi} \cdot \vec{\phi}\right) = 1$ we get

$$\delta\left(\vec{\phi} \cdot \vec{\phi}\right) = 2\vec{\phi} \cdot \delta\vec{\phi} = 0. \tag{9.35}$$

Then

$$\begin{aligned}
\delta\left(\vec{\phi} \cdot (\partial_j\vec{\phi} \times \partial_k\vec{\phi})\right) &= \delta\vec{\phi} \cdot \left(\partial_j\vec{\phi} \times \partial_k\vec{\phi}\right) + \vec{\phi} \cdot \left(\partial_j\delta\vec{\phi} \times \partial_k\vec{\phi}\right) + \vec{\phi} \cdot \left(\partial_j\vec{\phi} \times \partial_k\delta\vec{\phi}\right) \\
&= \delta\vec{\phi} \cdot \left(\partial_j\vec{\phi} \times \partial_k\vec{\phi}\right) + \partial_j\left(\vec{\phi} \cdot (\delta\vec{\phi} \times \partial_k\vec{\phi})\right) - \partial_j\vec{\phi} \cdot \left(\delta\vec{\phi} \times \partial_k\vec{\phi}\right) \\
&\quad \vec{\phi} \cdot \left(\delta\vec{\phi} \times \partial_j\partial_k\vec{\phi}\right) + \partial_k\left(\vec{\phi} \cdot (\partial_j\vec{\phi} \times \delta\vec{\phi})\right) \\
&\quad \partial_j\vec{\phi} \cdot \left(\delta\vec{\phi} \times \partial_k\vec{\phi}\right) - \vec{\phi} \cdot \left(\partial_j\partial_k\vec{\phi} \times \delta\vec{\phi}\right) \\
&= 3\,\delta\vec{\phi} \cdot \left(\partial_j\vec{\phi} \times \partial_k\vec{\phi}\right) + 2\partial_j\left(\vec{\phi} \cdot (\delta\vec{\phi} \times \partial_k\vec{\phi})\right),
\end{aligned} \tag{9.36}$$

where, in the last step, we use that the expression is contracted with ϵ^{ijk}. The total derivative terms give no contribution to any integral since $\delta\phi$ is of compact support. Now $\partial_j\phi$ and $\partial_k\phi$ are both orthogonal to ϕ, thus $\partial_j\vec{\phi} \times \partial_k\vec{\phi}$ is parallel to ϕ, giving

$$\delta\vec{\phi} \cdot \left(\partial_j\vec{\phi} \times \partial_k\vec{\phi}\right) = 0 \tag{9.37}$$

hence

$$\delta\left(\vec{\phi} \cdot (\partial_j\vec{\phi} \times \partial_k\vec{\phi})\right) = 2\partial_j\left(\vec{\phi} \cdot (\delta\vec{\phi} \times \partial_k\vec{\phi})\right). \tag{9.38}$$

Therefore the integral, Equation (9.34), is invariant under arbitrary continuous deformation of ϕ, since these are built up from a sequence of infinitesimal deformations of compact support. A continuous deformation of the surface over which the field is defined can also be interpreted as a continuous deformation of the ϕ field, thus g is also invariant under continuous deformations of the integration surface (remember that we are only in the Higgs vacuum).

Finally we can calculate g for

$$\phi^a = a\hat{x}^a = a\frac{x^a}{r}, \tag{9.39}$$

asymptotically, which corresponds to the winding number equal to one map. Then

$$\partial^i \phi^a = a\left(\frac{\delta^{ai}}{r} - \frac{x^a x^i}{r^3}\right) = \frac{a}{r}\left(\delta^{ai} - \hat{x}^a \hat{x}^i\right), \tag{9.40}$$

which gives

$$\epsilon^{ijk}\epsilon_{abc}\phi^a\partial_j\phi^b\partial_k\phi^c = a^3\epsilon_{ijk}\epsilon_{abc}\frac{x^a}{r^3}\left(\delta^{jb} - \hat{x}^j\hat{x}^b\right)\left(\delta^{kc} - \hat{x}^k\hat{x}^c\right) = \frac{2a^3}{r^2}\hat{x}^i. \quad (9.41)$$

Hence

$$g = \frac{1}{2ea^3}\oint_\Sigma dS^i\frac{2a^3}{r^2}\hat{x}^i = \frac{1}{2ea^3}8\pi a^3 = \frac{4\pi}{e}. \quad (9.42)$$

This answer is robust, in that it does not change for any infinitesimal changes and hence for any continuous change in the Higgs field. If we use the winding number 2 map, the answer for the integral will be $2 \times 4\pi/e$, and so on. If we write $\phi = R\phi_0$, then the winding number N map is obtained by taking $\phi = R^N\phi_0$.

If we transform $\phi^a \to \tilde{\phi}^a = \delta^{a3}a$, we cannot define the gauge transformation globally over any surface containing the core. We get the usual Dirac string singularity,

$$A_i^a = \delta^{a3}\frac{1}{er}\frac{(1-\cos\theta)}{\sin\theta}\hat{\varphi}_i, \quad (9.43)$$

where $\hat{\varphi}$ is the unit vector in the azimuthal direction.

9.3 Monopole Ansatz with Maximal Symmetry

The solution follows from the most general ansatz

$$\phi^a = H(aer)\frac{x^a}{er^2}$$

$$A_i^a = -\epsilon^{aij}\frac{x^j}{e^2r^2}(1 - K(aer)) + \frac{r^2\delta^{ai} - x^ix^a}{e^2r^3}B(aer) + \frac{x^ix^a}{e^2r^3}C(aer), \quad (9.44)$$

which is symmetric under the diagonal subgroup of the group $SO(3)_{\text{rot.}} \times SO(3)_{\text{iso-rot.}}$ of rotations and iso-rotations. If we had imposed invariance only under the $SO(3)_{\text{rot.}}$, the rotation subgroup alone, we would have to impose that ϕ^a is a constant on each spatial sphere, giving trivial asymptotic topology. On the other hand, the configuration that is invariant only under $SO(3)_{\text{iso-rot.}}$, the iso-rotational group, has the only possibility $\phi^a = 0$, which also has trivial topology. However, we can impose invariance under the next subgroup available, $SO(3)_{\text{diagonal}}$, the diagonal subgroup of rotations and iso-rotations, which the fields in Equation (9.44) satisfy.

Parity corresponds to the transformation

$$P: \phi^a(x^j,t) \to \phi^a(-x^j,t), \quad A_i^a(x^j,t) \to -A_i^a(-x^j,t) \quad (9.45)$$

and there is also the discrete transformation

$$Z: \phi^a(x^j,t) \to -\phi^a(x^j,t), \quad A_i^a(x^j,t) \to A_i^a(x^j,t). \quad (9.46)$$

P and Z individually reverse the magnetic charge, thus we cannot impose invariance under each separately. However, their product leaves the magnetic

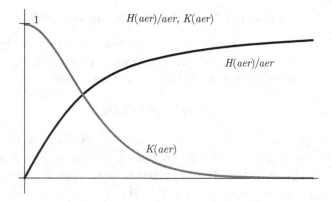

Figure 9.3. The curves of $H(aer)/aer$ and $K(aer)$

charge invariant. Hence, in the spirit of imposing the maximum symmetry on the solution without making it trivial, we impose that the ansatz be invariant under PZ. This implies $B(aer) = C(aer) = 0$.

9.3.1 Monopole Equations

We find, then, that $H(aer)$ and $K(aer)$ satisfy the system of equations

$$r^2 \frac{d^2}{dr^2} K(r) = K(r)H^2(r) + K(r)(K^2(r) - 1)$$
$$r^2 \frac{d^2}{dr^2} H(r) = 2K(r)^2 H(r) + \frac{\lambda}{e^2} H(r)(H^2(r) - a^2 r^2). \tag{9.47}$$

They have numerical solutions as depicted in Figure 9.3. In the Prasad–Sommerfield limit [104], $\lambda \to 0$, we know the exact solution

$$H(aer) = aer \coth(aer) - 1$$

$$K(aer) = \frac{aer}{\sinh(aer)}. \tag{9.48}$$

This solution corresponds to the famous 't Hooft–Polyakov magnetic monopole. In $3+1$ dimensions it is a static, stable, finite-energy solution to the equations of motion. In $2+1$ dimensions, but Euclideanized, it serves equally well as a finite-action, Euclidean space instanton, where it mediates tunnelling between different classical vacua, as we will see below.

9.4 Non-Abelian Gauge Field Theories

We must examine in some more detail what it means to have a quantum non-Abelian gauge theory.

9.4.1 Classical Non-Abelian Gauge Invariance

First we will consider non-Abelian gauge invariance more generally, and then apply it to our specific case. A non-Abelian gauge theory admits fields which transform according to given representations of a non-Abelian group,

$$\phi \to \mathcal{U}(g)\phi \quad \mathcal{U}(g) \in G, \tag{9.49}$$

where $\mathcal{U}(g)\mathcal{U}^\dagger(g) = \mathcal{U}^\dagger(g)\mathcal{U}(g) = 1$. If g does not depend on the spacetime point, we call the gauge transformation global, otherwise it is a local gauge transformation. However, the allowed variation of the gauge transformation is restricted to a region of compact support. It is easy to write a Lagrangian that is invariant under global gauge transformations, we simply construct it out of invariant polynomials of the fields. Spacetime derivatives commute with the gauge-transforming field $\mathcal{U}(g)$ and hence cause no problems. Now if we want to generalize the invariance to include local gauge transformations, we must introduce new fields. For our case

$$\phi^a \to (\mathcal{U}(g)\phi)^a$$
$$(\partial^\mu \phi)^a \to \partial^\mu (\mathcal{U}(g)\phi)^a$$
$$= (\mathcal{U}(g)\partial^\mu \phi)^a + ((\partial^\mu \mathcal{U}(g))\phi)^a. \tag{9.50}$$

That is, if $\mathcal{U}(g)$ depends on the spacetime point, the derivative does not commute with it. We must introduce a new field, the gauge field A_μ^a, with an inhomogeneous transformation property which will exactly cancel the extra term generated by the derivative. We replace all derivatives by

$$\partial_\mu \to \partial_\mu + A_\mu, \tag{9.51}$$

where A_μ is a vector field with values in the Lie algebra of the representation under which ϕ transforms. In our case

$$A_\mu = A_\mu^b \left(-\epsilon^{bac}\right), \tag{9.52}$$

thus

$$(D_\mu \phi)^a = \partial_\mu \phi^a - A_\mu^b \epsilon^{bac} \phi^c$$
$$= \partial_\mu \phi^a + \epsilon^{abc} A_\mu^b \phi^c. \tag{9.53}$$

A_μ is given the transformation property such that the covariant derivative transform covariantly:

$$D_\mu \phi \to \mathcal{U}(g) D_\mu \phi. \tag{9.54}$$

This is satisfied if

$$A_\mu \to \mathcal{U}(g) \left(A_\mu + \partial_\mu\right) \mathcal{U}^\dagger(g). \tag{9.55}$$

Evidently

$$
\begin{aligned}
D_\mu \phi = (\partial_\mu + A_\mu)\phi &\rightarrow (\partial_\mu + \mathcal{U}(g)(A_\mu + \partial_\mu)\mathcal{U}^\dagger(g))\mathcal{U}(g)\phi \\
&= (\partial_\mu \mathcal{U}(g))\phi + \mathcal{U}(g)\partial_\mu \phi + \mathcal{U}(g)A_\mu \phi + \mathcal{U}(g)(\partial_\mu \mathcal{U}^\dagger(g))\mathcal{U}(g)\phi \\
&= \mathcal{U}(g)(\partial_\mu + A_\mu)\phi + (\partial_\mu \mathcal{U}(g) + \mathcal{U}(g)(\partial_\mu \mathcal{U}^\dagger(g))\mathcal{U}(g))\phi \\
&= \mathcal{U}(g)(\partial_\mu + A_\mu)\phi \\
&= \mathcal{U}(g)D_\mu \phi.
\end{aligned}
\tag{9.56}
$$

The covariant derivative has the geometrical interpretation as the parallel transport in a fibre bundle with connection A_μ. For each infinitesimal path, $x^\mu \rightarrow x^\mu + dx^\mu$, we introduce the gauge field $A_\mu(x^\nu)$ and an element of the group,

$$
g(x + dx, A_\mu) = 1 + dx^\mu A_\mu.
\tag{9.57}
$$

Then for a finite path $\mathcal{C}$ we integrate this as

$$
g(\mathcal{C}, A) = P\left(\exp\left\{\int_\mathcal{C} dx^\mu A_\mu\right\}\right),
\tag{9.58}
$$

where the P symbol means the path-ordered integral. Intuitively this corresponds to the limit taken by multiplying the group elements of the form (9.57) for a finitely discretized approximation to the finite curve $\mathcal{C}$, in the order corresponding to the direction of the curve, and taking the limit that the discretization becomes infinitely fine. The other definition, which yields the same result, is to expand the exponential and then perform the multiple integral at each order, after applying the path-ordering to the integrand. A field is considered to have been transported in parallel in the connection A_μ if

$$
\begin{aligned}
\phi(x + dx) = \phi^{g(x+dx, A_\mu)}(x) &= \mathcal{U}(g(x + dx, A_\mu))\phi \\
&= \phi(x) + dx^\mu A_\mu \phi(x).
\end{aligned}
\tag{9.59}
$$

Then, in general,

$$
\begin{aligned}
\phi(x + dx) - \phi^{g(x+dx, A_\mu)}(x) &= dx^\mu(\partial_\mu + A_\mu(x))\phi(x) \\
&= dx^\mu D_\mu \phi(x)
\end{aligned}
\tag{9.60}
$$

defines the covariant derivative in the connection A_μ. Here $A_\mu = A_\mu^a t^a$, where t^a are the generators of the group in the representation that $\phi(x)$ transforms under.

9.4.2 The Field Strength

To construct the non-Abelian field strength we must consider a generalization of the Abelian version,

$$
F_{\mu\nu} = \partial_\mu A_\nu - \partial_\nu A_\mu.
\tag{9.61}
$$

This is invariant under Abelian gauge transformations

$$
A_\mu \rightarrow A_\mu + i\partial_\mu \Lambda
$$

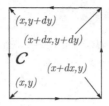

Figure 9.4. An infinitesimal closed loop $\mathcal{C}$

$$\delta F_{\mu\nu} = (\partial_\mu \partial_\nu - \partial_\nu \partial_\mu)\Lambda = 0. \tag{9.62}$$

We can write this as

$$F_{\mu\nu} \to e^{-i\Lambda} F_{\mu\nu} e^{i\Lambda} = F_{\mu\nu}. \tag{9.63}$$

For Abelian phases, $F_{\mu\nu}$ is invariant, but if we generalize this formula to non-Abelian groups, $F_{\mu\nu}$ does transform, but homogeneously. We construct $F_{\mu\nu}$ via parallel transport. The same construction works as in the abelian case. Consider a closed loop $\mathcal{C}$ drawn in Figure 9.4, and

$$g(\mathcal{C}, x, A) = P\left(\exp\left\{-\oint dx^\mu A_\mu\right\}\right)$$
$$= 1 - \oint dx^\mu A_\mu + \oint dx_1 \oint_{x_2 > x_1} dx_2^\mu A_\mu(x_2)^\nu A_\nu(x_1) + \cdots. \tag{9.64}$$

This group element transforms covariantly. Infinitesimally for each segment of the curve $\mathcal{C}$, we find

$$g(x + dx, A^g) = 1 - dx^\mu A_\mu^g$$
$$= 1 - dx^\mu \mathcal{U}(g)(A_\mu + \partial_\mu)\mathcal{U}^\dagger(g)$$
$$= \mathcal{U}(g)\left(1 - dx^\mu(A_\mu + (\partial_\mu \mathcal{U}^\dagger(g))\mathcal{U}(g))\right)\mathcal{U}^\dagger(g). \tag{9.65}$$

Now,

$$\mathcal{U}(g(x))\left(1 - dx^\mu\left(\partial_\mu \mathcal{U}^\dagger(g(x))\right)\mathcal{U}(g(x))\right)$$
$$= \mathcal{U}(g(x)) - dx^\mu \mathcal{U}(g(x))\partial_\mu \mathcal{U}^\dagger(g(x))\mathcal{U}(g(x))$$
$$= \mathcal{U}(g(x)) + dx^\mu \partial_\mu \mathcal{U}(g(x)) = \mathcal{U}(g(x + dx)) \tag{9.66}$$

hence

$$g(x + dx, A^g) = \mathcal{U}(g(x + dx))\left(1 - dx^\mu A_\mu\right)\mathcal{U}^\dagger(g(x))$$
$$= \mathcal{U}(g(x + dx))g(x + dx, A)\mathcal{U}^\dagger(g(x)). \tag{9.67}$$

Thus for the infinitesimal closed loop, as in Figure 9.4, starting and ending at x

$$g(\mathcal{C}, x, A^g) = \mathcal{U}(g(x))g(\mathcal{C}, x, A)\mathcal{U}^\dagger(g(x)), \tag{9.68}$$

which is exactly the covariant transformation property. Considering the second-order term in the expansion in Equation (9.64), we have for each straight line path part of the contour of direction l_μ

$$\int dx^\mu A_\mu = \int_0^1 dt l^\mu A_\mu(x^\nu + l^\nu t) = \int_0^1 dt l^\mu \left(A_\mu(x^\nu) + l^\sigma t \partial_\sigma A_\mu(x^\nu) \right) + o(l^3)$$
$$= l^\mu A_\mu(x^\nu) + \frac{1}{2} l^\mu l^\sigma \partial_\sigma A_\mu(x^\nu) + \cdots . \qquad (9.69)$$

Thus for the closed path we get to second-order contribution

$$\oint dx^\mu A_\mu(x^\nu) = \left\{ \left(dx^\mu A_\mu(x^\nu) + \frac{1}{2} dx^\mu dx^\sigma \partial_\sigma A_\mu(x^\nu) \right) \right.$$
$$+ \left(dy^\mu A_\mu(x^\nu + dx^\nu) + \frac{1}{2} dy^\mu dy^\sigma \partial_\sigma A_\mu(x^\nu) \right)$$
$$+ \left(-dx^\mu A_\mu(x^\nu + dx^\nu + dy^\nu) + \frac{1}{2} dx^\mu dx^\sigma \partial_\sigma A_\mu(x^\nu) \right)$$
$$\left. + \left(-dy^\mu A_\mu(x^\nu + dy^\nu) + \frac{1}{2} dy^\mu dy^\sigma \partial_\sigma A_\mu(x^\nu) \right) \right\}$$
$$= \left\{ \left(dx^\mu A_\mu(x^\nu) + \frac{1}{2} dx^\mu dx^\sigma \partial_\sigma A_\mu(x^\nu) \right) \right.$$
$$+ \left(dy^\mu [A_\mu(x^\nu)] + dx^\sigma \partial_\sigma A_\mu(x^\nu)] + \frac{1}{2} dy^\mu dy^\sigma \partial_\sigma A_\mu(x^\nu) \right)$$
$$+ (-dx^\mu [A_\mu(x^\nu) + dx^\sigma \partial_\sigma A_\mu(x^\nu) + dy^\sigma \partial_\sigma A_\mu(x^\nu)] + \frac{1}{2} dx^\mu dx^\sigma \partial_\sigma A_\mu(x^\nu))$$
$$\left. + \left(-dy^\mu [A_\mu(x^\nu) + dy^\sigma \partial_\sigma A_\mu(x^\nu)] + \frac{1}{2} dy^\mu dy^\sigma \partial_\sigma A_\mu(x^\nu) \right) \right\}$$
$$= dx^\sigma dy^\mu (\partial_\sigma A_\mu(x^\nu) - \partial_\mu A_\sigma(x^\nu)) . \qquad (9.70)$$

Notice that this term contributes with a minus sign in Equation (9.64). When integrating along one side in Equation (9.64), the second-order term gives directly

$$\int_x^{x+dx} dx_2^\mu \int_x^{x_2} dx_1^\mu A_\mu(x_2^\nu) A_\mu(x_1^\nu) = \int_0^1 dt \left(l^\mu A_\mu(x^\nu + l^\nu t) \int_0^t ds\, l^\sigma A_\sigma(x^\nu + l^\nu s) \right)$$
$$= \int_0^1 dt \left(l^\mu A_\mu(x^\nu) \int_0^t ds\, l^\sigma A_\sigma(x^\nu) \right)$$
$$= \int_0^1 dt \left(l^\mu A_\mu(x^\nu) t l^\sigma A_\sigma(x^\nu) \right)$$
$$= \frac{1}{2} l^\mu l^\sigma A_\mu(x^\nu) A_\sigma(x^\nu) . \qquad (9.71)$$

The two integrals simply factorize when the integrations are on two different segments and no factor of one half is generated. Hence adding up the contributions around the loop, substituting for l^μ with dx^μ or dy^μ gives

$$\oint dx_1^\nu \oint_{x_2 > x_1} dx_2^\mu A_\mu(x_2) A_\nu(x_1) = \Big\{ -dy^\mu A_\mu(x^\nu + dy^\nu)$$

$$\times \Big[\frac{1}{2}(-dy^\sigma) A_\sigma(x^\nu + dy^\nu) - dx^\sigma A_\sigma(x^\nu + dx^\nu + dy^\nu)$$

$$+ dy^\sigma A_\sigma(x^\nu + dx^\nu) + dx^\sigma A_\sigma(x^\nu) \Big]$$

$$- dx^\mu A_\mu(x^\nu + dx^\nu + dy^\nu) \Big[\frac{1}{2}(-dx^\sigma) A_\sigma(x^\nu + dx^\nu + dy^\nu)$$

$$+ dy^\sigma A_\sigma(x^\nu + dx^\nu) + dx^\sigma A_\sigma(x^\nu) \Big]$$

$$+ dy^\mu A_\mu(x^\nu + dx^\nu) \Big[\frac{1}{2} dy^\sigma A_\sigma(x^\nu + dx^\nu) + dx^\sigma A_\sigma(x^\nu) \Big]$$

$$+ dx^\mu A_\mu(x^\nu) \Big[\frac{1}{2} dx^\sigma A_\sigma(x^\nu) \Big] \Big\}$$

$$+ \Big\{ -\frac{1}{2} dy^\mu A_\mu(x^\nu) dy^\sigma A_\sigma(x^\nu) - \frac{1}{2} dx^\mu A_\mu(x^\nu) dx^\sigma A_\sigma(x^\nu)$$

$$- dx^\mu A_\mu(x^\nu) dy^\sigma A_\sigma(x^\nu)$$

$$+ \frac{1}{2} dy^\mu A_\mu(x^\nu) dy^\sigma A_\sigma(x^\nu) + dy^\mu A_\mu(x^\nu) dx^\sigma A_\sigma(x^\nu) + \frac{1}{2} dx^\mu A_\mu(x^\nu) dx^\sigma A_\sigma(x^\nu) \Big\}$$

$$= -dx^\sigma dy^\mu [A_\sigma(x^\nu) A_\mu(x^\nu) - A_\mu(x^\nu) A_\sigma(x^\nu)]. \qquad (9.72)$$

Adding up the two contributions, Equations (9.72) and (9.70), simply gives

$$P \exp\{dx^\mu A_\mu(x^\nu)\} = -dx^\sigma dy^\mu \left(\partial_\sigma A_\mu(x^\nu) - \partial_\mu A_\sigma(x^\nu) \right.$$

$$+ [A_\sigma(x^\nu), A_\mu(x^\nu)]) + o(dx)^3$$

$$\equiv -dx^\sigma dy^\mu F_{\sigma\mu} + o(dx)^3, \qquad (9.73)$$

which must transform covariantly. Actually we can write $F_{\mu\nu}$ as the commutator of two covariant derivatives,

$$F_{\mu\nu} = [D_\mu, D_\nu] = [\partial_\mu + A_\mu, \partial_\nu + A_\nu]$$

$$= [\partial_\mu, A_\nu] + [A_\mu, \partial_\nu] + [A_\mu, A_\nu]$$

$$= \partial_\mu A_\nu - \partial_\nu A_\mu + [A_\mu, A_\nu]. \qquad (9.74)$$

Then, due to the algebraic structure of $F_{\mu\nu}$, we immediately know that the Jacobi identity will be satisfied,

$$[D_\mu, [D_\nu, D_\sigma]] + [D_\sigma, [D_\mu, D_\nu]] + [D_\nu, [D_\sigma, D_\mu]] = 0$$

$$\Rightarrow [D_\mu, F_{\nu\sigma}] + [D_\sigma, F_{\mu\nu}] + [D_\nu, F_{\sigma\mu}] = 0, \qquad (9.75)$$

which in four dimensions is exactly the Bianchi identity,

$$\partial_\mu \epsilon^{\mu\nu\sigma\tau} F_{\sigma\tau} + [A_\mu, \epsilon^{\mu\nu\sigma\tau} F_{\sigma\tau}] = 0. \qquad (9.76)$$

Thus $F_{\mu\nu}$ is the appropriate covariant generalization of the usual Abelian definition of the field strength.

9.5 Quantizing Gauge Field Configurations

The physical (non-gauge) zero modes of the action come from translations of the positions of the monopoles and rotations of the monopoles in iso-space. This gives simply the volume of spacetime and the volume of the gauge group as a Jacobian factor. However, things are not so simple, since in a gauge theory there are lots of unphysical zero modes associated with gauge-equivalent configurations. The naive functional integral for a gauge theory is not well-defined, even in Euclidean space.

The Lagrangian of a gauge theory is called a singular Lagrangian, the equations of motion do not give rise to a well-defined initial value problem for the gauge fields. Obviously, if we fix the initial data, and find a solution of the equations of motion, there actually exist an infinite number of solutions of the equations of motion that satisfy the initial conditions, which are simply gauge transforms of the original solutions. The gauge transformations, of course, must be time-dependent, so that they do nothing to the gauge fields on the initial hyper-surface, but they do modify the gauge fields afterwards. The freedom to do time-dependent gauge transformations allow for this, and the solution of the initial value problem is not unique. Thus fixing the gauge becomes essential to define even the classical dynamics. Correspondingly, the quantum dynamics also requires gauge fixing in order to be well-defined. The important point is that, because of the gauge invariance, the actual physical content of the theory does not depend on the choice of gauge fixing.

The action is invariant under the infinite dimensional group of gauge transformations, $\mathcal{G}$. Thus

$$\mathcal{N} \int \mathcal{D}(A,\phi)\, e^{\frac{-S_E}{\hbar}} = (\text{volume}\,(\mathcal{G})) \left(\mathcal{N} \int_{\substack{\text{gauge} \\ \text{inequivalent}}} \mathcal{D}(A,\phi)\, e^{\frac{-S_E}{\hbar}} \right), \qquad (9.77)$$

as geometrically drawn in Figure 9.5. The volume $\mathcal{G}$ is, of course, infinite, it is not just a few zero modes which arise as in the propagator, but an infinity of zero modes due to arbitrary local gauge transformations. This infinite volume should cancel between numerator and denominator; however, we must realize how to define

$$\mathcal{N} \int_{\substack{\text{gauge} \\ \text{inequivalent}}} \mathcal{D}(A,\phi)\, e^{\frac{-S_E}{\hbar}} \qquad (9.78)$$

properly, *i.e.* in a gauge-invariant manner. The method for defining this integral is to begin in a canonical gauge, where the quantization is understood and well-defined, and then transform to any other gauge in an invariant way. This procedure was first spelled out by Faddeev and Popov [44].

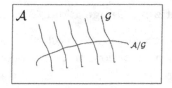

Figure 9.5. The space of all gauge fields, corresponding to the space $\mathcal{A}/G$ with leaves, foliated by the group of gauge transformation G

9.5.1 The Faddeev–Popov Determinant

We will start with the gauge choice

$$A^3 = 0. \tag{9.79}$$

This gauge condition is complete, which means that we may not make any further gauge transformations whose derivatives are of compact support. These are the so-called local gauge transformations, those that go sufficiently fast (often taken to be of compact support), to a constant at infinity. This constant is usually taken to be the identity. We insist on the gauge choice, that is, $A^3 = 0$, then any gauge transformation must satisfy

$$A^3 \to g^{-1}\partial_3 g = 0 \quad \Rightarrow \partial_3 g = 0. \tag{9.80}$$

But then g must be a global constant, everywhere equal to its value at infinity, chosen to be the identity. It is easy to convince ourselves that no local gauge transformation can be non-trivial and still be independent of x^3. Hence Equation (9.79) is a complete gauge-fixing condition as far as the group of local gauge transformations is concerned. We define

$$\mathcal{I} = \mathcal{N} \int \mathcal{D}(A, \phi)\, \delta\left(A_3\right) e^{\frac{-S_E(A,\phi)}{\hbar}}. \tag{9.81}$$

For any other gauge choice such that $F(A_i) = 0$ there must exist a gauge transformation $g_0(A)$ such that

$$(A_3)^{g_0(A)} = 0, \tag{9.82}$$

since it is understood that the set of gauge orbits of a given gauge slice must span the entire space of gauge fields at least locally.[1]

We define $\Delta(A)$ by

$$1 = \Delta(A) \int \mathcal{D}g \delta\left(F(A_i^g)\right), \tag{9.83}$$

---

[1] The Gribov ambiguity maintains that this is not exactly true. There do exist multiple gauge field configurations that respect the same gauge condition. However, these configurations are typically a finite distance away from each other. Thus the configurations that satisfy the gauge-fixing condition and their gauge orbits certainly give a complete foliation of the local neighbourhood of the space of gauge fields.

where $\mathcal{D}g$ corresponds to the integration measure for integration over the full group of local gauge transformations. This measure is defined in an invariant way, formally, the metric on the space of gauge transformations is defined as (in d dimensions)

$$(\delta g)^2 = -\int d^d x \operatorname{tr}\left((g^{-1}\delta g)(g^{-1}\delta g)\right). \tag{9.84}$$

Here δg corresponds to an element of the tangent space of the group of gauge transformations, this is called its Lie algebra. If h is an arbitrary fixed element of the group of gauge transformations, then the left multiplication by h in the group gives left multiplication of the algebra, $\delta(hg) = h\delta g$ and the 1-form $g^{-1}\delta g$ is left-invariant, as is the metric Equation (9.84). The metric is actually also invariant under right multiplication, since $\delta(gh) = (\delta g)h$, but then $\operatorname{tr}((gh)^{-1}\delta(gh)(gh)^{-1}\delta(gh)) = \operatorname{tr}((h^{-1}g^{-1}(\delta g)hh^{-1}g^{-1}(\delta g)h) = \operatorname{tr}\left((g^{-1}\delta g)(g^{-1}\delta g)\right)$. $\mathcal{D}g$ is then formally the corresponding volume form. We will mostly need to integrate over an infinitesimal neighbourhood of the identity. Here, with $g = 1 + \alpha$, where α is an infinitesimal element of the Lie algebra, we have, since $g^{-1}\delta g = \alpha$ to first order, and the analogue of the Euclidean geometry in the space of all α's

$$|\alpha|^2 = \int d^d x \operatorname{tr}\left(\alpha^2\right). \tag{9.85}$$

This then allows for the replacement $\mathcal{D}g \to \mathcal{D}\alpha$ with free, linear integration over α.

Notice that $\Delta(A)$ is gauge-invariant, for an arbitrary gauge transformation h,

$$\Delta(A^h) = \Delta(A). \tag{9.86}$$

This is because the integration measure over the group of gauge transformations is expected to be and can be defined to be gauge-invariant, that is,

$$\frac{1}{\Delta(A^h)} = \int \mathcal{D}g\, \delta\left(F((A_i^h)^g)\right) = \int \mathcal{D}(g)\, \delta\left(F(A_i^{gh})\right)$$
$$= \int \mathcal{D}(gh)\, \delta\left(F(A_i^{gh})\right) = \int \mathcal{D}g\, \delta(F(A_i^g))$$
$$= \frac{1}{\Delta(A)}. \tag{9.87}$$

$\Delta(A)$ is called the Faddeev–Popov factor. (We call to your attention that $(A_i^h)^g = A_i^{gh}$ as the group action works by left multiplication.) Then

$$\mathcal{I} = \mathcal{N}\int \mathcal{D}(A,\phi)\,\delta(A_3)e^{\frac{-S_E}{\hbar}}\left(\Delta(A)\int \mathcal{D}g\delta\left(F(A_i^g)\right)\right)$$
$$= \mathcal{N}\int \mathcal{D}g\int \mathcal{D}(A,\phi)\,\delta(A_3)e^{\frac{-S_E}{\hbar}}\Delta(A)\delta\left(F(A_i^g)\right)$$
$$= \mathcal{N}\int \mathcal{D}g\int \mathcal{D}(A,\phi)\,\delta(A_3^{g^{-1}})e^{\frac{-S_E}{\hbar}}\Delta(A^{g^{-1}})\delta\left(F(A_i)\right)$$
$$= \mathcal{N}\int \mathcal{D}(A,\phi)\,\delta\left(F(A_i)\right)e^{\frac{-S_E}{\hbar}}\Delta(A)\left(\int \mathcal{D}g\delta(A_3^{g^{-1}})\right). \tag{9.88}$$

Now let

$$g^{-1} = g'^{-1} g_0(A) \tag{9.89}$$

such that

$$(A_3)^{g_0(A)} = 0. \tag{9.90}$$

For a given g, g'^{-1} will depend on A; however, the integration over all g' will not, as the integration measure is invariant under left or right multiplication, as explained in the discussion after Equation (9.84). That is

$$\int \mathcal{D}g\, \delta\left(A_3^{g^{-1}}\right) = \int \mathcal{D}g'\, \delta\left((A_3^{g_0})^{g'^{-1}}\right) = \int \mathcal{D}g'\, \delta\left((0)^{g'^{-1}}\right) \tag{9.91}$$

is a constant, independent of A, and so we can absorb it into the normalization. Thus we get

$$\mathcal{I} = \mathcal{N}' \int \mathcal{D}(A, \phi) \delta\left(F(A_i)\right) \Delta(A) e^{\frac{-S_E}{\hbar}}. \tag{9.92}$$

We see how to change the gauge from the choice $A_3 = 0$ to an arbitrary gauge choice $F(A_i) = 0$, the integration measure must be appended with the Faddeev–Popov factor. The Faddeev–Popov factor,

$$\Delta^{-1}(A) = \int \mathcal{D}g\, \delta\left(F\left(A^g\right)\right) \tag{9.93}$$

will only get contributions from the infinitesimal neighbourhood of A around the point where $F(A) = 0$. Thus for A satisfying the gauge condition, we have, with $g = 1 + \alpha$, where α is an infinitesimal element of the the the Lie algebra,

$$F\left(A^{1+\alpha}\right) = F(A) + \int d^3y\, \frac{\delta F}{\delta A_i(y)} D_i \alpha(y) + o(\alpha^2), \tag{9.94}$$

since the change in the gauge field is exactly $\delta A_i(y) = D_i \alpha(y)$ and the integration is over α with measure $\mathcal{D}g \to \mathcal{D}\alpha$. Then generalizing the standard property of the integration over a delta function $\int d^n x \delta(M \cdot x) = (\det M)^{-1}$, we get

$$\Delta^{-1}(A) = \int \mathcal{D}\alpha \delta\left(\int d^3y\left(-D_i \frac{\delta F}{\delta A_i(y)}\right)\alpha(y)\right)$$
$$= \det^{-1}\left(-D_i \frac{\delta F}{\delta A_i(y)}\right)\left(\int \mathcal{D}\alpha \delta(\alpha(y))\right). \tag{9.95}$$

The last factor is 1, thus

$$\Delta(A) = \det\left(-D_i \frac{\delta F}{\delta A_i(y)}\right). \tag{9.96}$$

This expression is usually re-expressed as a fermionic functional integral over the so-called Faddeev–Popov ghost fields, which formally gives the determinant; however, for our analysis, we will not require or implement this step.

9.6 Monopoles in the Functional Integral

We want to calculate the functional integral

$$\langle 0| e^{-\frac{T\hat{H}}{\hbar}} |0\rangle = \mathcal{N} \int \mathcal{D}(A,\phi)\, e^{-\frac{S_E(A,\phi)}{\hbar}}. \tag{9.97}$$

We will calculate it in Gaussian approximation about the critical points of $S_E(A,\phi)$. This corresponds to integrating over the space of fields in the infinitesimal neighbourhood of the classical critical points, the monopole solutions. The usual understanding is that the contribution from the fields that are not in the infinitesimal neighbourhood of the monopole solutions will be suppressed by the exponential of the action. Knowing monopole solutions exist and are the critical points, we will get a result of the form

$$\langle 0| e^{-\frac{T\hat{H}}{\hbar}} |0\rangle = \mathcal{N} \sum_{n=-\infty}^{\infty} e^{-\frac{S_E(n\ \text{monopoles})}{\hbar}} \det^{-\frac{1}{2}} \left[\left(\frac{\delta^2 S_E}{\delta\phi_i^2} \right)\bigg|_{\text{crit.}} \right]. \tag{9.98}$$

To make this expression quantitative, we must do three further calculations:

1. Find the action for N instantons (n_1 monopoles and n_2 anti-monopoles with $n_1 + n_2 = N$).
2. Identify and separate the zero modes in the spectrum of Gaussian fluctuations.
3. Define the measure of functional integration to make the determinant in Equation (9.98) well-defined.

9.6.1 The Classical Action

As usual

$$\mathcal{N}\left(\frac{\delta^2 S_E}{\delta\phi_i^2} \right)\bigg|_{\text{crit.}} = \left(\frac{\left(\frac{\delta^2 S_E}{\delta\phi_i^2} \right)\big|_{\text{crit.}}}{\left(\frac{\delta^2 S_E}{\delta\phi_i^2} \right)\big|_{\text{vac.}}} \right) \mathcal{N}\left(\frac{\delta^2 S_E}{\delta\phi_i^2} \right)\bigg|_{\text{vac.}}$$

$$= K^n \cdot 1, \tag{9.99}$$

where "crit." stands for the critical point of n instantons, and "vac." stands for the vacuum configuration. The last factor is equal to 1 which serves to define $\mathcal{N}$

$$\mathcal{N}\left(\frac{\delta^2 S_E}{\delta\phi_i^2} \right)\bigg|_{\text{vac.}} \equiv 1. \tag{9.100}$$

The action for n widely separated instantons is n times that of one instanton. The number of such configurations behaves like

$$\sim \frac{(V\beta)^n}{n!}. \tag{9.101}$$

This "entropy" factor is, as usual, much larger than the corresponding factor when any subset of these n instantons are constrained to be close together, *i.e.* multi-monopole configurations. Even though the contribution of n widely separated instantons is suppressed by the exponential of its action $e^{-nS_E^0}$, the "entropy" factor can be big for a large finite spacetime volume $(V\beta)^n$, until eventually the $1/n!$ takes over as it will always eventually dominate.

The action for a single monopole is defined by a function $\epsilon\left(\frac{\lambda}{e^2}\right)$:

$$S_E^0 = \frac{m_W}{e^2}\,\epsilon\left(\frac{\lambda}{e^2}\right). \tag{9.102}$$

$m_W \sim a$ and the function ϵ can, in general, only be calculated numerically; however, in the Prasad–Sommerfield limit, $\epsilon(0) = 4\pi$, S_E^0 comes almost entirely from the integration over the core region

$$\int_{|\vec{x}|<R} d^3x \mathcal{L}^E = \frac{m_W}{e^2}\,\epsilon\left(\frac{\lambda}{e^2}\right)\left(1 + o\left(\frac{1}{m_W R}\right)\right). \tag{9.103}$$

The correction to the action from fields outside the core behaves like $\frac{1}{R}$, exactly the classical Coulomb self-energy of a magnetic charge.

For n well-separated monopoles of charge $\frac{4\pi q_a}{e}$, in addition to the Coulomb self-energy of each monopole, there is also a Coulomb interaction energy, a correction that is additive

$$S_E|_{\text{Coulomb}} = \frac{\pi}{2e^2}\sum_{a\neq b}\frac{q_a q_b}{|\vec{x}_a - \vec{x}_b|}, \tag{9.104}$$

with $q_a = \pm 1$. Then

$$S_E(n\text{ monopoles}) = \frac{m_W}{e^2}\,\epsilon\left(\frac{\lambda}{e^2}\right)\sum_a q_a^2 + \frac{\pi}{2e^2}\sum_{a\neq b}\frac{q_a q_b}{|\vec{x}_a - \vec{x}_b|} + o\left(\frac{1}{m_W R}\right), \tag{9.105}$$

where the small corrections exist because the monopoles are not point charges but spread out over regions of size $\frac{1}{m_W R}$. The additional Coulomb interaction energy term is non-negligible and has profound consequences.

9.6.2 Monopole Contribution: Zero Modes

Now we are in a position to analyse the zero-mode spectrum. If we write

$$A_i = A_i^{\text{cl}} + a_i \quad \phi = \phi^{\text{cl}} + \varphi, \tag{9.106}$$

where a_i and φ are quantum fluctuations about the classical values, we have the expansion of the action to second order in the fluctuations,

$$S_E = (S_E)_{\text{cl}} + (S_E)_2 + \cdots. \tag{9.107}$$

The first-order term vanishes because the equations of motion are satisfied for the classical fields, and $(S_E)_2$ is given by

$$(S_E)_2 = \int d^3x \, \text{tr} \left[\frac{1}{4e^2} \left(D_i^{\text{cl}} a_j - D_j^{\text{cl}} a_i \right)^2 + \frac{1}{2e^2} \left([a_i, a_j] F_{ij}^{\text{cl}} \right) + \frac{1}{2} [a_i, \phi^{\text{cl}}]^2 \right.$$
$$\left. + \frac{1}{2} \left(D_i^{\text{cl}} \varphi \right)^2 + \frac{1}{2} \varphi \mu^2 \left(\phi^{\text{cl}} \right) \varphi + \phi^{\text{cl}} [D_i^{\text{cl}} \varphi, a_i] + D_i^{\text{cl}} \phi^{\text{cl}} [a_i, \varphi] \right] \quad (9.108)$$

with

$$D_i^{\text{cl}} = \partial_i + [A_i^{\text{cl}}, . \quad (9.109)$$

This is a bilinear expression in a_i and φ, thus integration over these fields will give $\det^{-\frac{1}{2}}(\mathcal{O})$, where the operator $\mathcal{O}$ is the hermitean, linear, second-order differential operator appearing between these fields in Equation (9.108). We expect $\mathcal{O}$ to have eigenfunctions as (although they generally will be a continuous set)

$$\mathcal{O} \left(A^{\text{cl}}, \phi^{\text{cl}} \right) \begin{pmatrix} a_i^n \\ \phi^n \end{pmatrix} = \Omega_n^2 \begin{pmatrix} a_i^n \\ \phi^n \end{pmatrix}. \quad (9.110)$$

We expect the eigenvalues to be positive or zero, since the classical solution about which we expand the action is a minimum of the action. It is important to see that for any n such that $\Omega_n^2 > 0$ the corresponding eigenfunctions satisfy

$$D_i^{\text{cl}} a_i^n + [\phi^{\text{cl}}, \phi^n] = 0. \quad (9.111)$$

We will prove this from the hermiticity of the operator $\mathcal{O}$, and the evident fact that

$$a_i^0 = D_i^{\text{cl}} \alpha(x), \quad \phi^0 = [\phi^{\text{cl}}, \alpha(x)] \quad (9.112)$$

is a zero mode of $\mathcal{O}$ for every choice of $\alpha(x)$. a_i^0 and ϕ^0 are simply the changes induced by a gauge transformation, hence $S_E \left(A_i^{\text{cl}} + a_i^0, \phi^{\text{cl}} + \phi^0 \right) = S_E \left(A_i^{\text{cl}}, \phi^{\text{cl}} \right)$, which is valid order by order. This implies

$$S_2^E = \int dx (a_i^0, \phi^0) \mathcal{O} \begin{pmatrix} a_i^0 \\ \phi^0 \end{pmatrix} = 0. \quad (9.113)$$

Since $\mathcal{O}$ is hermitean, the modes for $\Omega_n^2 > 0$ are orthogonal to the zero modes hence

$$0 = \int d^3x \, \text{tr} \left(a_i^n a_i^0 + \phi^n \phi^0 \right)$$
$$= \int d^3x \, \text{tr} \left(a_i^n D_i^{\text{cl}} \alpha(x) + \phi^n [\phi^{\text{cl}}, \alpha(x)] \right)$$
$$= \int d^3x \, \text{tr} \left(\partial_i (a_i^n \alpha(x)) - (D_i^{\text{cl}} a_i^n) \alpha(x) + [\phi^n, \phi^{\text{cl}}] \alpha(x) \right)$$
$$= - \int d^3x \, \text{tr} \left((D_i^{\text{cl}} a_i^n + [\phi^{\text{cl}}, \phi^n]) \alpha(x) \right)$$
$$\Rightarrow D_i^{\text{cl}} a_i^n + [\phi^{\text{cl}}, \phi^n] = 0. \quad (9.114)$$

The conclusion in the last equation is reached since the integral must vanish for any choice of $\alpha(x)$. The $\alpha(x)$ zero modes in Equation (9.112) are not physical zero modes, they arise from the gauge invariance. If we impose the gauge choice

$$D_i^{\mathrm{cl}} A_i + \left[\phi^{\mathrm{cl}}, \phi\right] = 0 \tag{9.115}$$

with the understanding that the classical fields are assumed to satisfy this gauge condition, we can show that the unphysical gauge zero modes simply do not exist. Indeed, the gauge condition implies

$$
\begin{aligned}
0 = D_i^{\mathrm{cl}} A_i + \left[\phi^{\mathrm{cl}}, \phi\right] &= D_i^{\mathrm{cl}}(A_i^{\mathrm{cl}} + a_i) + \left[\phi^{\mathrm{cl}}, \phi^{\mathrm{cl}} + \varphi\right] \\
&= D_i^{\mathrm{cl}} A_i^{\mathrm{cl}} + \left[\phi^{\mathrm{cl}}, \phi^{\mathrm{cl}}\right] + D_i^{\mathrm{cl}} a_i + \left[\phi^{\mathrm{cl}}, \varphi\right] \\
&= D_i^{\mathrm{cl}} a_i + \left[\phi^{\mathrm{cl}}, \varphi\right].
\end{aligned}
\tag{9.116}
$$

Then we see that the norm of the putative zero mode that satisfies the gauge condition Equation (9.115), that is $D_i^{\mathrm{cl}} a_i^0 + \left[\phi^{\mathrm{cl}}, \phi^0\right] = 0$, simply vanishes:

$$
\begin{aligned}
\int d^3x \operatorname{tr}\left(a_i^0 a_i^0 + \phi^0 \phi^0\right) &= \int d^3x \operatorname{tr}\left(D_i^{\mathrm{cl}} \alpha(x) D_i^{\mathrm{cl}} \alpha(x) + (\left[\phi^{\mathrm{cl}}, \alpha(x)\right])^2\right) \\
&= -\int d^3x \operatorname{tr}\left(\left(D_i^{\mathrm{cl}} D_i^{\mathrm{cl}} \alpha(x) + \left[\phi^{\mathrm{cl}}, \left[\phi^{\mathrm{cl}}, \alpha(x)\right]\right]\right) \alpha(x)\right) \\
&= -\int d^3x \operatorname{tr}\left(\left(D_i^{\mathrm{cl}} a_i^0 + \left[\phi^{\mathrm{cl}}, \phi^0\right]\right) \alpha(x)\right) = 0.
\end{aligned}
\tag{9.117}
$$

This requires $a_i^0 = \phi^0 = 0$, that is, the pure gauge zero mode that satisfies the gauge condition simply does not exist.

The Faddeev–Popov factor comes from the gauge-fixing condition

$$F(A, \phi) = D_i^{\mathrm{cl}}(A_i) + \left[\phi^{\mathrm{cl}}, \phi\right] = 0. \tag{9.118}$$

Then following Equation (9.95) we have

$$
\begin{aligned}
F(A^{1+\alpha}, \phi^{1+\alpha}) &= D_i^{\mathrm{cl}}(A_i + D_i^A \alpha(x)) + \left[\phi^{\mathrm{cl}}, \phi + \left[\phi, \alpha(x)\right]\right] \\
&= D_i^{\mathrm{cl}} A_i + D_i^{\mathrm{cl}} D_i^A \alpha(x) + \left[\phi^{\mathrm{cl}}, \phi\right] + \left[\phi^{\mathrm{cl}}, \left[\phi, \alpha(x)\right]\right] \\
&= D_i^{\mathrm{cl}} A_i + \left[\phi^{\mathrm{cl}}, \phi\right] + D_i^{\mathrm{cl}} D_i^A \alpha(x) + \left[\phi^{\mathrm{cl}}, \left[\phi, \alpha(x)\right]\right].
\end{aligned}
\tag{9.119}
$$

Thus from Equation (9.96)

$$
\begin{aligned}
\Delta(A, \phi) &= \det\left(D_i^{\mathrm{cl}} D_i^A + \left[\phi^{\mathrm{cl}}, \left[\phi, \right.\right.\right) \\
&= \det\left(D_i^{\mathrm{cl}} D_i^{\mathrm{cl}} + \left[\phi^{\mathrm{cl}}, \left[\phi^{\mathrm{cl}}, \right.\right.\right)(1 + o(a_i, \varphi)).
\end{aligned}
\tag{9.120}
$$

9.6.3 Defining the Integration Measure

We can go further by defining the metric and integration measure on function space. We will integrate over an infinitesimal neighbourhood of the classical fields.

With $\delta A_i \equiv a_i = A_i - A_i^{\mathrm{cl}}$ and $\delta \phi \equiv \varphi = \phi - \phi^{\mathrm{cl}}$ to emphasize that we are in an infinitesimal neighbourhood of the classical fields, we can write the metric as

$$(\delta l)^2 = -\int d^3x \, \mathrm{tr}\left((\delta A_i)^2 + (\delta \phi)^2 \right). \qquad (9.121)$$

The minus sign is to take into account the anti-hermitean generators of the Lie algebra of the gauge group taken in the definition of the gauge fields and scalar fields. This metric is gauge-invariant since the infinitesimal change in the fields transform homogeneously under gauge transformations, and hence the gauge transformation cancels out due to the cyclicity of the trace. We parametrize the space of all gauge fields as a sub-manifold which corresponds to those gauge fields that satisfy the gauge condition, which is called the gauge slice, and orthogonal directions which correspond to gauge transformations. These lead to those gauge fields that do not satisfy the gauge condition but lie along the gauge orbit of the gauge fields on the gauge slice. We can expand the variations δA_i and $\delta \phi$ in terms of an arbitrary, linear combination of the eigenmodes of the operator $\mathcal{O}$, which respect the gauge condition, plus an arbitrary linearized gauge transformation. The eigenmodes translate us along the gauge slice while an arbitrary deformation off the gauge slice corresponds to a gauge transformation. Hence expanding to first order in ξ^n and $\alpha(x)$ gives

$$A_i = A_i^{\mathrm{cl}} + \sum_n \xi^n a_i^n + D_i^{\mathrm{cl}} \alpha(x)$$

$$\phi = \phi^{\mathrm{cl}} + \sum_n \xi^n \phi^n + [\phi^{\mathrm{cl}}, \alpha(x)] \qquad (9.122)$$

hence

$$
\begin{aligned}
(\delta l)^2 &= \sum_n (\xi_n)^2 - \int d^3x \, \mathrm{tr}\left(\left(D_i^{\mathrm{cl}} \alpha(x) \right)^2 + [\phi^{\mathrm{cl}}, \alpha(x)]^2 \right) \\
&= \sum_n (\xi_n)^2 - \int d^3x \, \mathrm{tr}\left(\alpha(x) \left(-D_i^{\mathrm{cl}} D_i^{\mathrm{cl}} - , \phi^{\mathrm{cl}}] [\phi^{\mathrm{cl}},) \alpha(x) \right) \right. \\
&= \sum_n (\xi_n)^2 - \int d^3x \, \mathrm{tr}\left(\alpha(x) \left(-D_i^{\mathrm{cl}} D_i^{\mathrm{cl}} - [\phi^{\mathrm{cl}}, [\phi^{\mathrm{cl}},) \alpha(x) \right) \right). \quad (9.123)
\end{aligned}
$$

Thus the measure is given by

$$
\begin{aligned}
\mathcal{D}(A_i, \phi) = \prod_x \mathcal{D}A_i(x) \mathcal{D}\phi(x) &= \prod_n d\xi_n \prod_x d\alpha(x) \det^{\frac{1}{2}}\left((D_i^{\mathrm{cl}} D_i^{\mathrm{cl}} + [\phi^{\mathrm{cl}}, [\phi^{\mathrm{cl}},)) \right) \\
&\equiv \prod_n d\xi_n \mathcal{D}\alpha(x) \det^{\frac{1}{2}}\left((D_i^{\mathrm{cl}} D_i^{\mathrm{cl}} + [\phi^{\mathrm{cl}}, [\phi^{\mathrm{cl}},)) \right)
\end{aligned}
$$

$$(9.124)$$

using a direct generalization of the corresponding volume measure for a finite dimensional system, if $ds^2 = \sum_{ij} g_{ij} dx^i dx^j$ then the volume measure is $dV = d^n x \sqrt{g}$, where $g = \det[g_{ij}]$. Then the integration giving rise to the Euclidean generating functional Equation (9.92) is given by

$$\mathcal{I} = \mathcal{N}' \int \mathcal{D}(A,\phi)\delta\left(F(A_i,\phi)\right)\Delta(A,\phi)e^{\frac{-S_E}{\hbar}}$$
$$= \mathcal{N}' \int \prod_n d\xi_n \mathcal{D}\alpha(x)\det^{\frac{1}{2}}\left(-\left(D_i^{\mathrm{cl}}D_i^{\mathrm{cl}} + [\phi^{\mathrm{cl}}, [\phi^{\mathrm{cl}},)\right)\delta\left(F(A_i,\phi)\right)\Delta(A,\phi)e^{\frac{-S_E}{\hbar}}.$$

$$(9.125)$$

But $\delta\left(F(A_i,\phi)\right) = \delta\left(D_i^{\mathrm{cl}}(A_i) + [\phi^{\mathrm{cl}},\phi]\right)$ and then using the expansion Equation (9.122) gives

$$\int \mathcal{D}\alpha(x)\delta\left(F(A_i,\phi)\right) = \int \mathcal{D}\alpha(x)\delta\left(-\left(D_i^{\mathrm{cl}}D_i^{\mathrm{cl}} + [\phi^{\mathrm{cl}}, [\phi^{\mathrm{cl}},)\right)\alpha(x)\right)$$
$$= \int \mathcal{D}\alpha(x)\det^{-1}\left(-\left(D_i^{\mathrm{cl}}D_i^{\mathrm{cl}} + [\phi^{\mathrm{cl}}, [\phi^{\mathrm{cl}},)\right)\right)(\delta(\alpha(x)))$$
$$= \det^{-1}\left(-\left(D_i^{\mathrm{cl}}D_i^{\mathrm{cl}} + [\phi^{\mathrm{cl}}, [\phi^{\mathrm{cl}},)\right)\right). \qquad (9.126)$$

We notice that this factor will actually neatly cancel out the Faddeev–Popov determinant. Indeed, we get

$$\mathcal{D}(A_i,\phi)\Delta(A_i,\phi) = \prod_n d\xi_n \det\left(D_i^{\mathrm{cl}}D_i^A + [\phi^{\mathrm{cl}}, [\phi,)\right) \frac{\det^{\frac{1}{2}}\left(D_i^{\mathrm{cl}}D_i^{\mathrm{cl}} + [\phi^{\mathrm{cl}}, [\phi^{\mathrm{cl}},)\right)}{\det\left(D_i^{\mathrm{cl}}D_i^{\mathrm{cl}} + [\phi^{\mathrm{cl}}, [\phi^{\mathrm{cl}},)\right)}$$
$$\approx \prod_n d\xi_n \det^{\frac{1}{2}}\left(D_i^{\mathrm{cl}}D_i^{\mathrm{cl}} + [\phi^{\mathrm{cl}}, [\phi^{\mathrm{cl}},)\right), \qquad (9.127)$$

where in the first line we have retained the full Faddeev–Popov factor multiplied by the factor coming from the measure and the integration over the gauge-fixing delta function.

There are still the physical zero modes corresponding to translation and internal rotational symmetries. The rotations give a finite constant volume factor which eventually cancels. Naively these are for translations

$$\tilde{a}_i^{(k,0)} = N^{-\frac{1}{2}}\partial_k A_i^{\mathrm{cl}}$$
$$\tilde{\phi}^{(k,0)} = N^{-\frac{1}{2}}\partial_k \phi^{\mathrm{cl}} \qquad (9.128)$$

however, these expressions do not satisfy the gauge condition. Augmenting by a gauge transformation gives (with $\alpha^k = -A_k^{\mathrm{cl}}$)

$$a_i^{(k,0)} = N^{-\frac{1}{2}}\left(\partial_k A_i^{\mathrm{cl}} - D_i^{\mathrm{cl}}A_k^{\mathrm{cl}}\right) = N^{-\frac{1}{2}}F_{ki}^{\mathrm{cl}}$$
$$\phi^{(k,0)} = N^{-\frac{1}{2}}\left(\partial_k \phi^{\mathrm{cl}} + [A_k^{\mathrm{cl}}, \phi^{\mathrm{cl}}]\right) = N^{-\frac{1}{2}}D_k^{\mathrm{cl}}\phi^{\mathrm{cl}} \qquad (9.129)$$

with $N = -\int d^3x\,\mathrm{tr}\left((F_{ki}^{\mathrm{cl}})^2 + (D_k^{\mathrm{cl}}\phi^{\mathrm{cl}})^2\right)$. The gauge condition

$$D_i^{\mathrm{cl}}F_{ki} + [\phi^{\mathrm{cl}}, D_k\phi^{\mathrm{cl}}] = 0 \qquad (9.130)$$

is just the equation of motion. Under a translation

$$\delta A_i = A_i^{\mathrm{cl}}(x + \delta R) - D_i^{\mathrm{cl}}(\delta R_j A_j) = \delta R_k F_{ki} = N^{\frac{1}{2}}\delta R_k a_i^{(k,0)} \qquad (9.131)$$

thus

$$d\xi_0^k = N^{\frac{1}{2}} dR_k \tag{9.132}$$

and

$$d^3\xi_0^k = N^{\frac{3}{2}} d^3\vec{R}. \tag{9.133}$$

So finally the integration measure is

$$\mathcal{D}(A_i,\phi)\Delta = N^{\frac{3}{2}} d^3\vec{R} \prod_{n\neq 0} d\xi_n \det^{\frac{1}{2}}\left(D_i^{\mathrm{cl}} D_i^{\mathrm{cl}} + [\phi^{\mathrm{cl}},[\phi^{\mathrm{cl}},\,)\right. . \tag{9.134}$$

For one monopole we have

$$Z_1 = \int N^{\frac{3}{2}} d^3\vec{R}\, \det^{\frac{1}{2}}\left(D_i^{\mathrm{cl}} D_i^{\mathrm{cl}} + [\phi^{\mathrm{cl}},[\phi^{\mathrm{cl}},\,)\right.\prod_{n\neq 0}\left(\frac{\Omega_n^0}{\Omega_n}\right) e^{-\frac{(S_E)_0}{\hbar}}$$

$$= \int \frac{m_W^{\frac{7}{2}}}{e}\alpha\left(\frac{\lambda}{e^2}\right) e^{-\epsilon\left(\frac{\lambda}{e^2}\right)\frac{m_W}{e^2}} d^3\vec{R} \tag{9.135}$$

from dimensional analysis and α is a function that can, in principle, be calculated. For N (not to be confused with the normalization above) instantons, n_1 monopoles and n_2 anti-monopoles,

$$Z_N = \frac{\zeta^N}{N!} \int \prod_{j=1}^N d^3\vec{R}_j \sum_{q_a=\pm 1} e^{-\frac{\pi}{2e^2}\sum_{a\neq b}\frac{q_a q_b}{|\vec{R}_a - \vec{R}_b|}} \tag{9.136}$$

and

$$Z = \sum_{N,q_a=\pm 1} \frac{\zeta^N}{N!} \int \prod_{j=1}^N d^3\vec{R}_j e^{-\frac{\pi}{2e^2}\sum_{a\neq b}\frac{q_a q_b}{|\vec{R}_a - \vec{R}_b|}}, \tag{9.137}$$

where

$$\zeta = \frac{m_W^{\frac{7}{2}}}{e}\alpha\left(\frac{\lambda}{e^2}\right) e^{-\epsilon\left(\frac{\lambda}{e^2}\right)\frac{m_W}{e^2}}. \tag{9.138}$$

9.7 Coulomb Gas and Debye Screening

This is exactly the partition function of a Coulomb gas. We know that such a gas has the property of screening. This is the same as confinement. Any electric fields will be cancelled exactly by a complete rearrangement of the particles in the gas.

If we re-express Z as a functional integral

$$Z = \int \mathcal{D}\chi e^{-\frac{\pi e^2}{2}\int d^3x (\nabla\chi)^2} \sum_{N,q_a=\pm 1} \frac{\zeta^N}{N!} \int \prod_{j=1}^N d^3\vec{R}_j e^{i\sum_a q_a \chi(\vec{R}_a)}. \tag{9.139}$$

Indeed,

$$\int \mathcal{D}\chi e^{-\frac{\pi e^2}{2}\int d^3x (\nabla\chi)^2 + i\sum_a q_a \chi(\vec{R}_a)} =$$

$$= \int \mathcal{D}\chi e^{-\frac{\pi e^2}{2} \int d^3x \left(-\chi \nabla^2 \chi + i \frac{2}{\pi e^2} \sum_a q_a \delta(\vec{x}-\vec{R}_a)\chi(\vec{x}) \right)}$$

$$= \int \mathcal{D}\chi e^{-\frac{\pi e^2}{2} \int d^3x \left(\chi + \frac{i}{\pi e^2} \sum_a q_a \delta(\vec{x}-\vec{R}_a)\left(\frac{1}{-\nabla^2}\right) \right)(-\nabla^2)\left(\chi + \frac{i}{\pi e^2}\left(\frac{1}{-\nabla^2}\right) \sum_b q_b \delta(\vec{x}-\vec{R}_b) \right)}$$

$$\times e^{-\frac{\pi e^2}{2} \int d^3x \frac{1}{(\pi e^2)^2} \sum_a q_a \delta(\vec{x}-\vec{R}_a)\left(\frac{1}{-\nabla^2}\right)(-\nabla^2)\left(\frac{1}{-\nabla^2}\right) \sum_b q_b \delta(\vec{x}-\vec{R}_b)}$$

$$= Ce^{-\frac{1}{2\pi e^2} \int d^3x \sum_a q_a \delta(\vec{x}-\vec{R}_a)\left(\frac{1}{-\nabla^2}\right) \sum_b q_b \delta(\vec{x}-\vec{R}_b)}$$

$$= Ce^{-\frac{1}{2\pi e^2} \int d^3x \sum_a q_a \delta(\vec{x}-\vec{R}_a) \frac{1}{4\pi} \int d^3y \frac{1}{|\vec{x}-\vec{y}|} \sum_b q_b \delta(\vec{y}-\vec{R}_b)}$$

$$= Ce^{-\frac{1}{2\pi e^2} \int d^3x \sum_a q_a \delta(\vec{x}-\vec{R}_a) \sum_b q_b \frac{1}{4\pi|\vec{x}-\vec{R}_b|}}$$

$$= Ce^{-\frac{1}{8\pi e^2} \sum_{a,b,a\neq b} q_a q_b \frac{1}{|\vec{R}_a-\vec{R}_b|}}, \tag{9.140}$$

where we absorb a harmless divergence at $a = b$ into the constant.[2] Thus (using $e \to e/2\pi$ in Equation (9.139)) we have

$$Z = \int \mathcal{D}\chi e^{-\frac{e^2}{8\pi} \int d^3x (\nabla\chi)^2} \sum_N \frac{\zeta^N}{N!} \int \prod_{j=1}^N d^3\vec{R}_j \left(e^{i\chi(\vec{R}_j)} + e^{-i\chi(\vec{R}_j)} \right)$$

$$= \int \mathcal{D}\chi e^{-\frac{e^2}{8\pi} \int d^3x (\nabla\chi)^2} \sum_N \frac{\zeta^N}{N!} \left(\int d^3\vec{R} \, 2\cos(\chi(\vec{R})) \right)^N$$

$$= \int \mathcal{D}\chi e^{-\frac{e^2}{8\pi} \frac{\pi e^2}{2} \int d^3x (\nabla\chi)^2} e^{2\zeta \int d^3x \cos(\chi(x))}$$

$$= \int \mathcal{D}\chi e^{-\frac{e^2}{8\pi} \int d^3x \left((\nabla\chi)^2 - M^2 \cos(\chi(x)) \right)} \tag{9.141}$$

with $M^2 = \frac{16\pi\zeta}{e^2}$.

There are no massless modes. The coupling constant, nominally taken as ζ, satisfies $\zeta \propto e^{-\left(\frac{m_W}{e^2}\right)\epsilon\left(\frac{\lambda}{e^2}\right)} << 1$ as $e \to 0$. This means that there are no massless gauge bosons, the low-energy Abelian theory is confined due to the effects of instantons. This is an incredible result; the theory is confining. Unfortunately, the result will not go over to four dimensions. However, in three dimensions, where the general arguments concerning the flux subtended by a large Wilson loop are critical, we find that the theory nevertheless favours confinement.

---

[2] We have a slight discrepancy with respect to Polyakov's paper [103]. We find that in Equation (9.139) we should replace $e \to e/2\pi$. This does not change the behaviour of the theory. We implement the change from now on.

10
Monopole Pair Production

In this chapter we will study the analysis by Affleck and Manton [4] of the decay of constant, external magnetic fields due to the production of magnetic monopole–anti-monopole pairs. The calculation is analogous to a calculation of the decay of external electric fields by Schwinger [109] due to the production of electron–positron pairs. In both cases the effect is due to non-perturbative tunnelling transitions.

10.1 't Hooft–Polyakov Magnetic Monopoles

In Chapter 9, we saw the solutions that correspond to magnetic monopoles, in the Georgi–Glashow model [54]; however, as we were in 2+1 dimensions these solutions were instantons in Euclidean three dimensions. Clearly the same solutions in 3+1 dimensions correspond to static soliton solutions and correspond to particle states of the 3+1-dimensional theory. There is a perturbative spectrum of particles corresponding to quantization of the small oscillations about the trivial vacuum. These particles correspond to a massless photon, a charged massive vector boson, and a neutral scalar from the Higgs field. We will consider the limit that the Higgs field mass and the vector gauge boson masses are very heavy while the photon remains massless. In this limit the monopoles are heavy, essentially point particles. We will see that in the presence of a constant external magnetic field, the Euclidean equations of motion admit instanton solutions that describe the production of monopole–anti-monopole pairs. The form of the instanton is surprisingly simple.

10.2 The Euclidean Equations of Motion

The solutions to the Euclidean equations of motion for a 't Hooft–Polyakov magnetic monopole in a constant external magnetic field must exist in general, as the initial value problem for the corresponding set of non-linear differential equations is well-defined. The solutions must be well-approximated by the solutions to the equations for point-like monopoles, certainly in the limit that the masses of the Higgs field and the massive vector bosons are taken to be very large. Then, apart from the self-action of each monopole being very large, the additional contribution to the action from the Euclidean trajectories of the monopoles will not diverge. The state of the system in the presence of a constant magnetic field should correspond to a meta-stable state, similar in principle to a false vacuum. There will be a finite probability for the creation of a monopole–anti-monopole pair. Creation of the pair of course costs energy; however, separating the monopoles in an external magnetic field gives back energy. After a separation to a critical radius, it is energetically favourable for the monopoles to separate to infinity. Thus the analogy to the decay of a meta-stable state is quite apt. The result is an exact analogy to the Schwinger calculation [109] of the decay of a constant electric field due to the creation of charged boson–anti-boson pairs. Schwinger found the amplitude

$$\Gamma = \frac{e^2 E^2}{8\pi^3} \sum_{n=1}^{\infty} \frac{(-1)^{n+1} e^{-n\pi m^2/eE}}{n^2} (1 + o(e^2)), \tag{10.1}$$

where E is the amplitude of the external electric field and m is the boson mass.

Manton and Affleck [4] found the result

$$\Gamma = \frac{g^2 B^2}{8\pi^3} e^{-\left(\pi M^2/gB + g^2/4\right)} \left(1 + o\left(\frac{g^3 B}{M^2}\right) + o(e^2)\right) \tag{10.2}$$

with g the magnetic charge, B the amplitude of the magnetic field, and M the mass of the monopole, which corresponds to the first term in the expansion found by Schwinger, interchanging electric charge and field with magnetic charge and field.

To find this amplitude, we will look for a solution to the classical Euclidean equations of motion that interpolate between the false vacuum in the presence of the constant background magnetic field, and the configuration containing a monopole–anti-monopole pair which are separating to infinity in the background magnetic field. The Euclidean solution will actually be a bounce-type instanton, thus we expect the pair will move apart up to a critical separation and then bounce back and return to each other and annihilate. The bounce point will correspond to the point at which the tunnelling occurs in Minkowski spacetime, and after the appearance of the physical monopoles in Minkowski spacetime, the magnetic field will pull them apart to infinite separation. The bounce should have one negative mode and all the rest positive. The negative mode will give rise to the imaginary part of the functional integral, with the appropriate analytical continuation. Effectively, the imaginary part of the functional integral is given by

$$\Im \left(T V K e^{-S_E} \right), \tag{10.3}$$

where

$$K = \frac{det^{-1/2}\left(\left.\frac{\delta^2 S_E}{\delta \phi_i^2}\right|_{\text{inst.}}\right)}{det^{-1/2}\left(\left.\frac{\delta^2 S_E}{\delta \phi_i^2}\right|_0\right)}. \tag{10.4}$$

There are also some zero modes that give the usual complications, which we will deal with using the Faddeev–Popov method. Our conventions will be the following for an $SU(2)$ gauge field $A_\mu = A_\mu^a T^a$ and a scalar field in the triplet representation, $\phi = \phi^a T^a$, where $T_{bc}^a = \epsilon^{abc}$ are the anti-symmetric 3×3 matrix representation of $SU(2)$,

$$\mathcal{L} = \frac{1}{e^2}\left(\frac{1}{4}F_{\mu\nu}^a F_{\mu\nu}^a + \frac{1}{2}(D_\mu\phi)^a(D_\mu\phi)^a + \frac{\lambda}{4e^2}\left(|\phi|^2 - M_W^2\right)^2\right), \tag{10.5}$$

where $[T^a, T^b] = \epsilon^{abc}T^c$, $|\phi|^2 = \phi^a\phi^a$, $F_{\mu\nu} = \partial_\mu A_\nu - \partial_\nu A_\mu - [A_\mu, A_\nu]$ and $D_\mu\phi = \partial_\mu\phi - [A_\mu, \phi]$, and M_W provides the mass scale. The equations of motion are

$$D_\mu F_{\mu\nu} = [D_\nu\phi, \phi]$$
$$D_\mu D_\mu\phi = \frac{\lambda}{e^2}\left(|\phi|^2 - M_W^2\right)\phi. \tag{10.6}$$

If we take $A_4 = 0$ and all fields independent of x_4, the equations of motion reduce to the static, Euclidean three-dimensional equations that we have already studied in Chapter 9, and there is a finite energy, stable, static non-trivial solution of the equations corresponding to the magnetic monopole. The action of the monopole is, of course, not finite as the solution is independent of x_4. The mass is

$$M = \frac{4\pi M_W}{e^2}k\left(\lambda/e^2\right) \quad \text{where } k \approx 1 \text{ for } \lambda/e^2 \leq 1 \tag{10.7}$$

and the magnetic charge is $g = 4\pi/e$, the core radius is $r_{cl} = g^2/M$ and the "Abelian" field strength can be defined as $f_{\mu\nu} = F_{\mu\nu}^a \phi^a/eM_W$. The Abelian field strength satisfies the Maxwell equation if $|\phi|^2 = M_W^2$ and $D_\mu\phi = 0$. In the limit of $\lambda \to \infty$, $e^2 \to \infty$ but λ/e^2 remaining finite, the monopole core size goes to zero and it looks very much like a point monopole.

10.3 The Point Monopole Approximation

Then in an external, constant magnetic field, the monopole solution cannot remain static. In Euclidean time, it must respect the Euclideanized magnetic "Lorentz" force law

$$M\frac{d^2 z_\mu}{ds^2} = -g\tilde{f}_{\mu\nu}\frac{dz_\nu}{ds}, \tag{10.8}$$

where z_μ is the position of the magnetic charge, s is a world line parameter normalized so that $\frac{dz_\nu}{ds}\frac{dz_\nu}{ds} = 1$ and $\tilde{f}_{\mu\nu} = \frac{1}{2}\epsilon_{\mu\nu\sigma\tau}f_{\sigma\tau}$. This equation is simply the dual of the usual Euclidean "Lorentz" force for a charged particle in electric and magnetic fields

$$M\frac{d^2 z_\mu}{ds^2} = -e f_{\mu\nu} \frac{dz_\nu}{ds}.$$ (10.9)

For the magnetic field with constant magnitude B in the three-direction, $f_{12} = B$ which means $\tilde{f}_{34} = B$. Then a solution of the equation of motion (10.8) is simply $z_1 = z_2 = 0$ and

$$z_3 = \frac{M}{gB} \cos\left(\frac{gB}{M}s\right) \quad z_4 = \frac{M}{gB} \sin\left(\frac{gB}{M}s\right).$$ (10.10)

The solution is obviously a circle. This is the analytic continuation of the corresponding Minkowski space solution, which would be a hyperbola.

10.4 The Euclidean Action

The point monopole equations of motion are, of course, approximative, but we can derive them in the limit of a weak external magnetic field [4]. This circular Euclidean solution is exactly the bounce solution that we are looking for. We can equally well think of the solution in the (x_3, x_4) plane as the creation of a monopole–anti-monopole pair, the two separating to a finite critical distance and then bouncing back together and annihilating. The diameter of the circle is the critical separation and corresponds to the point to which the pair separates in the Euclidean solution, but also the separation at which the pair appears in the tunnelling process, in Minkowski space. The circular solution neglects the Coulomb attraction between the monopole–anti-monopole pair. We will see that the Coulomb interaction does not greatly affect the instanton. To analyse the corrections, we consider the following decomposition of the action

$$S_E = \int d^4x \left(\mathcal{L} - \frac{1}{4}\tilde{f}_{\mu\nu}\tilde{f}_{\mu\nu} \right) + \int d^4x \frac{1}{4} \left(\tilde{f}_{\mu\nu}\tilde{f}_{\mu\nu} - f_{\mu\nu}^{ext.} f_{\mu\nu}^{ext.} \right),$$ (10.11)

where we have separated the Lagrangian into the first term that governs the dynamics above the Abelian gauge field and subtracted the action of the external gauge field. We define the dual Abelian gauge field into the core of the monopole as

$$\partial_\mu \tilde{f}_{\mu\nu} = \tilde{j}_\nu$$
$$\partial_\mu f_{\mu\nu} = 0$$ (10.12)

where $\tilde{j}_\nu$ is an appropriate, conserved, Abelian definition of the dual current into the core. Outside the core, $j_\nu = 0$ and the source-free Maxwell equations are perfectly valid. Equation (10.12) are just the Euclidean, dual, Abelian Maxwell equations with magnetic sources. As these are just the dual Maxwell equations, there exists a gauge potential $\tilde{a}_\mu$ such that

$$\tilde{f}_{\mu\nu} = \partial_\mu \tilde{a}_\nu - \partial_\nu \tilde{a}_\mu.$$ (10.13)

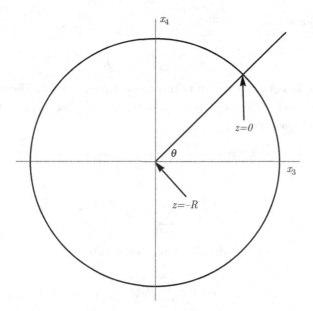

Figure 10.1. Circularly symmetric monopole–anti-monopole instanton

Exploiting the circular symmetry of the point-like solution we write

$$\tilde{a}_\mu = (0,0,-\sin\theta,\cos\theta)\psi(x,y.z)$$
$$\tilde{j}_\mu = (0,0,-\sin\theta,\cos\theta)\rho(x,y.z), \tag{10.14}$$

where x,y are the normal cartesian coordinates, but z,θ are polar coordinates in the x_3,x_4 plane, with the radius shifted so that $z=0$ corresponds to the radius of the circular point-like monopole instanton, *i.e.* the usual radial coordinate is $r=z+R$, as shown in Figure 10.1. Thus $z=-R$ is the origin, and we will expand the action about $z=0$. Then for the first term of the decomposition in Equation (10.11) we write

$$S_E^1 = 2\pi \int dxdydz(R+z)\,(\mathcal{L}-\mathcal{L}_{Abelian}), \tag{10.15}$$

where $\mathcal{L}_{Abelian} = \frac{1}{4}\tilde{f}_{\mu\nu}\tilde{f}_{\mu\nu}$ which can be evaluated from Equation (10.14)

$$\mathcal{L}_{Abelian} = \left(\frac{1}{2}(\partial_i\psi\partial_i\psi + \frac{1}{R+z}\psi\partial_z\psi + \frac{1}{2(R+z)^2}\psi^2\right), \tag{10.16}$$

where the index i goes over x,y,z and $\mathcal{L}$ is of course the full Lagrange density given in Equation (10.5). Away from $z=0$, we expect that the solution is exponentially zero, $D_i\phi \approx V(\phi) \approx e^{-M_W|\vec{x}|}$ and $F_{ij}^a F_{ij}^a \to f_{ij}f_{ij}$, exponentially fast, and consequently $(\mathcal{L}-\mathcal{L}_{Abelian})$ also vanishes exponentially.

We make no great error by changing the range of z from $-R \le z \le \infty$ to $-\infty \le z \le \infty$, as long as all fields and densities are exponentially small away from $z = 0$, thus we get

$$S_E^1 = 2\pi R \int d^3x \left(\mathcal{L} - \mathcal{L}_{Abelian}\right) + \frac{z}{R}\left(\mathcal{L} - \mathcal{L}_{Abelian}\right), \qquad (10.17)$$

where now the integral is over an entire three-dimensional Euclidean space. We expect that we can perform an expansion in powers of $1/R$. The Maxwell equation for the Abelian fields is

$$\partial_i \partial_i \psi + \frac{1}{R+z}\partial_z \psi - \frac{1}{(R+z)^2}\psi = \rho \qquad (10.18)$$

then if

$$\psi(\vec{x}) = \sum_{n=0}^{\infty} \psi_n(\vec{x})\frac{1}{R^n} \qquad (10.19)$$

the density $\rho(\vec{x})$ must also admit a similar expansion

$$\rho(\vec{x}) = \sum_{n=0}^{\infty} \rho_n(\vec{x})\frac{1}{R^n} \qquad (10.20)$$

as well as the Lagrange density $\mathcal{L}$. The terms in the expansion must be of alternating parity as $z \to -z$. The second term in Equation (10.17) vanishes to lowest order. The limit, as $R \to \infty$, *i.e.* $B \to 0$, which is the order $n = 0$ term, the solution is simply a static monopole at rest, the circle has infinite radius and thus becomes effectively a straight, world line. Then the first term of Equation (10.17) just gives

$$S_E^1 = 2\pi R(M - M_{Abelian}), \qquad (10.21)$$

where M is the mass of the monopole and $M_{Abelian}$ is just the contribution to the Coulomb energy from the zeroth order part of the current density $\rho_0(\vec{x})$, while the second term must give vanishing contribution due to parity. Thus, due to parity, the next correction only comes at $o\left(1/R^2\right)$.

10.5 The Coulomb Energy

The second term in the action, Equation (10.11), contains simply the energy in the Euclidean Abelian gauge fields, $\tilde{f}_{\mu\nu} = \tilde{f}_{\mu\nu}^{loop} + \tilde{f}_{\mu\nu}^{ext.}$, where $\tilde{f}_{\mu\nu}^{loop}$ comes from the monopole loop, and $\tilde{f}_{\mu\nu}^{ext.}$ comes from the fields outside of the loop. Then

$$S_E^2 = \frac{1}{4}\left(\tilde{f}_{\mu\nu}\tilde{f}_{\mu\nu} - f_{\mu\nu}^{ext.}f_{\mu\nu}^{ext.}\right) = \frac{1}{4}\tilde{f}_{\mu\nu}^{loop}\tilde{f}_{\mu\nu}^{loop} + \frac{1}{2}f_{\mu\nu}^{loop}f_{\mu\nu}^{ext.} \equiv S_E^{2,loop} + S_E^{2,int.}.$$
$$(10.22)$$

We will find

$$S_E^{2,loop} = \int d^4x \frac{1}{4}\tilde{f}_{\mu\nu}^{loop}\tilde{f}_{\mu\nu}^{loop} = \int d^4x d^4x' \frac{1}{8\pi^2}\frac{\tilde{j}_\mu(x)\tilde{j}_\mu(x')}{|x-x'|^2}. \qquad (10.23)$$

This can be shown by first observing that in the gauge $\partial_\mu \tilde{a}_\mu = 0$, we can solve the dual Maxwell field equation (10.12) for the dual gauge field simply as $\tilde{a}_\mu = \frac{1}{\Box}\tilde{j}_\mu$ where the Green's function is

$$\frac{1}{\Box} = -\frac{1}{4\pi^2}\frac{1}{|x-x'|^2} \tag{10.24}$$

and the dual field strength is as usual

$$\tilde{f}_{\mu\nu} = \partial_\mu \frac{1}{\Box}\tilde{j}_\nu - \partial_\nu \frac{1}{\Box}\tilde{j}_\mu. \tag{10.25}$$

Then it is straightforward to evaluate the contribution to the action

$$S_E^{2,loop} = \frac{1}{2}\int d^4x \left(\partial_\mu \frac{1}{\Box}\tilde{j}_\nu\right)\left(\partial_\mu \frac{1}{\Box}\tilde{j}_\nu\right) - \left(\partial_\mu \frac{1}{\Box}\tilde{j}_\nu\right)\left(\partial_\nu \frac{1}{\Box}\tilde{j}_\mu\right)$$

$$= \frac{1}{2}\int d^4x - \left(\frac{1}{\Box}\tilde{j}_\nu\right)(\tilde{j}_\nu) + \left(\frac{1}{\Box}\tilde{j}_\nu\right)\left(\partial_\mu\partial_\nu \frac{1}{\Box}\tilde{j}_\mu\right). \tag{10.26}$$

The second term in the first line vanishes after integration by parts, the second term in the last line vanishes since ∂_μ commutes with $1/\Box$ and $\partial_\mu\tilde{j}_\mu = 0$ by current conservation, which is necessary for the consistency of the dual Maxwell equations and is assumed to be verified by the current. Then

$$S_E^{2,loop} = \frac{1}{2}\int d^4x - \left(\frac{1}{\Box}\tilde{j}_\nu\right)(\tilde{j}_\nu) = \frac{1}{8\pi^2}\int d^4x d^4x' \frac{\tilde{j}_\mu(x)\tilde{j}_\mu(x')}{|x-x'|^2} \tag{10.27}$$

as desired. To calculate it explicitly is not too difficult. First of all, $\tilde{j}_\mu(x)\tilde{j}_\mu(x') = (\sin\theta\sin\theta' + \cos\theta\cos\theta')\rho(x)\rho(x') = \cos(\theta-\theta')\rho(x)\rho(x')$, thus we get, writing $d^2x = dx_1 dx_2$ and $d^2x' = dx'_1 dx'_2$

$$S_E^{2,loop} = \frac{1}{8\pi^2}\int d^2x d^2x \left(\frac{rr'\cos(\theta-\theta')\rho(x)\rho(x')drd\theta dr'd\theta'}{(x_1-x'_1)^2 + (x_1-x'_1)^2 + r^2 + r'^2 - 2rr'\cos(\theta-\theta')}\right). \tag{10.28}$$

The integral over θ and θ' can be done explicitly, we leave the reader to work it out or find it in tables, giving

$$S_E^{2,loop} = \int d^2x d^2x dr dr' \frac{1}{4}\rho(x)\rho(x')\left(\frac{W}{\sqrt{W^2-1}} - 1\right), \tag{10.29}$$

where, writing $(x_1-x'_1)^2 + (x_1-x'_1)^2 + (z-z')^2 = |\vec{x}-\vec{x}'|^2$

$$W = \frac{(x_1-x'_1)^2 + (x_1-x'_1)^2 + r^2 + r'^2}{2rr'}$$

$$= \frac{(x_1-x'_1)^2 + (x_1-x'_1)^2 + (R+z)^2 + (R+z')^2}{2(R+z)(R+z')}$$

$$= 1 + \frac{|\vec{x}-\vec{x}'|^2}{2R^2} - \frac{|\vec{x}-\vec{x}'|^2(z+z')}{2R^3} + o\left(\frac{1}{R^4}\right) \tag{10.30}$$

and intriguingly the terms $1/R$ exactly cancel. Then expanding carefully

$$\frac{W}{\sqrt{W^2-1}} - 1 = \frac{R}{|\vec{x}-\vec{x}'|} - 1 + \frac{z+z'}{2|\vec{x}-\vec{x}'|} + o\left(\frac{|\vec{x}|,z,z'}{R}\right) \tag{10.31}$$

and we note that actually the numerator only contributes to the the terms that have been neglected. Then in the evaluation of the contribution of this term to the action, the third term in Equation (10.31) vanishes because of parity when the lowest-order, spherically symmetric monopole charge density is put in for ρ, and the net remaining is simply

$$S_E^{2,loop} = \int d^2x d^2x dr dr' \frac{1}{4} \rho_0(x) \rho_0(x') \left(\frac{R}{|\vec{x} - \vec{x}'|} - 1 \right) + o\left(\frac{1}{R} \right). \tag{10.32}$$

The first term is exactly the Coulomb energy in the magnetic field while the second is proportion to the magnetic charge squared,

$$S_E^{2,loop} = 2\pi R M_{Abelian} - \frac{1}{4} g^2 + o\left(\frac{1}{R} \right), \tag{10.33}$$

where g is the magnetic charge. The first term exactly cancels against the identical term found in S_E^1, which is expected, since it arises solely because of the somewhat artificial Abelian magnetic charge density that was invented to extend the Abelian integration into the core. No physical phenomenon should depend on it. Thus

$$S_E^1 + S_E^{2,loop} = -\frac{1}{4} g^2 + 2\pi R M. \tag{10.34}$$

The interaction part of S_E^2, which we will call $S_E^{2,int.}$, is, integrating by parts and using the equation of motion,

$$S_E^{2,int.} = \int d^4x \frac{1}{2} \tilde{f}_{\mu\nu}^{loop} \tilde{f}_{\mu\nu}^{ext.} = -\int d^4x \tilde{j}_\mu \tilde{a}_\mu^{ext.}. \tag{10.35}$$

The external gauge potential can be taken with circular symmetry as

$$a_\mu^{ext.} = \left(0, 0, -\frac{1}{2} B(R+z) \sin\theta, \frac{1}{2} B(R+z) \cos\theta \right) \tag{10.36}$$

and the current is

$$j_\mu = (0, 0, -\sin\theta, \cos\theta) \rho(\vec{x}). \tag{10.37}$$

Then using $d^4x = dx_1 dx_2 dr d\theta r = dx_1 dx_2 dz d\theta (R+z) = d^3x (R+z) d\theta$ and integrating over θ gives a factor of 2π so that we get

$$S_E^{2,int.} = -\int d^3x \pi B(R+z)^2 \rho(\vec{x}) = -g\pi B R^2 + \cdots. \tag{10.38}$$

Thus the total action is

$$S_E = 2\pi M - g B \pi R^2 - \frac{1}{4} g^2 + o\left(\frac{1}{R^2} \right). \tag{10.39}$$

We vary the action with respect to R and demand that it be stationary to find the radius of the loop,

$$0 = \frac{\delta S_E}{\delta R} = 2\pi M - 2g\pi B R, \tag{10.40}$$

which gives $R = M/gB$. This is exactly the same value as in the case of the point-like monopoles, therefore we see that the inclusion of the Coulomb energy does not affect the radius of the loop. Inserting the value of R back into the action yields

$$S_E = \frac{\pi M^2}{g^2 B} - \frac{1}{4}g^2, \qquad (10.41)$$

and we observe that the Coulomb energy is $\sim 1/R$ integrated over a circle of circumference $2\pi R$, which yields $g^2/4$ which is independent of R. Finally, if we take the second variation we find

$$\frac{\delta^2 S_E}{\delta R^2} = -2gB\pi < 0, \qquad (10.42)$$

which means that the action has at least one negative mode and hence is at a saddle point. The negative mode is expected and gives rise to the decay width of the magnetic field.

10.6 The Fluctuation Determinant

We must now take into account the Gaussian integration over the fluctuations around the instanton

$$K = \frac{1}{2} \frac{\left| det \left(\frac{\delta^2 S_E}{\delta \phi_i^2} \Big|_{\text{inst.}} \right) \right|^{-1/2}}{\left(det \left(\frac{\delta^2 S_E}{\delta \phi_i^2} \Big|_0 \right) \right)^{-1/2}}. \qquad (10.43)$$

The factor of one-half occurs since we integrate over only half the Gaussian peak for the negative mode and any Faddeev–Popov factors are assumed to be included in the determinant. We have put the numerator in absolute value signs so that the negative mode does not give an imaginary value when we take the square root, as we explicitly put the i in by hand, in that the energy obtains an imaginary part $E = \mathcal{E} + i\Gamma$ with

$$\Gamma = VKe^{-S_E/\hbar}(1 + o(\hbar)). \qquad (10.44)$$

We must separate the zero modes, there are five, coming from four translations and one from internal rotation. The translation modes will give a familiar factor of the square root of the normalization

$$K \to \frac{1}{2} \prod_{\mu=1}^{4} \left(\frac{N_\mu}{2\pi e^2} \right)^{1/2} \frac{\left| det' \left(\frac{\delta^2 S_E}{\delta \phi_i^2} \Big|_{\text{inst.}} \right) \right|^{-1/2}}{\left(det \left(\frac{\delta^2 S_E}{\delta \phi_i^2} \Big|_0 \right) \right)^{-1/2}}. \qquad (10.45)$$

The internal rotation actually corresponds to the dyonic degree of freedom, internal rotation at a given angular frequency gives rise to a magnetically and electrically charged state, called the dyon. The full rate of pair production and consequent decay of the magnetic field must include the production of pairs of dyons. But for the lowest order, we can restrict ourselves to the case of a simple monopole pair production. The internal rotation is intimately connected with gauge fixing and the Faddev–Popov factor.

The translation zero modes naively are not gauge-invariant and must be made so by an accompanying gauge transformation, we find

$$(\delta A_\mu)_\nu = \partial_\nu A_\mu - D_\mu A_\nu = -F_{\mu\nu}$$
$$(\delta\phi)_\nu = \partial_\nu\phi - [A_\nu,\phi] = D_\nu\phi \tag{10.46}$$

and the normalization is (no sum on ν, sum on a assumed)

$$N_\nu = \int d^4x \left(\sum_\mu F^a_{\mu\nu} F^a_{\mu\nu} + (D_\nu)^a (D_\nu)^a \right). \tag{10.47}$$

The calculation of the determinant is possible in the limit $R \to \infty$ ($B \to 0$). In this limit, the fluctuations separate into those that change the shape of the monopole and those that change the shape of the loop.

Using the circular symmetry and the gauge $A_\theta = 0$, we have

$$\left.\frac{\delta^2 S_E}{\delta\phi_i^2}\right|_{inst.} = \left.\frac{\delta^2 S_E}{\delta\phi_i^2}\right|_{3,inst.} - \frac{1}{r^2}\frac{\partial^2}{\partial\theta^2}, \tag{10.48}$$

where the first term depends on x, y, r and is essentially a three-dimensional operator, while the second term comes from the kinetic energy, for example,

$$D_\mu D_\mu = D_1 D_1 + D_2 D_2 + D_r D_r + \frac{1}{r} D_r + \frac{1}{r^2}\frac{\partial^2}{\partial\theta^2}. \tag{10.49}$$

Eigenfunctions admit a separation of variables as

$$\Psi(x_1, x_2, r, \theta) = \psi(x_1, x_2, r) \begin{cases} \cos(n\theta) & n = 0, 1, 2, \cdots \\ \sin(n\theta) & n = 1, 2, 3, \cdots \end{cases} \tag{10.50}$$

and then in the sector of angular momentum n we get

$$\left.\frac{\delta^2 S_E}{\delta\phi_i^2}\right|_{inst.} = \left.\frac{\delta^2 S_E}{\delta\phi_i^2}\right|_{3,inst.} + \frac{n^2}{r^2}. \tag{10.51}$$

Now we make an expansion in $1/R$, with $z = r - R$, then, for example,

$$(D_\mu D_\mu)_3 = D_{x_1} D_{x_1} + D_{x_2} D_{x_2} + D_z D_z + \frac{1}{R+z} D_z$$
$$= D_i D_i + \left(\frac{1}{R} - \frac{z}{R^2} + \cdots \right) D_z. \tag{10.52}$$

To lowest order $(1/R)^0$ we just get the operator corresponding to the second variation of the Hamiltonian with a static monopole at $\vec{x} = 0$

$$\left.\frac{\delta^2 S_E}{\delta\phi_i^2}\right|_{3,inst.} = \left.\frac{\delta^2 H}{\delta\phi_i^2}\right|_{3,mono.} + o\left(\frac{1}{R}\right). \tag{10.53}$$

The angular momentum term also admits an expansion

$$\frac{n^2}{r^2} = \frac{n^2}{R^2}\left(1 - \frac{2z}{R} + \cdots\right) \qquad (10.54)$$

so that to lowest order we have

$$\left(\left.\frac{\delta^2 H}{\delta\phi_i^2}\right|_{3,mono.} + o\left(\frac{1}{R}\right) + \frac{n^2}{R^2} + o\left(\frac{1}{R}^3\right)\right)\psi_i^{(n)} = \lambda_i^{(n)}\psi_i^{(n)} \qquad (10.55)$$

and we note that the angular momentum term is a constant. The eigenvalues are then simply

$$\lambda_i^{(n)} = \omega_i^2 + \frac{n^2}{R^2}, \qquad (10.56)$$

where ω_i^2 are the eigenvalues of $\left.\frac{\delta^2 H}{\delta\phi_i^2}\right|_{3,mono.}$. The $\lambda_i^{(n)}$ admit an expansion in $1/R$ as

$$\lambda_i^{(n)} = \omega_i^2 + \frac{n^2}{R^2} + \frac{a_i}{R^2} + \frac{b_i n^2 + c_i}{R^4} + \cdots, \qquad (10.57)$$

where the odd powers vanish as the order zero eigenfunctions have definite parity under $z \to -z$. The correction a_i is difficult to compute, but it is expected to give a small correction for the non-zero eigenmodes. To calculate them in principle, we must find the correction to the instanton to order $o(1/R^2)$ and then compute the correction to the eigenvalues to second order in perturbation theory. However, for the zero modes the correction is important, but easily calculable.

There are three translational zero modes; first, consider the modes for translation in the x_1 and x_2 directions. These are out of the plane of the loop and correspond to $\omega_{x_1}^2 = 0$ and $\omega_{x_2}^2 = 0$. For these $n = 0$ and $\lambda_{x_1}^{(0)} = 0 = \lambda_{x_2}^{(0)}$. Thus for these to remain zero modes to order $1/R$ we must have $a_{x_1} = a_{x_2} = 0$. For translation in the z direction, we see these are translational zero modes of the monopole in the plane of the loop. These must come with multiplicity two as there are two independent directions for the translation. Furthermore, they must deform the loop, hence they must correspond to $n \neq 0$. Indeed, the first deformation of the loop occurs for $n = 1$ and the two independent angular eigenmodes give the two independent directions of the deformation. Thus we require that $\lambda_z^{(1)} \equiv \lambda_{x_3}^{(1)} = \lambda_{x_4}^{(1)} = 0$. For the zero-order Hamiltonian, we already have $\omega_{x_3}^2 = 0$ and $\omega_{x_4}^2 = 0$, hence to order $1/R$ we must have

$$\lambda_z^{(1)} = 0 = 0 + \left.\frac{n^2}{R^2}\right|_{n=1} + \frac{a_z}{R^2} + \cdots = \frac{1}{R^2} + \frac{a_z}{R^2} \quad \Rightarrow \quad a_z = -1. \qquad (10.58)$$

We perform exactly the same separation of variables and analysis for the denominator in Equation (10.45)

$$\left.\frac{\delta^2 S_E}{\delta\phi_i^2}\right|_{3,0} = \left.\frac{\delta^2 H^0}{\delta\phi_i^2}\right|_3 + o\left(\frac{1}{R}\right), \qquad (10.59)$$

which gives

$$\lambda_{i,0}^{(n)} = \omega_{i,0}^2 + \frac{n^2}{R^2} + \frac{a_{i,0}}{R^2} + \cdots. \tag{10.60}$$

The determinant corresponds to the product of the eigenvalues, thus the angular momentum family corresponding to eigenmode i contributes as

$$\ln K_i = -\frac{1}{2}\left(\ln \lambda_i^{(0)} - \ln \lambda_{i,0}^{(0)} + 2\sum_{n=1}^{\infty}\left(\ln \lambda_i^{(n)} - \ln \lambda_{i,0}^{(n)}\right)\right), \tag{10.61}$$

where the factor of 2 is because all the $n \neq 0$ modes come with multiplicity two while the mode $n = 0$ is solitary. To perform the summation we use the Euler–Maclaurin formula [2]

$$f(0) + 2\sum_{n=1}^{N} f(n) = 2\left(\int_0^N dx\, f(x)\right) + f(N) + B_1(f'(N) - f'(0))$$
$$- \frac{1}{12}B_2(f'''(N) - f'''(0)) + \cdots, \tag{10.62}$$

where the B_is are the Bernoulli numbers and $f(n) = \ln(\omega_i^2 + \frac{n^2}{R^2} + \frac{a_i}{R^2} + \cdots) - \ln(\omega_{i,0}^2 + \frac{n^2}{R^2} + \frac{a_{i,0}}{R^2} + \cdots)$. For large n, we expect that $\lambda_i^{(n)} \to \lambda_{i,0}^{(n)}$, hence $f(N), f'(N), f'''(N), \cdots$ all vanish. Also since $\lambda_i^{(n)}$ is actually a function of n^2 the odd derivatives vanish at $n = 0$, and only the first term contributes, giving (letting $Ry = x$)

$$\ln K_i = -R\int_0^{\infty} dy\left(\ln(\omega_i^2 + y^2 + \frac{a_i}{R^2} + \cdots) - \ln(\omega_{i,0}^2 + y^2 + \frac{a_{i,0}}{R^2} + \cdots)\right)$$
$$= -R\int_0^{\infty} dy\left(\ln\left(\frac{\omega_i^2 + y^2}{\omega_{i,0}^2 + y^2}\right) + o\left(\frac{1}{R^2}\right)\right)$$
$$= -R\pi(\omega_i^2 - \omega_{i,0}^2) + o\left(\frac{1}{R}\right). \tag{10.63}$$

This follows from using the integral,

$$= R\int_0^{N/R} dy \ln(\omega^2 + y^2)\, Ry\ln(\omega^2 + y^2) - 2Ry + 2R\omega\arctan\frac{y}{\omega}\Big|_0^{N/R}$$
$$= N\ln\left(\omega^2 + \frac{N^2}{R^2}\right) - 2N + 2R\omega\arctan\left(\frac{N}{\omega R}\right)$$
$$= N\ln\left(\omega^2 + \frac{N^2}{R^2}\right) - 2N + R\pi\omega \tag{10.64}$$

taking $N \to \infty$. This approximation is fine for all the angular momentum families that do not have exact zero modes. For $n = 0, 1$ we would get a vanishing result and singularities in the amplitude.

We can, of course, still apply the method to the comparison theory of the true vacuum without the monopole. Here we get

$$= \frac{1}{2}\ln\lambda_{i,0}^{(0)} + \sum_{n=1}^{N}\ln\lambda_{i,0}^{(n)}\frac{1}{2}\ln\omega_{i,0}^2 + \sum_{n=1}^{N}\left(\omega_{i,0}^2 + \frac{n^2}{R^2}\right)$$

$$= R\int_0^{N/R} dy\ln(\omega_{i,0}^2 + y^2) + \frac{1}{2}\ln\left(\omega_{i,0}^2 + \frac{N^2}{R^2}\right) + o\left(\frac{1}{N}\right)$$

$$= R\pi\omega_{i,0} + (2N+1)\ln\left(\frac{N}{R}\right) - 2N + o\left(\frac{1}{N}\right). \tag{10.65}$$

This follows from the integral Equation (10.64) after adding $\frac{1}{2}\ln\left(\omega_{i,0}^2 + \frac{N^2}{R^2}\right)$ and expanding for large N.

For the three zero modes, the sum over $\lambda_a^{(n)}$ for $a = x_1, x_2, z$ is done explicitly excluding $\lambda_{x_1}^{(0)}$, $\lambda_{x_2}^{(0)}$ and $\lambda_z^{(1)}$ (with multiplicity two). We will use the Stirling approximation $\ln N! \approx N\ln N - N + \frac{1}{2}\ln(2\pi N)$. For $a = x_1, x_2$ we get, noting $\omega_a^2 = 0$

$$-\sum_{n=1}^{N}\ln\lambda_a^{(n)} = -\sum_{n=1}^{N}\ln\left(\left(\frac{n^2}{R^2}\right) + o\left(\frac{1}{R}\right)\right)$$

$$\approx -2\ln\left(\frac{1}{R^N}\prod_{n=1}^{N}n\right) = -2\ln\left(\frac{N!}{R^N}\right)$$

$$= -2\left(N\ln N - N + \frac{1}{2}\ln(2\pi N)\right) + 2N\ln R$$

$$= -2N\ln\left(\frac{N}{R}\right) + 2N - \ln(2\pi N) + o\left(\frac{1}{N}\right). \tag{10.66}$$

Then subtracting the true vacuum result, Equation (10.65), from the result in the presence of the instanton, Equation (10.66), we get

$$\ln K_{x_1} = -R\pi(\omega_{x_1} - \omega_{x_1.0}) - \ln(2\pi R) \quad \text{and} \quad \ln K_{x_2} = -R\pi(\omega_{x_2} - \omega_{x_1.0}) - \ln(2\pi R) \tag{10.67}$$

keeping in mind that $\omega_{x_1}^2 = \omega_{x_1}^2 = 0$. For $a = z$ we have $\lambda_z^{(1)} = 0$, thus we must perform the sum

$$-\frac{1}{2}\ln|\lambda_z^{(0)}| - \sum_{n=2}^{N}\ln\lambda_z^{(n)}, \tag{10.68}$$

where we have put absolute value signs around $\lambda_z^{(0)}$ as it is negative (and the i is taken out explicitly in the Equations (10.43) and (10.44)). As $\omega_z^2 = 0$ and $a_1 = -1$, we get $\lambda_z^{(0)} = -1/R$. Furthermore, putting $\lambda_z^{(n)} = (n^2 - 1)/R$, we get

$$-\frac{1}{2}\ln|\lambda_z^{(0)}| - \sum_{n=2}^{N}\ln\lambda_z^{(n)} = -\frac{1}{2}\ln\left|\frac{-1}{R^2}\right| - \sum_{n=2}^{N}\ln\left(\frac{n^2 - 1}{R^2}\right). \tag{10.69}$$

We evaluate the sum as follows

$$= \sum_{n=2}^{N} \ln\left(\frac{n^2-1}{R^2}\right) \sum_{n=2}^{N} \ln\left(\frac{n^2}{R^2}\left(1-\frac{1}{n^2}\right)\right) = \sum_{n=2}^{N} \ln\frac{n^2}{R^2} + \sum_{n=2}^{N} \ln\left(1-\frac{1}{n^2}\right)$$

$$= \ln\prod_{n=2}^{N} \frac{n^2}{R^2} + \ln\prod_{n=2}^{N}\left(1-\frac{1}{n^2}\right) = 2\ln N! + \ln\prod_{n=2}^{N}\left(\frac{(n+1)(n-1)}{n^2}\right)$$

$$= 2\ln N! - 2(N-1)\ln R + \sum_{n=2}^{N}\left(\ln\left(\frac{n+1}{n}\right) - \ln\left(\frac{n}{n-1}\right)\right)$$

$$= 2N\ln\left(\frac{N}{R}\right) - 2N + \ln 2\pi N + 2\ln R - \ln(2) \tag{10.70}$$

as the final sum is telescopic and gives the $-\ln 2$. Adding the $-\frac{1}{2}\ln\left(\frac{|-1|}{R^2}\right) = \ln R$ gives

$$-\frac{1}{2}\ln|\lambda_z^{(0)}| - \sum_{n=2}^{N} \ln\lambda_z^{(n)} = \ln R - \left(2N\ln\left(\frac{N}{R}\right) - 2N + \ln 2\pi N + 2\ln R - \ln(2)\right)$$

$$= -2N\ln\left(\frac{N}{R}\right) + 2N - \ln(\pi N R). \tag{10.71}$$

Then subtracting the vacuum result, Equation (10.65), we get

$$\ln K_z = -R\pi(\omega_z - \omega_{z,0}) - \ln\pi R^2, \tag{10.72}$$

where of course $\omega_z = 0$. Thus finally adding all the three contributions together we get

$$\sum_i \ln K_i = -R\pi(\omega_i - \omega_{i,0}) - 2\ln 2\pi R - \ln\pi R^2 = R\pi(\omega_i - \omega_{i,0}) - \ln 4\pi^3 R^4 \tag{10.73}$$

or equally well

$$K = \frac{1}{4\pi^3 R^4} e^{-R\pi\sum_i(\omega_i - \omega_{i,0})}. \tag{10.74}$$

The sum $\frac{1}{2}\sum_i(\omega_i - \omega_{i,0})$ has a perfect physical interpretation as the renormalized energy of the magnetic monopole due to vacuum fluctuations about the monopole configuration. This energy is properly subtracted with the energy of the vacuum fluctuations about the true vacuum. Thus we write

$$\frac{1}{2}\sum_i(\omega_i - \omega_{i,0}) = \Delta M. \tag{10.75}$$

The Faddeev–Popov factors, which we have not explicitly dealt with, will also contribute; however, their contribution also simply contributes to the renormalization of the mass of the monopole.

10.7 The Final Amplitude for Decay

The final thing we must calculate are the normalization factors of the translation zero modes using the explicit expressions for the zero modes given by Equation (10.46). We use the coordinates x_1, x_2, r, θ, but will rather use $r = z + R$. First for the directions $i = x_1, x_2$, circular symmetry gives a factor of 2π. The field strength and covariant derivatives of the scalar field are independent of the θ direction, *i.e.* $F_{\theta,\mu} = 0, D_\theta \phi = 0$. The dominant contribution comes from the regions near $z = R$. We can use spherical symmetry in the three independent coordinates x_1, x_2, z. Then the normalization is given by,

$$
= N_i \, 2\pi \int dx_1 dx_2 dr r \left(\sum_\mu F_{\mu i}^a F_{\mu i}^a + (D_i \phi)^a (D_i \phi)^a \right)
$$

$$
\approx 2\pi R \int dx_1 dx_2 dr \left(\sum_\mu F_{\mu i}^a F_{\mu i}^a + (D_i \phi)^a (D_i \phi)^a \right)
$$

$$
= \frac{2\pi R}{3} \int d^3x \left(\sum_{ij} F_{ij}^a F_{ij}^a + (D_i \phi)^a (D_i \phi)^a \right) \tag{10.76}
$$

as, for example, $F_{21}^2 + F_{31}^2 = (2/3)(F_{21}^2 + F_{31}^2 + F_{32}^2) = (1/3)\sum_{jk} F_{jk}^2$.

For the mode $i = 3, 4$ we get a similar expression, but there is angular dependence. Then, for example, $D_3 = \cos\theta D_z$ and we get

$$
= N_3 \, 2\pi \int dx_1 dx_2 dr r \left(\sum_\mu F_{\mu 3}^a F_{\mu 3}^a + (D_3 \phi)^a (D_3 \phi)^a \right)
$$

$$
= R \int d^3x \int d\theta \left(\sum_{i=1,2} F_{iz}^a F_{iz}^a + (D_z \phi)^z (D_z \phi)^a \right) \cos^2\theta
$$

$$
= N_i \frac{1}{2\pi} \int_0^{2\pi} d\theta \cos^2\theta = \frac{1}{2} N_i. \tag{10.77}
$$

Thus we only have to evaluate the integral $\int d^3x \left(\sum_{ij} F_{ij}^a F_{ij}^a + (D_i \phi)^a (D_i \phi)^a \right)$, which can be related easily to the monopole mass. The monopole mass is given by

$$
M = \frac{1}{e^2} \int d^3x \left(\frac{1}{4} F_{jk}^a F_{jk}^a + \frac{1}{2} (D_j \phi)^a (D_j \phi)^a + V(\phi) \right). \tag{10.78}
$$

However, the expression for mass, which is the energy of the monopole, must be stationary with respect to arbitrary variations for the fields. Making a scale transformation $\phi(x) \to \phi(\alpha x)$ and $A(x) \to a A(\alpha x)$ and demanding the mass be stationary at $\alpha = 1$ gives

$$
\int d^3x \left(\left(\frac{1}{4} F_{jk}^a F_{jk}^a \right) - \frac{1}{2} (D_j \phi)^a (D_j \phi)^a - 3V(\phi) \right) = 0. \tag{10.79}
$$

Thus

$$\int d^3x V(\phi(x)) = \frac{1}{3}\int d^3x \frac{1}{4}f^2 - \frac{1}{3}(D\phi)^2, \tag{10.80}$$

which gives

$$M = \frac{1}{e^2}\int d^3x \left(\frac{1}{3}F^a_{jk}F^a_{jk} + \frac{1}{3}(D_j\phi)^a (D_j\phi)^a\right) = \frac{1}{e^2}\frac{N_i}{2\pi R}. \tag{10.81}$$

So $N_i = 2\pi Re^2 M$ and $N_3 = N_4 = \pi Re^2 M$. Thus

$$\left(\frac{N_i}{2\pi e^2}\right)^{1/2} = (RM)^{1/2} \quad i = 1,2, \quad \left(\frac{N_i}{2\pi e^2}\right)^{1/2} = (RM/2)^{1/2} \quad i = 3,4 \tag{10.82}$$

and

$$K = \frac{1}{2}\prod_{i=1}^{4}\left(\frac{N_i}{2\pi e^2}\right)^{1/2} K' = RM \times \frac{RM}{2}\frac{1}{4\pi^3 R^4}e^{-R\pi 2\Delta M} = \frac{M^2}{8\pi^3}e^{-R\pi 2\Delta M} \tag{10.83}$$

Then putting in the factor for the classical instanton action we get the final expression for the amplitude of the decay of the magnetic field

$$\Gamma = \frac{M^2}{8\pi^3 R^2}e^{-R\pi 2\Delta M}e^{-\left(\pi M^2/g^2 B - g^2/4\right)}. \tag{10.84}$$

Using $M/R = gB$, writing $M_{ren.} = M + \Delta M$ and assuming $\Delta M \ll M$

$$\Gamma = \frac{g^2 B^2}{8\pi^3}e^{-\left(\pi M_{ren.}^2/g^2 B - g^2/4\right)}. \tag{10.85}$$

We have not taken into account the zero mode corresponding to internal rotations. As we have mentioned, this mode corresponds to the dyonic excitation. Without the creation of dyonic pairs, the zero mode will give a factor of

$$\left(\frac{J}{Re^2}\right)^{1/2}, \tag{10.86}$$

where J/R is defined to be the normalization of this zero mode. J is calculable from the exact solutions for the dyons as is the mass of the dyon [66]. There is a whole family of dyon solutions with all possible charges, all of which can be produced in pairs. We will not treat the calculation in detail here and refer the reader to the original article [4]. We simply quote the final result, writing Γ_M for the pure monopole result Equation (10.85)

$$\Gamma = \Gamma_M\left(\frac{J}{Re^2}\right)^{1/2}\sum_{-\infty}^{\infty}e^{-(\pi J/Re^2)n^2} = \Gamma_M\sum_{-\infty}^{\infty}e^{-(\pi M/gB)(e^2 n^2/J)} \tag{10.87}$$

using the Poisson summation formula

$$\sum_m f(m) = \sum_m\left(\int dx e^{2\pi i m x}f(x)\right) \tag{10.88}$$

and performing the ensuing Gaussian integral and that $M/R = gB$.

11
Quantum Chromodynamics (QCD)

11.1 Definition of QCD

Quantum Chromodynamics is the theory of strong interactions. It is a non-Abelian gauge theory based on the gauge group $SU(3)$, which is called the colour gauge group. The gauge symmetry is preserved in this theory and, specifically, it is not spontaneously broken. The gauge bosons that carry the strong interaction are called gluons. The matter content of the theory consists of quarks, which are spin one-half fermions that transform according to the fundamental representation of $SU(3)$, that is a three-component, complex triplet. The quark model was proposed in the 1960s and 1970s and elaborated in its incorporation into the "standard model" of particle physics corresponding to a gauge-theoretic description of the strong, weak and electromagnetic interactions. This model is now at the level of a confirmed theory. An untold number of experimental data have shown the existence of quarks and gluons, in addition to the matter content corresponding to the non-strongly interacting particles, the leptons, and the corresponding gauge bosons of the weak and electromagnetic interactions, which are known as the W and Z gauge bosons, and the photon.

The strong interactions govern the interactions that give rise to nuclear forces. The matter that experiences these forces is generally called hadronic matter. The hadrons split into two categories: baryons, which correspond to the neutron, proton and atomic nuclei, which seem to be stable; and mesons, such as the pions, kaons and others, which all seem to be unstable. The fundamental building blocks of the hadrons are the quarks. The quarks interact directly with the gauge bosons of the colour $SU(3)$ gauge group, which are the gluons. The quarks have colour charges and couple directly to the gluons, which themselves have colour charges. However, it is believed that the QCD vacuum is such that colour charges are confined, that free colour charges correspond to states of infinite energy. Therefore, the observable hadrons must all be colour singlet states. The baryons correspond to the bound states of three quarks, and a colour singlet in the three-fold tensor product of the fundamental representation $3 \otimes 3 \otimes 3 = 10 \oplus 8 \oplus 8 \oplus 1$. The mesons correspond to bound states of quarks and anti-quarks, $3 \otimes \bar{3} = 8 \oplus 1$. There are many other possibilities for obtaining singlets, but these have not been experimentally observed.

11.1.1 The Quark Model and Chiral Symmetry

In the 1960s the quark model of hadrons was invented, with contributions from many different authors coming independently. It was understood that quarks come in many flavours, and these were named up, down, charm, strange, top, bottom, and more, if necessary. In our daily experience, we only encounter the up and down quarks. During the 1960s and 1970s, it was discovered how the quarks fit together to give rise to the observable hadrons, and also their interactions with the non-hadronic particles called generically leptons, the electron, muon, and taon, their neutrinos. The quark model seemed to indicate the existence of families of elementary particles, which bring together the strong, weak and electromagnetic interaction with gauge group $SU_c(3) \times SU(2) \times U(1)$, the gauge group of the standard model. Models of grand unification correspond to the inclusion of this group inside a single, semi-simple group, with symmetry breaking giving rise to the observed symmetry group of the standard model. The $SU_c(3)$ is the colour gauge group of QCD. The weak interactions are mediated by the $SU(2)$, while the $U(1)$ corresponds to what is called weak hypercharge. The weak $SU(2)$ is spontaneously broken to a $U(1)$ subgroup, the by now celebrated Higgs field and Higgs mechanism, and the actual electromagnetic $U(1)$ gauge group corresponds to a linear combination of this unbroken remnant of the weak $SU(2)$ and the $U(1)$ hypercharge gauge symmetry. We will not elaborate the full standard model here, it is out of our interest and there are many very good references that describe the standard model in all its detail. For us it will suffice to know that the left-handed quark fields and the leptons feel the weak interaction, which only acts on left-handed fields, and transform according to the doublet representation of the weak interaction gauge symmetry. All right-handed fields, quark or lepton, do not feel the weak interaction, and only feel the strong and electromagnetic interaction.

The first family comprises left-handed up and down quarks forming a doublet of the weak interactions based on the group $SU(2)$ and transforming individually according to a $U(1)$ charge called weak hypercharge, along with the left-handed electron and its neutrino, which also form a weak doublet with their respective weak hypercharges. The family is completed with the right-handed partners of the up and down quarks and the right-handed partner of the electron. The neutrino was not supposed to have a right-handed partner; however, this is no longer certain as it has been observed that the neutrinos must have mass. For the purposes of this book, we will not add a right-handed neutrino. The right-handed partners of all the particles did not experience the weak interaction but did experience the weak hypercharge, and each member had a corresponding value for the weak hypercharge. The second family comprises the charm and strange quarks and the muon and its neutrino; and the third family comprises the top and bottom quarks and the taon and its neutrino. Chiral symmetry corresponds to the notion that there is a complete symmetry under unitary rotation of the quarks amongst themselves. In principle, this would correspond to a "flavour" symmetry group of $SU_f(6)$.

Chiral symmetry is, explicitly, badly broken by the mass spectrum of the quarks. The best preserved subgroup is chiral $SU(2)$ (which is also, coincidentally, the weak interaction symmetry) corresponding to iso-rotations of the up and down quarks amongst themselves as these quarks have masses in the range of a few MeV, which is almost negligible at the scale of the strong interactions. Including the next lightest quark, the strange quarks gives rise to chiral $SU(3)$ symmetry, which is broken at a 10% level as the strange quark mass is around 100 MeV. This symmetry was named $SU_f(3)$, the three-dimensional unitary symmetry of flavour. Identification of this symmetry led to a great advance in the organization of the hadronic particle spectrum. This meant that the Lagrangian of the quarks was made up of three fermionic fields and it is invariant under the unitary rotation of the three fields into each other. The energy eigenstates then must form representations of this group of symmetry, much like the energy levels of the hydrogen atom form representations of the group of spatial rotations, $SO(3)$. Even though the $SU_f(3)$ is broken at the 10% level, the physical hadrons, which are the energy eigenstates of the theory, are easily identifiable as being members of various representations of this symmetry group. The baryons form the representations 8 and 10 of $SU_f(3)$, while the mesons fall into the 8, as shown in Figures 11.1, 11.2 and 11.3, which were created by [84, 83, 82].

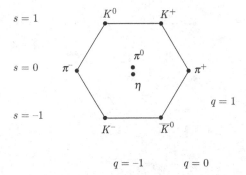

Figure 11.1. QCD flavour diagram of the meson octet

$SU_f(3)$, being a symmetry of the theory, cannot be responsible for the strong force between the hadrons. The strong force must be independent of the flavour symmetry, for the flavour symmetry to manifest itself as a symmetry of the mass spectrum. The charm quark mass is about $1.2\ GeV$ and the top and bottom masses are over $150\ GeV$, hence invoking chiral symmetry including these quarks is quite unrealistic. But what was holding the quarks together?

11.1.2 Problems with Chiral Symmetry

1. Chiral $SU(3)$ symmetry implies the existence of multiplets of hadronic particle states, which have all been observed, and brings order to the chaos of the zoo of observed hadronic particles. However, there is a problem, as chiral symmetry predicts hadronic states such as the Δ^{++} which is made of three up quarks or the Δ^- the corresponding states of three down quarks or the Ω^- that of three strange quarks, each of them in a spin 3/2 state. The problem has to do with their wave functions. The three quarks should be in a spatially symmetric state as there is no additional angular momentum, a spin-symmetric state

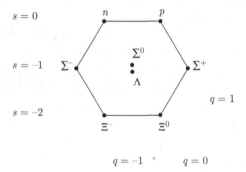

Figure 11.2. QCD flavour diagram of the baryon octet

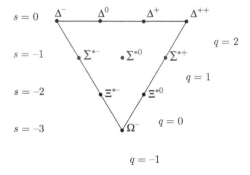

Figure 11.3. QCD flavour diagram of the baryon decouplet

giving rise to the spin 3/2 of the state and an iso-spin symmetric state as the iso-spin of each quark is identical. Such a state is not permitted for fermions by the Pauli exclusion principle, which requires that the wave function of identical fermions must be anti-symmetric under the exchange of any two. Therefore the quarks must have another, hidden quantum number, and the wave function of the state must be anti-symmetric under this hidden degree of freedom.

2. There exists a second experimental reason why the quarks should come in three colours. The ratio

$$R = \frac{\sigma(e^+e^- \to q\bar{q})}{\sigma(e^+e^- \to \mu^+\mu^-)} = \sum_i Q_i^2 \tag{11.1}$$

is simply predicted by perturbation theory, where Q_i is the electrical charge of the quark. This ratio is measured experimentally and gives a rising function of the incoming energy, with a few isolated peaks corresponding to resonances at the positions of particles. However, it reaches a first plateau with a numerical value of 2 when it crosses the threshold for production of the strange quark. Now the sum over the charges of the lightest quarks, up, down and strange, which are, respectively, $\frac{2}{3}$, $-\frac{1}{3}$, $-\frac{1}{3}$, is given by

$$\sum_{\text{lightest quarks}} Q_i^2 = \left(\frac{2}{3}\right)^2 + \left(-\frac{1}{3}\right)^2 + \left(-\frac{1}{3}\right)^2 = \frac{2}{3}. \tag{11.2}$$

Clearly if each quark came three times with three colours we get the required value 2. Increasing the energy of the scattering, once we pass the charm threshold at about 1.2 GeV, the value of R increases to a second plateau at $3\frac{1}{3}$. This corresponds exactly to the addition of the charge of the charm quark squared, $\left(\frac{2}{3}\right)^2 \times 3$. Finally after crossing the bottom quark threshold at an energy of about 4.2 GeV, the value of R again increases to a plateau at $3\frac{2}{3}$ corresponding to the charge of the bottom quark, appropriately $\left(\frac{1}{3}\right)^2 \times 3$.

3. Another experimental reason for three colours has to do with the decay rate of the neutral pion to two photons, $\pi^0 \to 2\gamma$. This decay is mediated by the so-called anomaly diagram. The amplitude for the decay predicted if only one quark is circulating in the triangle is exactly three times too small from the observed amplitude.

4. "Anomaly cancellation" gives another reason to believe that there must be three colours. As mentioned, part of the flavour symmetry group is actually also gauged and gives rise to the weak and electromagnetic interaction. Sometimes gauge symmetries are broken by quantization of chiral fermions. A gauged symmetry must be respected at the quantum level; it is necessary to prove the renormalizability of the theory. Invariance under gauge transformations for the quantum theory is used in an essential manner to prove renormalizability. Therefore, it is imperative that the weak and

electromagnetic gauge symmetries are anomaly-free. The anomalies of the corresponding gauge group all potentially reside in the weak hypercharge $U(1)$ symmetry. The weak hypercharge of the left-handed up and down quarks is $\frac{1}{3}$ while that of the right-handed up quark is $\frac{4}{3}$ and that of the right-handed down quark is $-\frac{2}{3}$. The left-handed leptons, the electron and its neutrino, have weak hypercharge -1 while the right-handed electron has weak hypercharge -2, and we are assuming that the right-handed neutrino does not exist. The anomaly is proportional to the sum of the cubes of all the left-handed hypercharges minus the same for the right-handed charges. We must not forget that the quarks each come in three colours, giving an additional factor of three, and then this gives

$$3 \times \left(\frac{1}{3}\right)^3 + 3 \times \left(\frac{1}{3}\right)^3 + (-1)^3 + (-1)^3 - \left(3 \times \left(\frac{4}{3}\right)^3 + 3 \times \left(-\frac{2}{3}\right)^3 + (-2)^3\right)$$

$$= \left(\frac{1}{9}\right) + \left(\frac{1}{9}\right) - 2 - \left(\frac{64}{9}\right) + \left(\frac{8}{9}\right) + 8 = 0. \tag{11.3}$$

5. Finally, there has to be some mechanism by which the colour degree of freedom is not seen in hadronic states, and has to be confined. There is a good theoretical indication why a non-Abelian gauge theory could supply the correct interaction. First of all, the colour degree of freedom is flavour-blind, it is identical for each flavour. However, QCD being a renormalizable theory, we can perturbatively calculate the renormalization of the coupling constant. Non-trivial renormalization means that naive calculations of, say, the perturbative corrections to the coupling constant give infinite answers. However, by scaling the bare coupling constants of the theory appropriately, all the infinities can be absorbed into these inobservable, infinite, bare coupling constants, while the physically observed coupling constants are finite and defined at a chosen energy scale. However, then the value of the coupling constant at different energy scales is predicted by finite scaling, which is called the renormalization group. Perturbative calculations indicate that as the energy scale is increased the value of the coupling constant decreases (rendering, in fact, the perturbative calculations, which are valid for a small coupling constant, more and more precise). Evidently for lower and lower energies the coupling constant must increase. These properties are called asymptotic freedom at high energies and infrared slavery at low energies. Of course, the perturbative calculation becomes less and less reliable as the coupling constant increases, and hence actually only indications of infrared slavery are predictable via the perturbation theory. Nevertheless, the picture for quarks emerges, that when they are close together, at short distances which correspond to high energies, they are essentially free and non-interacting. However, as they try to separate from one another, at long distances, the force between them increases and, in principle, it would require infinite energy to

separate them infinitely far. The theoretical prediction of asymptotic freedom has been observed experimentally. When very high electrons impinge on a hadronic target and suffer deep inelastic scattering, they scatter off the individual quarks, which, because of the high energy of the electrons, are being probed at very short distances. The quarks then should behave as free particles. This is exactly what is observed. The deep inelastic scattering cross-section for electrons on hadrons exhibits the property of scaling, that the cross-section is simply that of an electron scattering from a free quark of momentum $x \times p_H$, where p_H is the total momentum of the hadron and x is the fraction of that momentum carried by the quark, multiplied by a factor that corresponds to the probability of finding a quark with momentum fraction x.

Thus the colour degree of freedom arose, and making it a local gauge degree of freedom gave the added bonus that it provided a means for obtaining interactions between the quarks that would in principle bind them together.

11.1.3 The Lagrangian of QCD

The Lagrangian density of N flavours of free quarks is given by

$$\mathcal{L} = \sum_{a=1}^{N} \bar{\psi}_\alpha^a \left(i\gamma^\mu \partial_\mu - m^a \right) \psi_\alpha^a. \tag{11.4}$$

The label a corresponds to the different flavours, while the label α corresponds to the colour and the summation over repeated colour, flavour and Lorentz indices is assumed.[1] Interaction terms involving just the fields ψ_α themselves, such as $\left(\bar{\psi}_\alpha^a \psi_\alpha^a \right)^2$ or $\left(\bar{\psi}_\alpha^a \gamma^\mu \psi_\alpha^a \right) \left(\bar{\psi}_\beta^b \gamma_\mu \psi_\beta^b \right)$ and any others, are not renormalizable. To have interactions between the quarks, we must add other fields such as gauge fields or scalar fields with which the quarks interact, and then with each other through the exchange of the additional particles. We will consider the idea of gauging the added $SU(3)$ colour symmetry, the symmetry in any case seems to be required for the existence of fermionic statistics of the quarks in some of the hadronic states.

The colour degree of freedom corresponds to the index α, which goes from 1 to 3, and we will now add gauge fields corresponding to making the gauge symmetry $SU(3)$ local,

$$\mathcal{L} = \sum_{i=a}^{N} \bar{\psi}_\alpha^a \left(i\gamma^\mu (\partial_\mu + A_\mu) - m^a \right) \psi_\alpha^a. \tag{11.5}$$

---

[1] The colour metric or the flavour metric are both simply the identity matrix so we will write the indices above or below depending on convenience. The Lorentz indices are summed with the Minkowski metric, thus for these we will rigorously only sum a raised index with a repeated lowered index.

The covariant derivative $D_\mu = \partial_\mu + A_\mu$ now appears with $A_\mu = iA_\mu^i \lambda_i$, $i = 1, \cdots, 8$, and the λ_i correspond to the 3×3 Gell-Mann matrices

$$\lambda_1 = \begin{pmatrix} 0 & 1 & 0 \\ 1 & 0 & 0 \\ 0 & 0 & 0 \end{pmatrix}, \ \lambda_2 = \begin{pmatrix} 0 & -i & 0 \\ i & 0 & 0 \\ 0 & 0 & 0 \end{pmatrix}, \ \lambda_3 = \begin{pmatrix} 1 & 0 & 0 \\ 0 & -1 & 0 \\ 0 & 0 & 0 \end{pmatrix},$$

$$\lambda_4 = \begin{pmatrix} 0 & 0 & 1 \\ 0 & 0 & 0 \\ 1 & 0 & 0 \end{pmatrix}, \ \lambda_5 = \begin{pmatrix} 0 & 0 & -i \\ 0 & 0 & 0 \\ i & 0 & 0 \end{pmatrix}, \ \lambda_6 = \begin{pmatrix} 0 & 0 & 0 \\ 0 & 0 & 1 \\ 0 & 1 & 0 \end{pmatrix},$$

$$\lambda_7 = \begin{pmatrix} 0 & 0 & 0 \\ 0 & 0 & -i \\ 0 & i & 0 \end{pmatrix}, \ \lambda_8 = \frac{1}{\sqrt{3}} \begin{pmatrix} 1 & 0 & 0 \\ 0 & 1 & 0 \\ 0 & 0 & -2 \end{pmatrix}. \tag{11.6}$$

The Gell-Mann matrices satisfy the Lie algebra of $SU(3)$,

$$[\lambda_i, \lambda_j] = i f^{ijk} \lambda_k, \tag{11.7}$$

where f^{ijk} are the structure constants of $SU(3)$. The structure constants are completely anti-symmetric, $f^{ijk} = -f^{jik} = -f^{ikj}$ with $f^{123} = 1, f^{147} = -f^{156} = f^{246} = f^{257} = f^{345} = -f^{367} = 1/2, f^{458} = f^{678} = \sqrt{3}/2$. To this action, we add the Lagrangian for the gauge fields

$$\mathcal{L}_{gauge} = -\frac{1}{4g^2} F_{\mu\nu}^i F^{i\mu\nu}, \tag{11.8}$$

where, as previously defined, $F_{\mu\nu}^i$ is obtained from

$$[D_\mu, D_\nu] = i F_{\mu\nu}^i \lambda_i, \tag{11.9}$$

explicitly

$$F_{\mu\nu}^i = \partial_\mu A_\nu^i - \partial_\mu A_\nu^i - f^{ijk} A_\mu^j A_\nu^k. \tag{11.10}$$

Our aim in this book is to consider the importance of the classical solutions to the Euclidean equations of motion, the instantons. Thus we will write the Euclidean Lagrangian density as

$$\mathcal{L}_E = \frac{1}{4} F_{\mu\nu}^i F_{\mu\nu}^i = -\frac{1}{2} Tr \left(F_{\mu\nu} F_{\mu\nu} \right), \tag{11.11}$$

where now the Lorentz index becomes a Euclidean vectorial index and the metric in Euclidean space is just the identity, hence we change the sign in the first equality, and

$$F_{\mu\nu} = \partial_\mu A_\nu - \partial_\nu A_\mu + [A_\mu, A_\nu], \tag{11.12}$$

which is an anti-hermitean matrix-valued field.

11.2 Topology of the Gauge Fields

We shall look for configurations of finite Euclidean action

$$S_E = \int d^4x \mathcal{L}_E. \tag{11.13}$$

We assume that for large radius r in four-dimensional Euclidean space, the gauge fields can be expanded in powers of $1/r$. For finite action then, $F_{\mu\nu}$ must decrease as $o(1/r^2)$, where $o(1/r^2)$ means faster than $1/r^2$. This implies that the gauge field must decrease at least as $o(1/r)$ up to pure gauge terms

$$A_\mu = o\left(\frac{1}{r}\right) + g(\Omega)\partial_\mu g^{-1}(\Omega), \tag{11.14}$$

where $g(\Omega)$ is a function only of the angular variables Ω at infinity. Then $g(\Omega)\partial_\mu g^{-1}(\Omega) \sim 1/r$, and this yields the required behaviour for $F_{\mu\nu}$.

But $g(\Omega)$ is defined essentially at infinity of Euclidean spacetime, $\mathbf{R}^4$, which is topologically the three-sphere S^3. Thus $g(\Omega)$ defines a mapping of the three-sphere at infinity into the gauge group $SU(3)$,

$$g(\Omega) : S^3 \to SU(3). \tag{11.15}$$

These fall into the homotopy classes of mappings which define the homotopy group $\Pi_3(SU(3))$. Gauge group configurations $g_1(\Omega)$ and $g_2(\Omega)$ can be continuously deformed one into the other only if they fall into the same homotopy class. We write $g_1(\Omega) \sim g_2(\Omega)$ if they are in the same homotopy class. The homotopy group is well known,

$$\Pi_3(SU(3)) = \mathbb{Z}, \tag{11.16}$$

where $\mathbb{Z}$ corresponds to the integers, and an integer corresponding to a homotopy class is called the winding number. This means that each configuration can be associated with a class of homotopically equivalent configurations, which have the same winding number. Configurations with different winding numbers cannot be continuously deformed one into another, since the winding number can only change discretely. Continuous changes cannot change the winding number. Consequently, different gauge field configurations of finite Euclidean action must also fall into topologically distinct homotopy classes. A gauge field configuration $A_1(x)$ with a limiting value defined by the asymptotic gauge group configuration $g_1(\Omega)$ cannot be continuously deformed into another gauge field configuration $A_2(x)$ with a limiting value defined by the asymptotic gauge group configuration $g_2(\Omega)$ unless $g_1(\Omega) \sim g_2(\Omega)$. If the asymptotic gauge group configurations are in different homotopy classes, the existence of a deformation of the gauge fields into each other continuously keeping the Euclidean action finite would be a contradiction, as it would provide a deformation of one asymptotic gauge group configuration into the other.

We might imagine that as the theory is invariant under local gauge transformations, we might be able to remove the asymptotic gauge dependence. Suppose we make a gauge transformation at infinity,

$$g \to hg \qquad (11.17)$$

for some group element h. Then the gauge field transforms as

$$
\begin{aligned}
A_\mu &\to h(A_\mu + \partial_\mu)h^{-1} \\
&= h\left(g\partial_\mu g^{-1} + o(1/r) + \partial_\mu\right)h^{-1} \\
&= hg(\partial_\mu g^{-1})h^{-1} + h\partial_\mu h^{-1} + o(1/r) \\
&= hg(\partial_\mu (hg)^{-1}) + o(1/r).
\end{aligned} \qquad (11.18)
$$

Thus if we chose $h = g^{-1}$, we could eliminate g. But this is impossible because the gauge transformation h should be a differentiable function defined over the whole space $\mathbf{R}^4$. At least h should be a continuous function over all of $\mathbf{R}^4$. Thus if we define $h = g^{-1}$ at infinity, we must be capable of continuing the definition of h throughout space, including the origin. This is clearly impossible since the origin is a degenerate sphere on which the mapping must be trivial. This implies that the gauge transformation h must be in a class of gauge transformations that can be continuously deformed to the trivial mapping. Hence h cannot satisfy $h = g^{-1}$ at infinity, as g is not in the class of trivial mappings. Thus any gauge transformation h can modify g at infinity, but only within its homotopy class, $g \to hg \sim g$. The integer invariant corresponding to the homotopy class of g is seen to be exactly the Chern number of the gauge field configuration.

We can explicitly construct the gauge transformations that give rise to the different classes of gauge fields

$$
g^{(0)}(x) = 1
$$

$$
g^{(1)}(x) = \frac{x_4 + i\vec{x}\cdot\vec{\sigma}}{(x_4 + |\vec{x}|^2)^{1/2}}
$$

$$
\cdot
$$
$$
\cdot
$$
$$
\cdot
$$

$$
g^{(\nu)}(x) = \left(g^{(1)}\right)^{\nu}
$$

$$
\cdot
$$
$$
\cdot
$$
$$
\cdot \qquad (11.19)
$$

defined over each S^3 that contains the origin. The gauge transformations are singular at the origin (except $g^{(0)}$).

11.2.1 Topological Winding Number

We can explicitly calculate the winding number of the gauge field configuration through the following analysis. Consider the integral

$$\nu = \frac{-1}{24\pi^2}\int d^3\theta \epsilon^{ijk} Tr\left((g\partial_i g^{-1})(g\partial_j g^{-1})(g\partial_k g^{-1})\right),\qquad (11.20)$$

where the integral is over a three-sphere with local coordinates θ_i. For any local, infinitesimal transformation of g, $g \to g(1+\delta T)$ and $g^{-1} \to (1-\delta T)g^{-1}$ so that $gg^{-1}=1$ is unchanged. This means that with $\delta g = g\delta T$ and $\delta g^{-1} = -\delta T g^{-1}$ we will show that ν is unchanged. We then find

$$\begin{aligned}
\delta(g\partial_k g^{-1}) &= g\delta T\partial_k g^{-1} - g\partial_k(\delta T g^{-1})\\
&= g\delta T\partial_k g^{-1} - g(\partial_k \delta T)g^{-1} - g\delta T\partial_k g^{-1}\\
&= -g(\partial_k \delta T)g^{-1}.
\end{aligned}\qquad (11.21)$$

Thus the change in ν is

$$\begin{aligned}
\delta\nu &= \frac{1}{24\pi^2}\int d^3 x\,\epsilon^{ijk} Tr\left(g\partial_i\delta T)g^{-1}(g\partial_j g^{-1})(g\partial_k g^{-1})\right) \times 3\\
&= \frac{1}{8\pi^2}\int d^3\theta\,\epsilon^{ijk} Tr\left((\partial_i\delta T)(\partial_j g^{-1})g(\partial_k g^{-1})g\right)\\
&= \frac{-1}{8\pi^2}\int d^3\theta\,\epsilon^{ijk} Tr\left((\partial_i\delta T)(\partial_j g^{-1})(\partial_k g)\right)\\
&= \frac{-1}{8\pi^2}\int d^3\theta\,\epsilon^{ijk}\partial_i Tr\left((\delta T)(\partial_j g^{-1})(\partial_k g)\right) = 0,
\end{aligned}\qquad (11.22)$$

where in the first line the factor of 3 comes because the contribution from each of the three factors is the same, in the third line we use $g(\partial_k g^{-1})g = -\partial_k g$ and the last line vanishes as the integral is of a total derivate over a three-sphere that has no boundary.

We can evaluate ν explicitly for $g^{(1)}$. At the "north pole", $x^4=1, x^i \approx 0$ then we can take $\theta^i = x^i$

$$\begin{aligned}
g^{(1)}\partial_i\left(g^{(1)}\right)^{-1}\Big|_{north\ pole} &= \left(\frac{x^4+i\vec{x}\cdot\vec{\sigma}}{(x^4+|\vec{x}|^2)^{1/2}}\right)\partial_i\left(\frac{x^4-i\vec{x}\cdot\vec{\sigma}}{(x^4+|\vec{x}|^2)^{1/2}}\right)\Big|_{x^4=1,x^i=0}\\
&= -i\sigma_i - \left(\frac{(x^4-i\vec{x}\cdot\vec{\sigma})x^i}{(x^4+|\vec{x}|^2)^{3/2}}\right)\Big|_{x^4=1,x^i=0} = -i\sigma_i. (11.23)
\end{aligned}$$

However, the symmetry of the configuration means that the integrand is the same at all points on the sphere. Hence,

$$\begin{aligned}
\epsilon^{ijk} Tr\left((g\partial_i g^{-1})(g\partial_j g^{-1})(g\partial_k g^{-1})\right) &= i\epsilon^{ijk} Tr\left(\sigma_i\sigma_j\sigma_k\right)\\
&= i\epsilon^{ijk} Tr\left(i\epsilon_{ijl}\sigma_l\sigma_k\right)\\
&= -\epsilon^{ijk}\epsilon_{ijl}2\delta_{lk} = -2\cdot 6 = -12\ (11.24)
\end{aligned}$$

using $\sigma_i\sigma_j = i\epsilon^{ijk}\sigma_k + \delta_{ij}$. Thus

$$\nu = -\frac{1}{24\pi^2}(-12)\int d^3\theta = \frac{1}{2\pi^2}\int d^3\theta = 1, \tag{11.25}$$

since the volume of the unit three-sphere is exactly $2\pi^2$. This is obtainable by integrating over the angular variables in $\mathbf{R}^4$ in the generalization of spherical coordinates. It is easy to see the $\nu(g_1g_2) = \nu(g_1) + \nu(g_2)$, indeed, using form notation

$$\begin{aligned}
\nu(g_1g_2) &= \frac{-1}{24\pi^2}\int Tr\left(g_1g_2 d(g_1g_2)^{-1}\right)^3 \\
&= \frac{-1}{24\pi^2}\int Tr\left(g_1g_2(dg_2^{-1})g_1^{-1} + g_1 dg_1^{-1}\right)^3 \\
&= \nu(g_1) + \nu(g_2) + 3\frac{-1}{24\pi^2} \\
&\quad \times \int Tr\left(g_1g_2(dg_2^{-1})g_1^{-1}g_1(dg_1^{-1})(g_1g_2(dg_2^{-1})g_1^{-1} + g_1 dg_1^{-1})\right) \\
&= \nu(g_1) + \nu(g_2) + \frac{-1}{8\pi^2} \\
&\quad \times \int Tr\left(g_2(dg_2^{-1})(dg_1^{-1})g_1g_2(dg_2^{-1}) + g_2(dg_2^{-1})(dg_1^{-1})g_1(dg_1^{-1})g_1\right) \\
&= \nu(g_1) + \nu(g_2) + \frac{-1}{8\pi^2}\int d\left(Tr\left(g_2(dg_2^{-1})(dg_1^{-1})g_1\right)\right) \\
&= \nu(g_1) + \nu(g_2)
\end{aligned} \tag{11.26}$$

where we have used $d(gd(g^{-1})) = -gd(g^{-1})gd(g^{-1})$.

We can define

$$G_\mu = 4\epsilon_{\mu\nu\lambda\sigma}Tr\left(A_\nu\partial_\lambda A_\sigma + \frac{2}{3}A_\nu A_\lambda A_\sigma\right) \tag{11.27}$$

then

$$\begin{aligned}
\partial_\mu G_\mu &= 4\epsilon_{\mu\nu\lambda\sigma}Tr\left(\partial_\mu A_\nu\partial_\lambda A_\sigma + \frac{2}{3}(\partial_\mu A_\nu A_\lambda A_\sigma + A_\nu\partial_\mu A_\lambda A_\sigma + A_\nu A_\lambda\partial_\mu A_\sigma)\right) \\
&= 4\epsilon_{\mu\nu\lambda\sigma}Tr\left(\partial_\mu A_\nu\partial_\lambda A_\sigma + 2(\partial_\mu A_\nu A_\lambda A_\sigma)\right) \\
&= 4\epsilon_{\mu\nu\lambda\sigma}Tr\left((\partial_\mu A_\nu + A_\mu A_\nu)(\partial_\lambda A_\sigma + A_\lambda A_\sigma)\right) \\
&= \epsilon_{\mu\nu\lambda\sigma}Tr\left((\partial_\mu A_\nu - \partial_\nu A_\mu + [A_\mu, A_\nu])(\partial_\lambda A_\sigma - \partial_\sigma A_\lambda + [A_\lambda, A_\sigma])\right) \\
&= \epsilon_{\mu\nu\lambda\sigma}Tr\left(F_{\mu\nu}F_{\lambda\sigma}\right). \tag{11.28}
\end{aligned}$$

But

$$\int d^4x\,\partial_\mu G_\mu = \oint_{r\to\infty} dS_\mu G_\mu = \oint_{r\to\infty} dS_\mu 4\epsilon_{\mu\nu\lambda\sigma}Tr\left(A_\nu F_{\lambda\sigma} - \frac{1}{3}A_\nu A_\lambda A_\sigma\right) \tag{11.29}$$

and since

$$F_{\lambda\sigma} = o\left(\frac{1}{r^2}\right) \Rightarrow A_\nu = g\partial_\nu(g^{-1)} + o\left(\frac{1}{r}\right),\tag{11.30}$$

then the first term in Equation (11.29) falls off too fast to contribute while the second term gives exactly the expression for ν in Equation (11.20). Hence

$$\nu = \frac{1}{32\pi^2}\int d^4x Tr\left(\epsilon_{\mu\nu\lambda\sigma}F_{\mu\nu}F_{\lambda\sigma}\right)$$

$$= \frac{1}{16\pi^2}\int d^4x Tr\left(F_{\mu\nu}\tilde{F}_{\mu\nu}\right)\tag{11.31}$$

where the dual field strength is defined as $\tilde{F}_{\mu\nu} = \frac{1}{2}\epsilon_{\mu\nu\lambda\sigma}F_{\lambda\sigma}$.

We can summarize our findings as follows.

1. Each gauge field configuration of finite Euclidean action is associated with an integer, called its Pontryagin number.
2. It is impossible to continuously deform one gauge field configuration into another with different Pontryagin numbers, keeping the Euclidean action finite.

For any other gauge group, $SU(3)$ in particular, there is a theorem by Bott [18] that says that any mapping of S^3 into a semi-simple Lie group G can be continuously deformed to a mapping into a $SU(2)$ subgroup of G. Hence everything that we have shown for $SU(2)$ is actually valid for any semi-simple Lie group G. The only thing that changes is the normalization in the formulae for the winding number. However, if we use the notion of the Cartan scalar product on the Lie algebra of G, defining

$$\langle T^a T^b\rangle = \delta^{ab} = \alpha Tr\left(T^a T^b\right)\tag{11.32}$$

then α depends on the representation of the T^a, but the formula for the Pontryagin number is universal

$$\nu = \frac{1}{32\pi^2}\int d^4x\langle F_{\mu\nu}\tilde{F}_{\mu\nu}\rangle.\tag{11.33}$$

Now with the possibility of many inequivalent classical sectors in the space of field configurations, we expect the existence of the many different vacuum configurations, and of course the possibility of quantum tunnelling between them.

11.3 The Yang–Mills Functional Integral

We begin with the functional integral

$$\mathcal{I} = \mathcal{N}\int \mathcal{D}A e^{\int d^4x \frac{1}{4g^2}\langle F_{\mu\nu}F_{\mu\nu}\rangle}.\tag{11.34}$$

We must fix the gauge, we will choose $A_3 = 0$. We then have the following observations:

1. It is easy to see that all gauge field configurations may be put into this gauge, simply take

$$h = \mathcal{P}\left(\exp \int_{-\infty}^{x_3} dx'^3 A_3(x^1, x^2, x'^3, x^4)\right). \tag{11.35}$$

Then

$$h\partial_3 h^{-1} = -hA_3 h^{-1} \tag{11.36}$$

hence

$$A_3' = h(A_3 + \partial_3)h^{-1} = hA_3 h^{-1} - hA_3 h^{-1} = 0. \tag{11.37}$$

2. The Faddeev–Popov factor is just a constant.

11.3.1 Finite Action Gauge Fields in a Box

We will consider the theory in a finite spatial volume V, but always have in mind that $V \to \infty$ at the end. The same for the Euclidean time T. We must choose boundary conditions on the walls. We will choose the boundary conditions such that the bulk equations of motion are not modified because of them. The general variation of the action is

$$
\begin{aligned}
\delta S &= \int d^4x \frac{\partial \mathcal{L}}{\partial A_\mu} \delta A_\mu + \frac{\partial \mathcal{L}}{\partial \partial_\nu A_\mu} \delta \partial_\nu A_\mu \\
&= \int d^4x \left(\frac{\partial \mathcal{L}}{\partial A_\mu} - \frac{\partial \mathcal{L}}{\partial \partial_\nu A_\mu}\right) \delta A_\mu + \partial_\mu \left(\frac{\partial \mathcal{L}}{\partial \partial_\nu A_\mu} \delta A_\mu\right) \\
&= \int d^3s\, \hat{n}^\nu \frac{\partial \mathcal{L}}{\partial \partial_\nu A_\mu} \delta A_\mu + \int d^4x \left(\frac{\partial \mathcal{L}}{\partial A_\mu} - \frac{\partial \mathcal{L}}{\partial \partial_\nu A_\mu}\right) \delta A_\mu \\
&= \int d^3s\, \hat{n}^\nu F_{\nu\mu} \delta A_\mu + \int d^4x \left(\frac{\partial \mathcal{L}}{\partial A_\mu} - \frac{\partial \mathcal{L}}{\partial \partial_\nu A_\mu}\right) \delta A_\mu.
\end{aligned}
\tag{11.38}
$$

Therefore, to not have any contribution from the boundary we must impose

$$\hat{n}^\nu F_{\nu\mu} \delta A_\mu = 0 \tag{11.39}$$

on the boundary. We can decompose δA_μ into its normal and tangential components, $\delta A_\mu = (\delta A^{norm.})\hat{n}_\mu + \delta A_\mu^{tang.}$, where $\hat{n}_\mu \delta A_\mu^{tang.} = 0$. Then the boundary condition Equation (11.39) becomes

$$\hat{n}^\nu F_{\nu\mu}\left((\delta A^{norm.})\hat{n}_\mu + \delta A_\mu^{tang.}\right) = \hat{n}^\nu F_{\nu\mu} \delta A_\mu^{tang.} = 0, \tag{11.40}$$

since $F_{\nu\mu} = -F_{\mu\nu}$. Thus we are required to fix the tangential components of the gauge field on the boundary and, consequently, we impose that the tangential components may not be varied on the boundary, so that $\delta A_\mu^{tang.} = 0$. We must also respect the gauge-fixing condition, $A_3 = 0$, and we are only interested in field configurations whose action remains finite as the box size is taken to infinity. We

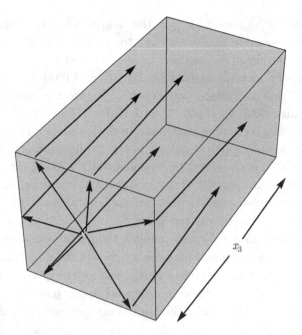

Figure 11.4. Paths over the boundary defining the gauge group element

will see that these conditions mean that the winding number inside the box must be a definite integer. We will show that the only vestige of the boundary conditions is that the winding number inside the box is a definite integer.

Indeed, inside a large box of dimensions (L_1, L_2, L_3, T), gauge fields that remain of finite action when the box is taken to infinite size must have the behaviour

$$A_\mu = g \partial_\mu g^{-1} + o\left(\frac{1}{r}\right) \tag{11.41}$$

on the boundary. $g \partial_\mu g^{-1}$ is obtained from the limiting values of the gauge field configuration, and hence must be continuously defined over the entire boundary. g is extracted by performing the path-ordered exponential integral, as shown in Figure 11.4, along a nest of paths that start at an initial point x_0^μ on the boundary at $x^3 = -L_3/2$ and move along and cover the boundary to all other points on the boundary x^μ

$$g(x^\mu) = \mathcal{P} \exp\left(-\int_{x_0^\mu}^{x^\mu} A_\nu(x'^\mu) dx'^\nu\right). \tag{11.42}$$

The integrability condition that the gauge group element obtained from the path-ordered exponential integral from two different paths is the same, and is exactly that the field strength vanishes on a surface whose boundary comprises the two paths. This condition can be easily verified for an infinitesimal loop. The field strength does indeed vanish for $A_\mu = g \partial_\mu g^{-1}$. Thus g is continuously defined over the entire boundary. g is the unique solution of the linear, first-order differential equation $\partial_\mu g^{-1} = g^{-1} A_\mu$ (or equivalently $\partial_\mu g = -A_\mu g$), up to an irrelevant, multiplicative, constant gauge group element. Equivalently, the actual gauge group element defined by Equation (11.41) is also ambiguously defined up to a constant gauge group element g_0; we can simply take $g \to g g_0$ as then the gauge field $A_\mu = g \partial_\mu g^{-1}$ is invariant. The constant gauge group element is irrelevant, it does not contribute to the action or any winding number. The integration paths are perpendicular to the x^3 direction on the two faces at $x^3 = \pm L_3/2$, hence g is necessarily independent of x^3. Along the other surfaces, we integrate along lines parallel to the x^3 direction, but since $A_3 = 0$ the gauge group element is unchanged. On the two surfaces at $x_3 = \pm L_3/2$, the gauge transformation is not necessarily the same.

Specifying g on the boundary fixes only the tangential components of A_μ since g only varies along the boundary that corresponds to the directions tangential to the boundary. The normal component of A_μ must also be given by the form given in Equation (11.41). However, these then depend on how g varies as we move away from the boundary into the bulk. The normal components of A_μ do not need to be specified, since all we insist on is that the boundary values do not contribute to the equations of motion. Thus we do not have to specify the variation of g as we move away from the boundary into the bulk. One thing is important, since the surface of the box is topologically S^3, the gauge group element g defined on the boundary can perfectly well be in a non-trivial homotopy class of $\Pi_3(G)$, and hence may not necessarily be continuously defined throughout the entire box. Indeed, g is only defined by the asymptotic behaviour of the gauge field on and near the boundary.

On the surfaces at $x^3 = \pm L_3/2$, the gauge group element depends, in principle, non-trivially on the three coordinates (x^1, x^2, x^4) and $g(x^1, x^2, -L_3/2, x^4) \neq g(x^1, x^2, +L_3/2, x^4)$ as in Figure 11.5. But on the surfaces that connect the boundaries of these two ends, since $A_3 = g \partial_3 g^{-1} = 0$ from the gauge condition, we must have that g is independent of x^3. Thus the values of g on the boundaries of the two end surfaces at $x^3 = \pm L_3/2$, *i.e.* for at least one of: $x^1 = \pm L_1/2$, $x^2 = \pm L_2/2$ or $x^4 = \pm T/2$, and $x^3 = \pm L_3/2$, are the same. Now we will perform a gauge transformation by $h(x^1, x^2, x^3, x^4)$, which is actually independent of x^3 and defined by the value of the gauge group element at the surface $x^3 = -L_3/2$, *i.e.*

$$h(x^1, x^2, x^3, x^4) = g^{-1}(x^1, x^2, -L_3/2, x^4). \tag{11.43}$$

Then

$$A_\mu \to h(A_\mu + \partial_\mu)h^{-1} = g^{-1}(x^1, x^2, -L_3/2, x^4)(A_\mu + \partial_\mu)g(x^1, x^2, -L_3/2, x^4). \tag{11.44}$$

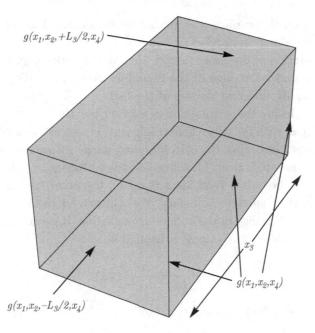

$g(x_1,x_2,+L_3/2,x_4)$

x_3

$g(x_1,x_2,x_4)$

$g(x_1,x_2,-L_3/2,x_4)$

Figure 11.5. Boundary with the gauge group element

But since $A_\mu = g\partial_\mu g^{-1} + o\left(\frac{1}{r}\right)$ for large r, we get

$$A_\mu \to \left(g^{-1}\big|_{x^3=-L_3/2}\right)\left(g\partial_\mu g^{-1} + o\left(\frac{1}{r}\right) + \partial_\mu\right)\left(g\big|_{x^3=-L_3/2}\right). \qquad (11.45)$$

We emphasize that the g appearing inside the middle bracket depends on x^3 while that on the outside is independent of x_3 and is equal to its value at $x^3 = -L_3/2$. Evidently we can then write

$$A_\mu = g_1 \partial_\mu g_1^{-1} + o\left(\frac{1}{r}\right), \qquad (11.46)$$

where

$$g_1(x^\mu) = \left(g^{-1}\big|_{x^3=-L_3/2}\right)g(x^\mu). \qquad (11.47)$$

Evidently, g_1 is equal to the identity on the surface at $x^3 = -L_3/2$ and also on the surfaces where x^3 changes, $-L_3/2 \to L_3/2$. On the surface at $x^3 = L_3/2$, g_1 must then be the identity on the boundary of this surface (for $x^1 = \pm L_1/2$ and so on), but for interior values of x^1, x^2 and x^4 generally $g\big|_{x^3=L_3/2}(x^1,x^2,x^4)$, need not be equal to $g\big|_{x^3=-L_3/2}(x^1,x^2,x^4)$. Indeed, if the instanton number of the gauge field configuration is non-trivial, then $g_1\big|_{x^3=L_3/2} \neq \mathbb{I}$ since the gauge transformation h of Equation (11.43) cannot change the instanton number. The instanton number given by the integral Equation (11.31) is gauge-invariant. The surface at $x^3 = L_3/2$ with its boundary identified is also topologically S^3, and g_1 defined on this surface goes to the identity at its boundary. This means that g_1

is also defined on this surface with its boundary identified as topologically S^3. Thus g_1 defines a map from $S^3 \to G$, which contains all the topological winding number information of the gauge field defined in the entire box. We want to see what happens under a change of the boundary conditions. We will implement this by fractionally changing the size of the box.

We imagine placing the original box in a larger box that is extended along the x^3 direction by Δ with the gauge field configuration also extended into the larger box. In the larger box, after the corresponding gauge transformation, there will also be a gauge group element, g_2, defined on the surface at $x^3 = L_3/2 + \Delta$, which is identity on its boundary (and identity on all the other surfaces of the box), just like g_1 on its respective boundary. We will extend the box in such a way that the fractional change in the volume is negligible. If we choose $\Delta = (L_3)^{1/2}$, the fractional change in the volume is negligible,

$$\frac{\delta V}{V} = \frac{L_1 L_2 L_4 \Delta}{L_1 L_2 L_3 L_4} = \frac{\Delta}{L_3} = (L_3)^{-1/2} \to 0. \qquad (11.48)$$

Alternatively, the volume of the larger box is

$$V + \delta V = L_1 L_2 L_3 L_4 \left(1 + \frac{1}{\Delta}\right) = V \left(1 + \frac{1}{L_3^{1/2}}\right) \to V \quad \text{when} \quad L_3 \to \infty. \ (11.49)$$

If g_1 and g_2 are in the same homotopy class, we will show that all gauge field configurations defined in the smaller box can be extended to gauge field configurations in the larger box, with negligible change in the action. If g_1 and g_2 are not in the same homotopy class, this is not the case: there has to be at least one more instanton outside the smaller box which implies an increase the action by at least $8\pi^2/g^2$, which is the minimum action of one instanton, as we will see in the next section[2]. This increase in the action is independent of the amount of the extension of volume of the box, even if the volume is only extended fractionally, negligibly. Thus for g_1 and g_2 in the same homotopy class, the action changing negligibly means that extending the box is simply equivalent to a changing boundary condition $g_1 \to g_2$. The action is invariant, but the only vestige of the boundary condition is the topological winding number encoded in g_1 or any other homotopically equivalent boundary gauge group element and the corresponding boundary condition.

Let $g(x^1, x^2, s, x^4)$ with $s \in [0, 1]$ be a homotopy from g_1 to g_2:

$$g(s = 0) = g_1, \ g(s = 1) = g_2. \qquad (11.50)$$

Then for

$$h = h(x^1, x^2, x^3, x^4) = g(x^1, x^2, (x^3 - L_3/2)/\Delta, x^4) \qquad (11.51)$$

[2] See Equation (11.71).

for $x^3 \in [+L_3/2, L_3/2 + \Delta]$, and with the gauge field extended as $A_\mu = h\partial_\mu h^{-1}$, evidently the action does not change. However, the gauge condition $A_3 = 0$ is not respected. The choice

$$A_\mu = \begin{cases} h\partial_\mu h^{-1} & \mu \neq 3 \\ 0 & \mu = 3 \end{cases} \tag{11.52}$$

does satisfy the gauge condition, but the action is slightly changed. The increase in the action is easily calculated, using gauge invariance. We transform the gauge field of Equation (11.52) by h^{-1}, which gives

$$A_\mu = \begin{cases} 0 & \mu \neq 3 \\ h^{-1}\partial_3 h & \mu = 3 \end{cases}. \tag{11.53}$$

Then, integrating only over the extension,

$$S_E = -\int d^4x \frac{1}{4g^2} \langle F_{\mu\nu} F_{\mu\nu} \rangle = -\int d^4x \frac{1}{4g^2} \langle F_{\mu3} F_{\mu3} \rangle. \tag{11.54}$$

Then $A_3 = g^{-1}\partial_s g \partial_3 \left(\frac{x^3 - L_3/2}{\Delta} \right) = (g^{-1}\partial_s g) \frac{1}{\Delta} \sim \frac{1}{\Delta}$, and consequently

$$F_{\mu3} = \partial_\mu A_3 - \partial_3 A_\mu = \partial_\mu A_3 \sim \frac{1}{\Delta} \tag{11.55}$$

as the commutator term vanishes. Thus $\langle F_{\mu\nu} F_{\mu\nu} \rangle \sim \frac{1}{\Delta^2}$. But the integral

$$\int_{L_3/2}^{L_3/2+\Delta} d^4x \sim \Delta, \tag{11.56}$$

which implies that the action also changes by a negligible amount

$$\delta S_E \sim \frac{1}{\Delta} = (L_3)^{-1/2} \to 0. \tag{11.57}$$

Hence a change in the boundary conditions that preserves the winding number is just a surface effect, not a volume effect. The action is invariant. Therefore, only the winding number remains, which is defined by the gauge group element $g(x^1, x^2, L_3/2, x^4)$ which defines a map $S^3 \to G$.

Now suppose we decided to choose a different boundary condition, not the one that fixes the tangential components of the gauge field on the boundary but some arbitrary, other boundary condition. We will still work with the gauge condition $A_3 = 0$. The condition that the action be finite still imposes that

$$A_\mu = g\partial_\mu g^{-1} + o\left(\frac{1}{r}\right) \tag{11.58}$$

and nothing obstructs from performing the x^3 independent gauge transformation that gives $\tilde{g}(x^1, x^2, x^4)$ on the end surface at $x^2 = L_3/2$, in the same way as before. We will compare gauge field configurations in this gauge. But now $\tilde{g}$ will not correspond to any such gauge group element obtained when the boundary conditions fixed the tangential components of the gauge field, although the homotopy class of $\tilde{g}$ is fixed by the winding number. Thus gauge field configurations giving rise to $\tilde{g}$ would not be included in the subset of configurations that satisfy the boundary conditions on the tangential components of the gauge field that we have considered. However, the arguments given above show that, although we do not get the exact gauge field configurations with different boundary conditions, we do get configurations that are arbitrarily close, by small deformations near the end at $x^3 = L_3/2$. Indeed, any given $\tilde{g}$ can be obtained by making changes to the gauge field configuration only in the extended part of the box, as we did when defining the homotopy from g_1 to g_2, now we will simply consider the homotopy from g_1 to $\tilde{g}$. Thus all gauge field configurations apart from a small difference in the extended portion of the box are permitted by our boundary conditions, and this difference gives negligible change for a large enough box.

11.3.2 The Theta Vacua

Therefore, for a large enough box, we can simply forget the boundary conditions, but impose that all configurations in the functional integration correspond to those of a fixed winding number n.

$$F(V,T,n) \equiv \mathcal{N} \int \mathcal{D}A e^{-S_E} \delta_{\nu n}, \qquad (11.59)$$

where $\mathcal{D}A = \mathcal{D}(A_1, A_2, A_4)$. $F(V,T,n)$ is a matrix element between an initial state and a final state that are determined by the boundary conditions. For T_1 and T_2 taken very large,

$$F(V,T_1 + T_2, n) = \sum_{n_1 + n_2 = n} F(V,T_1,n_1) F(V,T_2,n_2). \qquad (11.60)$$

This is because the winding number

$$\nu = \frac{1}{32\pi^2} \int d^4 x \langle F_{\mu\nu} \tilde{F}_{\mu\nu} \rangle \qquad (11.61)$$

is an integral of a local density $\langle F_{\mu\nu} \tilde{F}_{\mu\nu} \rangle$. This means that one way to put a configuration of winding number n into the box with Euclidean time length given by $T_1 + T_2$ is to put $\nu = n_1$ into the first part of the box and $\nu = n_2$ into the second part. Such configurations neglect configurations with significant action on the border between the two parts; however, we expect that this contribution is negligible for large T_1 and T_2. Normally a matrix element for $T = T_1 + T_2$ that gets a contribution from only one energy state follows a multiplicative law. The convolutive law of combination of the matrix elements above, Equation (11.61), can be simply disentangled into the more familiar multiplicative law by a simple

Fourier transformation. Defining

$$F(V,T,\theta) = \sum_n e^{in\theta} F(V,T,n) \equiv \mathcal{N} \int \mathcal{D}A e^{-S_E} e^{i\nu\theta} \qquad (11.62)$$

implies

$$F(V,T_1+T_2,\theta) = F(V,T_1,\theta)F(V,T_2,\theta). \qquad (11.63)$$

This implies the existence of states such that

$$F(V,T,\theta) = N'\langle\theta|e^{-HT}|\theta\rangle, \qquad (11.64)$$

where the states $|\theta\rangle$ are eigenstates of the Hamiltonian. Our field theory is now surprisingly separated into a family of sectors enumerated by θ. In each sector we use the same action except we add an extra term proportional to $\nu = \theta\langle F_{\mu\nu}\tilde{F}_{\mu\nu}\rangle$.

We can obtain all of these results from the functional integral and from the possible instanton solutions to the Euclidean equations of motion. If there is a solution for $\nu = 1$, all the results follow. Suppose such a solution exists with action S_0. Then translation invariance gives at least four zero modes, and

$$\langle\theta|e^{-HT}|\theta\rangle = N' \int \mathcal{D}A e^{-S_E} e^{i\nu\theta}$$

$$= \sum_{n,\bar{n}} \left(\left(Ke^{-S_0}\right)^{n+\bar{n}} (VT)^{n+\bar{n}} e^{i(n-\bar{n})\theta} \right) / n!\bar{n}!$$

$$= e^{2KVTe^{-S_0}\cos\theta} \qquad (11.65)$$

where K is the usual determinantal factor and a sum has been done over n instantons and $\bar{n}$ anti-instantons. Then we can read off the energy of the $|\theta\rangle$ states,

$$E(\theta) = -2VK\cos\theta e^{-S_0}. \qquad (11.66)$$

We can also compute the expectation value

$$\langle\theta|\langle F_{\mu\nu}\tilde{F}_{\mu\nu}\rangle|\theta\rangle = \frac{1}{VT} \int d^4x \langle\theta|\langle F_{\mu\nu}\tilde{F}_{\mu\nu}\rangle|\theta\rangle$$

$$= \frac{32\pi^2 \int \mathcal{D}A\, \nu e^{-S_E} e^{i\nu\theta}}{VT \int \mathcal{D}A e^{-S_E} e^{i\nu\theta}}$$

$$= \frac{-32\pi^2 i}{VT} \frac{d}{d\theta} \ln\left(\int \mathcal{D}A e^{-S_E} e^{i\nu\theta} \right)$$

$$= \frac{-32\pi^2 i}{VT} \left(-2K\cos\theta e^{-S_0} \right) VT$$

$$= -64\pi^2 iK e^{-S_0} \sin\theta. \qquad (11.67)$$

The answer is imaginary, but this is correct. Since $\langle F_{\mu\nu}\tilde{F}_{\mu\nu}\rangle = \langle F_{12}F_{34} + perm.\rangle$ in Euclidean space, analytic continuation to Minkowski space yields, for example, $F_{j4} \to iF_{j0}$. Hence the imaginary result in Euclidean space corresponds to the correct, real result in Minkowski space. Everything depends on θ, it is a physical parameter.

11.3.3 The Yang–Mills Instantons

The instantons do actually exist as solutions of the Euclidean equations of motion. We can prove that the action in the single instanton sector is bounded from below, and when the bound is saturated, the configuration must satisfy the equations of motion. Consider the inequality which is evidently satisfied

$$- \int d^4x \langle (F_{\mu\nu} \pm \tilde{F}_{\mu\nu})(F_{\mu\nu} \pm \tilde{F}_{\mu\nu}) \rangle \geq 0, \qquad (11.68)$$

note that the minus sign is there because our gauge fields and field strengths are anti-hermitean. This implies

$$- \int d^4x \left(\langle F_{\mu\nu} F_{\mu\nu} \rangle + \langle \tilde{F}_{\mu\nu} \tilde{F}_{\mu\nu} \rangle \pm 2 \langle F_{\mu\nu} \tilde{F}_{\mu\nu} \rangle \right) \geq 0. \qquad (11.69)$$

But the first two terms are equal, hence choosing the $\pm$ sign appropriately, we have

$$- \int d^4x \langle F_{\mu\nu} F_{\mu\nu} \rangle \geq \left| \int d^4x \langle F_{\mu\nu} \tilde{F}_{\mu\nu} \rangle \right|. \qquad (11.70)$$

But the right-hand side is just the instanton number while the left-hand side is proportional to the action, thus we find

$$S_E \geq \frac{8\pi^2}{g^2} |\nu| \qquad (11.71)$$

as we had promised to show earlier. The equality is attained for

$$F_{\mu\nu} = \pm \tilde{F}_{\mu\nu}, \qquad (11.72)$$

where we get the $+$ sign for $\nu \geq 0$ and the minus sign for $\nu \leq 0$. If we can find the solutions of Equation (11.72), we automatically get solutions of the full equations of motion, as the action is minimal for such configurations and hence, must be stationary. A bonus is that Equation (11.72) as a differential equation is now a first-order equation, instead of a second-order equation, and is consequently easier to solve.

For $\nu = 1$ we will look for a solution that asymptotically behaves as

$$A_\mu = g^{(1)} \partial_\mu \left(g^{(1)} \right)^{-1} + o\left(\frac{1}{r} \right), \qquad (11.73)$$

where $g^{(1)}$ is the gauge group element defined in Equation (11.19). $g^{(1)}$ is spherically symmetric, hence we make the ansatz

$$A_\mu = f(r) r^2 g^{(1)} \partial_\mu \left(g^{(1)} \right)^{-1} = -i A_\mu^i \sigma^i. \qquad (11.74)$$

Using a double index notation, for the gauge group $SU(2)$ seen as the diagonal subgroup of $SO(4)$, we can write

$$A_\mu^i = \frac{1}{2} \left(A_\mu^{0i} + \frac{1}{2} \epsilon^{ijk} A_\mu^{jk} \right) \qquad (11.75)$$

for anti-symmetric $SO(4)$-valued gauge fields $A_\mu^{\alpha\beta} = -A_\mu^{\beta\alpha}$. An easy calculation with g given in Equation (11.19) gives

$$A_\mu^{\alpha\beta} = f(r)(x_\alpha \delta_{\mu\beta} - x_\beta \delta_{\mu\alpha}). \tag{11.76}$$

Then a straightforward calculation gives

$$F_{\mu\nu}^{\alpha\beta} = (2f - r^2 f^2)(\delta_{\mu\alpha}\delta_{\nu\beta} - \delta_{\mu\beta}\delta_{\nu\alpha}) + \left(\frac{f'}{r} + f^2\right)$$

$$\times (x_\alpha x_\mu \delta_{\nu\beta} - x_\alpha x_\nu \delta_{\mu\beta} + x_\beta x_\nu \delta_{\mu\alpha} - x_\beta x_\mu \delta_{\nu\alpha}). \tag{11.77}$$

The condition of self duality, $F_{\mu\nu} = \tilde{F}_{\mu\nu}$, is automatically satisfied for the first term $(\delta_{\mu\alpha}\delta_{\nu\beta} - \delta_{\mu\beta}\delta_{\nu\alpha}) = \epsilon_{\mu\nu\sigma\tau}\epsilon_{\sigma\tau\alpha\beta}$

$$\frac{1}{2}\epsilon_{\lambda\rho\mu\nu}(\delta_{\mu\alpha}\delta_{\nu\beta} - \delta_{\mu\beta}\delta_{\nu\alpha}) = \frac{1}{2}\epsilon_{\lambda\rho\mu\nu}\epsilon_{\mu\nu\sigma\tau}\epsilon_{\sigma\tau\alpha\beta} = \epsilon_{\lambda\rho\sigma\tau}\epsilon_{\sigma\tau\alpha\beta} \tag{11.78}$$

but not so for the second term

$$\frac{1}{2}\epsilon_{\lambda\rho\mu\nu}(x_\alpha x_\mu \delta_{\nu\beta} - x_\alpha x_\nu \delta_{\mu\beta} + x_\beta x_\nu \delta_{\mu\alpha} - x_\beta x_\mu \delta_{\nu\alpha}) = \epsilon_{\lambda\rho\mu\beta}x_\alpha x_\mu - \epsilon_{\lambda\rho\mu\alpha}x_\beta x_\mu. \tag{11.79}$$

Thus we obtain a self dual field strength by imposing

$$\frac{f'}{r} + f^2 = 0. \tag{11.80}$$

This integrates trivially as

$$f(r) = \frac{1}{r^2 + \lambda^2}, \tag{11.81}$$

where λ is an arbitrary integration constant. Thus

$$A_\mu = \frac{r^2}{r^2 + \lambda^2} g^{(1)} \partial_\mu \left(g^{(1)}\right)^{-1}. \tag{11.82}$$

We will find that there exist eight parameters corresponding to symmetries of the action that are broken by the solution. These correspond in principle to scale transformations, rotations, translations, special conformal transformations and global gauge transformations. Scale transformations correspond to changing λ. Note that the global gauge transformations preserve the gauge-fixing conditions $A_3 = 0$. Rotations and global gauge transformations are tied together, the solution is invariant under the diagonal subgroup of simultaneous rotations and global gauge transformations by the same amount. Note that the rotation group $SO(3)$ and the global gauge group $SU(2)$ are essentially the same group. Special conformal transformations can be obtained by translations and gauge transformations and hence do not give rise to new solutions. This in the end leaves eight parameters, coming from one scale transformation, four translations and three rotations (or equivalently global gauge transformations). For a configuration on n instantons and $\bar{n}$ anti-instantons, the number of parameters is simply $8(n + \bar{n})$.

11.4 Theta Vacua in QCD

The existence of the instanton solutions means that there exist different, inequivalent classical ground states, between which the instantons mediate quantum tunnelling. We have not explicitly seen these vacuum configurations; to uncover them, it is more convenient to use the temporal gauge, *i.e.* $A_0 = 0$. As, in principle, everything we do does not depend on the gauge choice, we are free to take any gauge that we want. The dynamical variables in this gauge are just the spatial components of the gauge field A_i.

In Minkowski space, the Hamiltonian is given by

$$\mathcal{H} = \int d^3x \frac{1}{2} \left((E_i^a)^2 + (B_i^a)^2 \right), \tag{11.83}$$

where the electric and magnetic fields are given by

$$E_i^a = \dot{A}_i^a$$
$$B_i^a = \frac{1}{2}\epsilon_{ijk} \left(\partial_j A_k^a - \partial_k A_j^a + f^{abc} A_j^b A_k^c \right). \tag{11.84}$$

In this gauge, since there is no field A_0, the equation of motion that usually comes from varying with respect to it is missing. This is the Gauss law

$$\mathcal{G}^a = \partial_i \dot{A}_i^a + f^{abc} A_i^b \dot{A}_i^c \equiv (D_i E_i)^a = 0. \tag{11.85}$$

However, in this gauge the Hamiltonian is invariant under time-independent, spatial gauge transformations. The corresponding conserved charge is actually a local expression, exactly the Gauss law operator, $\dot{\mathcal{G}}^a = 0$, *i.e.*

$$[\mathcal{H}, \mathcal{G}^a] = 0. \tag{11.86}$$

Thus we must impose this constraint on the initial values of the fields, then since the time evolution preserves the constraint, the Gauss law operator will be preserved for all time.

Now in the quantum theory, eigenstates of the field operator $A_i^a(x)$ correspond to the states $|A_i^a(x)\rangle$ and the amplitude for a transition between two such states is given by the functional integral

$$\langle \tilde{A}_i^a(x)|e^{-i\mathcal{H}T}|A_i^a(x)\rangle = \int_{A_i^a(x)}^{\tilde{A}_i^a(x)} \mathcal{D}\left(A_i^a(\vec{x}(t))\right) e^{\frac{-i}{2g^2} \int_0^T d^t d^3x \left((E_i^a)^2 + (B_i^a)^2 \right)}, \tag{11.87}$$

where the functional integral is over all time histories $A_i^a(\vec{x}(t))$ that interpolate between the initial and final configurations. In the quantum theory, however, the Gauss law constraint is imposed as a constraint on the Hilbert space, a physical state in the Hilbert space of all states is one that is annihilated by the Gauss law operator

$$\mathcal{G}^a(x)|\Psi\rangle = (D_i E_i)^a|\Psi\rangle = 0, \tag{11.88}$$

the states $|A_i^a(x)\rangle$ do not satisfy this constraint. We wish to characterize the states that do satisfy the Gauss constraint.

Under a gauge transformation

$$A_i \to \mathcal{U}_\lambda A_i \mathcal{U}_\lambda^{-1} + i\mathcal{U}_\lambda \partial_i \mathcal{U}_\lambda^{-1} \tag{11.89}$$

with $\mathcal{U}_\lambda = e^{i\lambda^a T^a}$ where $\lambda(\vec{x})$ is independent of t. Defining the corresponding conserved charge

$$Q_\lambda = \int d^3x \dot{A}_i^a \left(\partial_i\lambda^a + F^{abc}A_i^b\lambda^c\right)$$

$$= -\int d^3x \lambda^a(\vec{x})(D_iE_i)^a \tag{11.90}$$

integrating by parts and assuming $\lambda^a \to 0$ for $|\vec{x}| \to \infty$. Then the gauge transformation can be effected by Q_λ as

$$A_i\vec{x},t) \to A_i'\vec{x},t) = e^{-iQ_\lambda}A_i\vec{x},t)e^{iQ_\lambda}$$
$$= A_i\vec{x},t) - i[Q_\lambda, A_i\vec{x},t)] \quad \text{for infinitesimal } \lambda$$
$$= A_i\vec{x},t) - i\int d^3y$$
$$\times \left[\dot{A}_i^a(\vec{y},t)\left(\partial_i\lambda^a(\vec{y},t) + f^{abc}A_i^b(\vec{y},t)\lambda^c(\vec{y},t)\right), A_i(\vec{x},t)\right]$$
$$= A_i\vec{x},t) - (D_i\lambda)^a\vec{x},t)T^a \tag{11.91}$$

using the equal time canonical commutator $[\dot{A}_i^a(\vec{y},t)), A_j^b(\vec{x},t))] = \delta^{ab}\delta^3(\vec{x}-\vec{y})$ and that the time variable in the integral expression for Q_λ can be chosen arbitrarily since it is in fact time-independent and here conveniently chosen equal to the time variable t in $A_i(\vec{x},t)$. Thus Q_λ generates the infinitesimal gauge transformation corresponding to λ, and physical states should be invariant under the action of this gauge transformation, *i.e.*

$$e^{iQ_\lambda}|\Psi\rangle = |\Psi\rangle \tag{11.92}$$

for λ falling off sufficiently fast as $|\vec{x}| \to \infty$. But

$$e^{iQ_\lambda}|A_i^a(x)\rangle = |\mathcal{U}_\lambda(A_i + i\partial_i)\mathcal{U}_\lambda^{-1}\rangle, \tag{11.93}$$

where $\mathcal{U}_\lambda = e^{i\lambda^a(\vec{x})T^a}$. Therefore, a physical state will be obtained if we sum over all states that are related by a gauge transformation

$$|A_i(\vec{x})\rangle_{physical} = \int \mathcal{D}\lambda'^a(\vec{x})e^{iQ_{\lambda'}}|A_i^a(x)\rangle \tag{11.94}$$

integrating over a dummy field variable λ'. This is obvious since the integration measure is invariant under translation, hence

$$e^{iQ_\lambda}|A_i(\vec{x})\rangle_{physical} = e^{iQ_\lambda}\int \mathcal{D}\lambda'^a(\vec{x})e^{iQ_{\lambda'}}|A_i^a(x)\rangle = \int \mathcal{D}\lambda'^a(\vec{x})e^{iQ_{\lambda+\lambda'}}|A_i^a(x)\rangle$$

$$= \int \mathcal{D}\lambda'^a(\vec{x})e^{iQ_{\lambda'}}|A_i^a(x)\rangle = |A_i(\vec{x})\rangle_{physical}. \tag{11.95}$$

Now what are the possible classical ground-state configurations? Such configurations must have $\mathcal{H} = 0$. This requires $E_i^a = 0$ and $B_i^a = 0$. A vanishing magnetic field means that the gauge field is pure gauge $A_i = g\partial_i g^{-1}$ and a vanishing electric field means $\dot{A}_i = 0$, which implies $\dot{g} = 0$. Thus for a classical vacuum the gauge field must be $A_i = g(\vec{x})\partial_i(g(\vec{x}))^{-1}$, and then we must implement gauge invariance as in Equation (11.94). Writing a corresponding state as $|g(\vec{x})\rangle$, for a gauge transformation

$$|g(\vec{x})\rangle \to e^{iQ_\lambda}|g(\vec{x})\rangle = |\mathcal{U}_\lambda g(\vec{x})\rangle = |\tilde{g}(\vec{x})\rangle. \tag{11.96}$$

Since $\lambda \to 0$ for $|\vec{x}| \to \infty$, $\tilde{g}$ and g must be homotopically equivalent, *i.e.* $\mathcal{U}_\lambda$ is homotopically trivial (evidently, just switch $\lambda \to 0$). Hence a potential vacuum state $|0\rangle$ is given by

$$|0\rangle = \sum_{g \,\in\, \text{one homotopy class}} |g(\vec{x})\rangle. \tag{11.97}$$

Without loss of generality, for the state $|0\rangle$ we choose the equivalence class corresponding to all gauge group elements that are homotopically trivial, *i.e.* in the same homotopy class as the constant, identity gauge transformation $g = \mathbb{I}$ and are generated by multiplication by $\mathcal{U}_\lambda$.

$$|0\rangle = \sum_{g \,\in\, \text{trivial homotopy class}} |g(\vec{x})\rangle. \tag{11.98}$$

But what if we define a different vacuum, obtained from $|\bar{g}(\vec{x})\rangle$

$$\overline{|0\rangle} = \sum_{g \,\in\, \text{ homotopy class of } \bar{g}} |g(\vec{x})\rangle \tag{11.99}$$

for some other gauge group element $\bar{g}$ which is not in the trivial homotopy class. $\bar{g}$ must go to identity at infinity. If $\bar{g}$ does not go to identity at infinity, and instead goes to some other constant gauge group element, g_0, then such a state is irrelevant. The matrix element between the state so defined and the state $|0\rangle$ must necessarily vanish since we must integrate over configurations, in Equation (11.87), which are spatially constant at infinity but change in time from $\mathbb{I}$ to g_0. Such configurations will have a non-zero $\dot{A}_i$ over an infinite spatial volume (at infinity), for which the action is infinite and consequently the transition amplitude vanishes. Thus $\bar{g}$ defines an element of the homotopy classes that need not be the trivial class. Evidently the state $\overline{|0\rangle}$ is also a vacuum state; the corresponding classical field configurations that we integrate over $A_i = g\partial_i g^{-1}$ in Equation (11.94), are all of zero energy.

All gauge group elements that we are considering here satisfy $\lim_{|\vec{x}|\to\infty} g(\vec{x}) \to \mathbb{I}$. Thus, topologically, all gauge group elements are defined on $\mathbf{R}^3$ with the point at infinity identified, topologically S^3. Each $g(\vec{x})$ defines a mapping from $S^3 \to G$, which fall into the homotopy classes of $\Pi_3(G) = \mathbb{Z}$. We can index the homotopy classes by an integer n,

$$g^{(0)} = \mathbb{I}$$
$$g^{(1)} = e^{i\pi \frac{\vec{x} \cdot \vec{\sigma}}{(|\vec{x}|^2 + \lambda^2)^{1/2}}}$$
$$g^{(n)} = \left(g^{(1)}\right)^n. \tag{11.100}$$

Correspondingly, we can enumerate the classical vacua with the integer n

$$|n\rangle = \sum_{g \text{ of winding number } n} |g\rangle \tag{11.101}$$

and the winding number is given by the formula

$$n = -\frac{1}{24\pi^2} \int d^3 x \epsilon_{ijk} Tr\left(g\partial_i g^{-1} g\partial_j g^{-1} g\partial_k g^{-1}\right). \tag{11.102}$$

If we denote by $\mathcal{R}$, the operator that implements the gauge transformation $g^{(1)}$, then

$$\mathcal{R}|n\rangle = |n+1\rangle. \tag{11.103}$$

Note that $g^{(1)} \neq e^{i\lambda^a T^a}$ with $\lim_{|\vec{x}| \to \infty} \lambda^a \to 0$, hence gauge invariance does not impose that the vacuum state be invariant under action of $\mathcal{R}$. But physically we would imagine that gauge invariance would require at least that

$$\mathcal{R}|\Psi\rangle = e^{i\theta}|\Psi\rangle, \tag{11.104}$$

since we cannot physically measure an overall phase factor. A vacuum state that satisfies Equation (11.104) is called a theta vacuum, and is denoted by $|\theta\rangle$.

There is no physical principle that can predict θ. However, θ must be time-independent since $[\mathcal{H}, \mathcal{R}] = 0$. θ must also be gauge-invariant. We say that θ labels the superselection sectors of the Hilbert space and the Hamiltonian is diagonal, block by block, for each superselection sector indexed by θ. We can explicitly construct the state labelled by θ by a simple Fourier sum

$$|\theta\rangle = \sum_{n=-\infty}^{\infty} e^{in\theta}|n\rangle. \tag{11.105}$$

These are the physical vacua of QCD, gauge invariant under trivial gauge transformations, and invariant up to an overall phase under topologically non-trivial gauge transformations. In the next section we will see how instantons give rise to these theta vacua.

11.4.1 Instantons: Specifics

In this section we will complete the specifics of the instanton solutions, some of which we have already used. The solution from Equation (11.75) is given by the gauge field configuration for gauge group $SU(2)$ (we will put all indices down for convenience)

$$A_\mu^{\alpha\beta} = \frac{1}{r^2 + \lambda^2}(x_\alpha \delta_{\mu\beta} - x_\beta \delta_{\mu\alpha}). \tag{11.106}$$

The corresponding field strength is

$$F_{\mu\nu}^{\alpha\beta} = \frac{r^2 + 2\lambda^2}{(r^2 + \lambda^2)^2}(\delta_{\mu\alpha}\delta_{\nu\beta} - \delta_{\mu\beta}\delta_{\nu\alpha}). \tag{11.107}$$

Equivalently in matrix form

$$A_\mu = \frac{r^2}{r^2 + \lambda^2}g^{(1)}\partial_\mu\left(g^{(1)}\right)^{-1}. \tag{11.108}$$

Obviously, as $r \to \infty$, the field strength vanishes quadratically as $\sim 1/r^2$. Thus the action of the instanton is located in an essentially compact region of Euclidean spacetime. This was the reason for the name "instanton"; if we scan up through Euclidean time, there is nothing at the beginning, then, for an instant, the instantons turn on and off in a localized spatial region, and then, again, there is nothing.

The instanton solution is rotationally invariant when compensated by a global gauge transformation. This is evident for the field strength $F_{\mu\nu}^{\alpha\beta}$. For the gauge field we must have the same. A rotation is defined by

$$x_\mu \to x'_\mu = R_{\mu\nu}x_\nu \tag{11.109}$$

with

$$R_{\mu\nu}R_{\mu\sigma} = \delta_{\nu\sigma}, \tag{11.110}$$

since $x_\mu x_\mu \to R_{\mu\nu}x_\nu R_{\mu\sigma}x_\sigma = R_{\mu\nu}R_{\mu\sigma}x_\nu x_\sigma$. For infinitesimal transformations, we have $R_{\mu\nu} = \delta_{\mu\nu} + \Lambda_{\mu\nu}$, where $\Lambda_{\mu\nu}$ is infinitesimal. Then

$$x_\mu \to x'_\mu = x_\mu + \Lambda_{\mu\nu}x_\nu \tag{11.111}$$

and

$$(\delta_{\mu\nu} + \Lambda_{\mu\nu})(\delta_{\mu\sigma} + \Lambda_{\mu\sigma}) = \delta_{\nu\sigma} + \Lambda_{\sigma\nu} + \Lambda_{\nu\sigma} = \delta_{\nu\sigma}, \tag{11.112}$$

which requires

$$\Lambda_{\sigma\nu} + \Lambda_{\nu\sigma} = 0 \tag{11.113}$$

or equivalently, $\Lambda_{\sigma\nu} = -\Lambda_{\nu\sigma}$. Then the gauge field transforms as

$$A_\mu(x) \to A'_\mu(x') = A_\mu(x) + (\Lambda_{\mu\nu} + \delta_{\mu\nu}\Lambda_{\sigma\tau}x_\sigma\partial_\tau)A_\nu(x) \tag{11.114}$$

thus the gauge field $A_\mu(x) = \frac{r^2}{r^2+\lambda^2}g\partial_\mu g^{-1}$ will transform as

$$A_\mu(x) \to A'_\mu(x') = A_\mu(x) + \frac{r^2}{r^2 + \lambda^2}\left(\Lambda_{\mu\nu}g\partial_\nu g^{-1} + (\Lambda_{\sigma\tau}x_\sigma\partial_\tau g)\partial_\mu g^{-1}\right.$$
$$\left. + g\partial_\mu(\Lambda_{\sigma\tau}x_\sigma\partial_\tau g^{-1}) - g\Lambda_{\mu\tau}\partial_\tau g^{-1}\right)$$
$$= \frac{r^2}{r^2 + \lambda^2}\left(g(x')\partial_\mu g^{-1}(x')\right) \tag{11.115}$$

using that r is invariant and $g(x') = g(x) + \Lambda_{\sigma\tau}x_\sigma\partial_\tau g$ to first order. Then explicitly with $g(x) = g^1(x) = \frac{1}{r}(x^4 + ix^i\sigma^i)$

$$g(x') = \frac{1}{r}(x^4 + ix^i\sigma^i) + \frac{1}{r}\Lambda_{\sigma\tau}x_\sigma\partial_\tau(x^4 + ix^i\sigma^i). \qquad (11.116)$$

Then with $\Lambda_{i4} = \lambda_i$ and $\Lambda_{ij} = \epsilon_{ijk}\alpha_k$ we get

$$g(x') = g(x) + \frac{1}{r}\lambda_i(x_i\partial_4 - x_4\partial_i)(x_4 + ix_j\sigma_j)$$
$$+ \frac{1}{r}\epsilon_{ijk}\alpha_k x_i\partial_j ix_l\sigma_l$$
$$= g(x) + \frac{1}{r}\lambda_i(x_i - ix_4\sigma_i) + \frac{i}{r}\epsilon_{ijk}\alpha_k x_i\sigma_j$$
$$= g(x) + \frac{1}{r}(x_4(-i\lambda_i\sigma_i) + x_i(\lambda_i + i\epsilon_{ijk}\sigma_j\alpha_k))$$
$$= \frac{1}{r}(x_4(1 - i\lambda_i\sigma_i) + x_i(i\sigma_i + \lambda_i + i\epsilon_{ijk}\sigma_j\alpha_k))$$
$$= \frac{1}{r}(1 - i\gamma_i\sigma_i)(x_4 + ix_j\sigma_j)(1 + i\beta_k\sigma_k)$$
$$(11.117)$$

where $\lambda_i = \gamma_i - \beta_i$ and $\alpha_i = \gamma_i + \beta_i$, since, continuing the algebra to first order we get

$$g(x') = = \frac{1}{r}(x_4(1 - i(\gamma_i - \beta_i)\sigma_i) + x_i(i\sigma_i + \gamma_j\sigma_j\sigma_i - \beta_j\sigma_i\sigma_j))$$
$$= \frac{1}{r}(x_4(1 - i(\gamma_i - \beta_i)\sigma_i) + x_i(i\sigma_i + (\gamma_i - \beta_i) + i\epsilon_{ijk}(\gamma_j + \beta_j)\sigma_k)) \quad (11.118)$$

confirming

$$g(x') = (1 - i\gamma_i\sigma_i)g(x)(1 + i\beta_i\sigma_i) = a^{-1}g(x)b, \qquad (11.119)$$

where $a = (1 + \gamma_i\sigma_i)$ while $b = (1 + i\beta_i\sigma_i)$ to first order. Then from Equation (11.115)

$$A'_\mu(x') = \frac{r^2}{r^2 + \lambda^2}a^{-1}gb\partial_\mu(b^{-1}g^{-1}a) = a^{-1}A_\mu(x)a. \qquad (11.120)$$

Thus the solution is clearly invariant under an arbitrary choice of b. Thus the instanton is invariant under an arbitrary choice of b, but it is not invariant under a.

This should give rise to three zero modes, it does, but in a slightly more complicated way. The important point is that rotations can be compensated by global gauge transformations. The rotation group $SO(4) = SO_a(3) \times SO_b(3)$ is explicitly broken to the $SO_a(3)$ subgroup. The instanton is invariant under the $SO_b(3)$ subgroup and it does not give rise to new solutions or equivalently to zero modes. In principle, this subgroup can be used to characterize the representations under which the physical states of the theory transform, exactly as, for example, the invariance of a physical system under spatial rotations tells us that the physical states of the system must transform according to representations of the rotation group. The broken subgroup is $SO_a(3)$, which should give rise to new solutions and zero modes, but its transformation can be exactly compensated by a transformation of the group of global gauge transformation $SO_{gl.}(r)$, which is also $SU(2) = SO_{gl.}(3)$. Under a global gauge transformation

$$A'_\mu(x') \to hA'_\mu(x')\, h^{-1} = ha^{-1}A_\mu(x)\, ah^{-1} \tag{11.121}$$

hence $h = a$ is a symmetry of the solution. Therefore, the symmetry group $SO_{gl.}(3) \times SO_a(3)$ is in fact broken to the diagonal subgroup $SO_d(3)$, corresponding to $h = a$, which remains a symmetry of the instantons. The anti-diagonal subgroup $SO_{ad}(3)$, with $h = a^{-1}$, is broken by the instanton configuration, and gives rise to exactly three zero modes. Thus the rotation group is broken to $SO_a(3)$

$$SO(4) = SO_a(3) \times SO_b(3) \to SO_a(3), \tag{11.122}$$

while the group of global gauge transformations $SU(2) = SO_{gl.}(3)$ is mixed with the rotation group

$$SO_{gl.}(3) \times SO_a(3) = SO_d(3) \times SO_{ad}(3) \tag{11.123}$$

and the diagonal subgroup remains an symmetry of the solution, while the anti-diagonal subgroup gives rise to three zero modes.

11.4.2 Transitions Between Vacua

The instantons are perfectly suited to describing quantum tunnelling transitions between the $|n\rangle$ vacua. The solution can be put into the gauge $A_4 = 0$ by the straightforward gauge transformation $A_\mu \to h(A_\mu + \partial_\mu)h^{-1}$ with

$$h = \mathcal{P}\left(e^{i\int_{-\infty}^{x_4} dx'_4 A_4(x'_4)}\right). \tag{11.124}$$

Then at $\tau = -\infty$ the gauge field is

$$A_i|_{\tau=-\infty} = 0 \tag{11.125}$$

but at $\tau = \infty$ we have

$$A_i|_{\tau=\infty} = g\partial_i g^{-1}, \tag{11.126}$$

where

$$g = e^{-i\pi\frac{\vec{x}\cdot\vec{\sigma}}{(x^2+\lambda^2)^{1/2}}}. \tag{11.127}$$

The gauge field is given in slightly different notation by 't Hooft [112],

$$A^a_\mu = 2\frac{x_\nu}{x^2+\lambda^2}\eta_{a\mu\nu}, \tag{11.128}$$

where $\eta_{a\mu\nu}$ is the 't Hooft tensor

$$\eta_{a\mu\nu} = \epsilon_{4a\mu\nu} + \frac{1}{2}\epsilon_{abc}\epsilon_{bc\mu\nu} \qquad (11.129)$$

with corresponding field strength

$$F_{\mu\nu}^a = 4\frac{\lambda^2}{(x^2+\lambda^2)^2}\eta_{a\mu\nu}. \qquad (11.130)$$

This configuration corresponds to a change in the winding number between $\tau = -\infty$ and $\tau = \infty$

$$\begin{aligned}
\Delta n &= \frac{-1}{24\pi^2}\int d^3x\,\epsilon_{ijk}Tr\left(g\partial_i g^{-1}g\partial_j g^{-1}g\partial_k g^{-1}\right)\Big|_{\tau=-\infty}^{\tau=\infty} \\
&= \frac{1}{32\pi^2}\int d^3x\,G_4\Big|_{\tau=-\infty}^{\tau=\infty} \\
&= \frac{1}{32\pi^2}\int d^4x\,\partial_\tau G_4 \\
&= \frac{1}{32\pi^2}\int d^4x\,(\partial_\mu G_\mu - \partial_i G_i) \qquad (11.131)
\end{aligned}$$

where G_μ was defined in Equation (11.27). The spatial components G_i are, using $A_4 = 0$,

$$G_i \sim \epsilon_{i\mu0\lambda}(A_\mu\partial_\nu A_\lambda + \frac{2}{3}A_\mu A_\nu A_\lambda) = \epsilon_{i\mu0\lambda}A_\mu\dot{A}_\lambda = \epsilon_{ijk}A_j\dot{A}_k, \qquad (11.132)$$

which vanish as $|\vec{x}| \to \infty$ since the electric field $\dot{A}_k = E_k \to 0$ so that the subtracted spatial divergence gives no contribution from the surface at infinity. Thus we find

$$\Delta n = \frac{1}{32\pi^2}\int d^4x\,\partial_\mu G_\mu = \frac{1}{32\pi^2}\int d^4x\,\langle F_{\mu\nu}\tilde{F}_{\mu\nu}\rangle = 1. \qquad (11.133)$$

Thus the instanton mediates transitions with the vacua $|0\rangle \to |1\rangle$.

11.5 Instantons and Confinement

We will now consider quantum corrections that come from the Gaussian functional integral that should be performed about the instanton configuration. We must first extract the zero modes. For the single instanton, there exist, in fact, eight. In principle the amplitude $\langle 1|e^{-HT}|0\rangle$ is given by

$$\langle 1|e^{-HT}|0\rangle \equiv Z(T) = e^{-S_0}\int \mathcal{D}(Q_\mu^a)e^{-\frac{1}{2}\int d^4x\,Q\cdot\left(\frac{\delta^2}{\delta Q^2}\mathcal{L}(A_\mu^a)\right)\cdot Q}, \qquad (11.134)$$

where Q_μ^a is the fluctuation that gives rise to the quantum corrections. The zero modes coming from translations and scale transformations can be eliminated by using a Faddeev–Popov method. We insert one into the integral

$$1 = S_0^5 \int d^4 R \delta^4 \left(\int d^4 x \left(\mathcal{L}(A_\nu^a(x))(x_\mu - R_\mu) \right) \right) \int_0^\infty d(\lambda^2)$$

$$\delta \left(\int d^4 x (\mathcal{L}(A_\nu^a(x))((x-R)^2 - \lambda^2)) \right). \tag{11.135}$$

The delta functions choose the values of R and λ as

$$\delta^4 \left(\int d^4 x \left(\mathcal{L}(A_\nu^a(x))(x_\mu - R_\mu) \right) \right) = \delta^4 \left(\left(\int d^4 x \mathcal{L}(A_\nu^a(x)) \right) (\bar{R}_\mu - R_\mu) + \cdots \right)$$

$$= \frac{\delta^4(R_\mu - \bar{R}_\mu)}{\left(\int d^4 x \mathcal{L}(A_\nu^a(x)) \right)^4} = \frac{\delta^4(R_\mu - \bar{R}_\mu)}{S_0^4}$$

$$\tag{11.136}$$

with the obvious definition

$$\bar{R}_\mu = \frac{\int d^4 x \left(\mathcal{L}(A_\nu^a(x)) x_\mu \right)}{\int d^4 x \mathcal{L}(x)}. \tag{11.137}$$

Furthermore,

$$\delta \left(\int d^4 x (\mathcal{L}(A_\nu^a(x))((x-R)^2 - \lambda^2)) \right) = \frac{\delta(\lambda^2 - \bar{\lambda}^2)}{\int d^4 x \mathcal{L}(x)} = \frac{\delta(\lambda^2 - \bar{\lambda}^2)}{S_0} \tag{11.138}$$

with

$$\bar{\lambda}^2 = \frac{\int d^4 x \mathcal{L}(A_\nu^a(x))(x-R)^2}{\int d^4 x \mathcal{L}(x)} \tag{11.139}$$

and here R_μ could be replaced with $\bar{R}_\mu$ because of the first delta function. Evidently, $\bar{R}$ depends on what gauge field $A_\nu^a(x)$ is chosen: it should correspond to an instanton, and it contains the data on where the instanton is and its scale, λ. Then starting with $Z(T)$ slightly differently

$$Z(T) = S_0^5 \int d(\lambda^2) \int d^4 R \int \mathcal{D}(Q_\mu^a) e^{-S_E} \times$$

$$\times \delta^4 \left(\int d^4 x \left(\mathcal{L}(Q_\mu^a(x))(x_\mu - R_\mu) \right) \right) \delta \left(\int d^4 x (\mathcal{L}(Q_\mu^a(x))((x-R)^2 - \lambda^2)) \right).$$

$$\tag{11.140}$$

First we perform a translation and a conformal scaling

$$x_\mu \to x'_\mu = \lambda x_\mu + R_\mu$$
$$Q_\mu^a(x_\nu) \to \lambda Q_\mu^a(x'_\nu). \tag{11.141}$$

The action is invariant under a translation and conformal scaling (actually under all special conformal transformations) with

$$\mathcal{L}(x) = -\frac{1}{4g^2} Tr \left(\partial_\mu Q_\nu(x) - \partial_\nu Q_\mu(x) + [Q_\mu(x), Q_\nu(x)] \right)^2 \tag{11.142}$$

we have

$$d^4x\mathcal{L}(\lambda Q(\lambda x - R)) = -\lambda^{-4}d^4x'Tr\left(\lambda\partial'_\mu\lambda Q_\nu(x') - \lambda\partial'_\nu\lambda Q_\mu(x')\right.$$
$$+\lambda^2[Q_\mu(x'),Q_\nu(x')])^2/4g^2$$
$$= -d^4x'Tr\left(\partial'_\mu Q_\nu(x') - \partial'_\nu Q_\mu(x') + [Q_\mu(x'),Q_\nu(x')]\right)^2/4g^2$$
$$= d^4x'\mathcal{L}(x') \tag{11.143}$$

and we will simply rename $x' \to x$. Then we will change the functional integration variable

$$Q^a_\mu \to (A^a_\mu + Q^a_\mu)(x_\nu), \tag{11.144}$$

where A^a_μ is a classical field, an instanton solution. Expanding the action to second order in Q^a_μ (the first-order variation vanishes after one integration by parts, as A^a_μ satisfies the equations of motion), we get

$$Z(T) = S_0^5 \int d(\lambda^2) \int d^4R \int \mathcal{D}(Q^a_\mu)e^{-\int d^4x\mathcal{L}(A^a_\mu(x))+\frac{1}{2}Q\cdot\frac{\delta^2}{\delta Q^2}\mathcal{L}(A^a_\mu(x))\cdot Q} \times$$
$$\times\delta^4\left(\int d^4x\left(\mathcal{L}(A_\mu + Q_\mu)(\lambda x_\mu)\right)\right)\delta\left(\int d^4x(\mathcal{L}(A_\mu + Q_\mu)(\lambda^2 x^2 - \lambda^2))\right)$$
$$= S_0^5 \int \frac{d(\lambda^2)}{\lambda^2} \int \frac{d^4R}{\lambda^4} \int \mathcal{D}(Q^a_\mu)e^{-\int d^4x\mathcal{L}(A^a_\mu(x))+\frac{1}{2}Q\cdot\frac{\delta^2}{\delta Q^2}\mathcal{L}(A^a_\mu(x))\cdot Q} \times$$
$$\times\delta^4\left(\int d^4x\left(\mathcal{L}(A_\mu + Q_\mu)(x_\mu)\right)\right)\delta\left(\int d^4x(\mathcal{L}(A_\mu + Q_\mu)(x^2 - 1))\right). \tag{11.145}$$

Now we choose the instanton configuration to be centred on the origin and of unit scale size, *i.e.*

$$\int d^4x\mathcal{L}(A_\nu)x_\mu = \int d^4x\mathcal{L}(A_\nu)(x^2 - 1) = 0. \tag{11.146}$$

Then expanding in the first-order Taylor expansion of the action in Q_ν, and using the notation D^A_σ to be the covariant derivative with respect to the classical field A_μ, $D^A_\sigma\cdot = \partial_\sigma\cdot + [A_\sigma,\cdot]$ and the corresponding field strength $F^A_{\sigma\tau}$, we have

$$\mathcal{L}(A_\nu + Q_\nu) = \mathcal{L}(A_\nu) - \frac{1}{g^2}Tr\left(F^A_{\sigma\tau}D^A_\sigma Q_\tau\right) \tag{11.147}$$

we get in the delta functions

$$\delta^4\left(\int d^4x\frac{1}{g^2}Tr\left(F^A_{\sigma\tau}D^A_\sigma Q_\tau x_\mu\right)\right) = \delta^4\left(\int d^4x\frac{1}{g^2}Tr\left(F^A_{\mu\tau}Q_\tau\right)\right) \tag{11.148}$$

and

$$\delta\left(\int d^4x\frac{1}{g^2}Tr\left(F^A_{\sigma\tau}D^A_\sigma Q_\tau(x^2 - 1)\right)\right) = \delta\left(-2\int d^4x\frac{1}{g^2}Tr\left(F^A_{\sigma\tau}Q_\tau x_\sigma\right)\right). \tag{11.149}$$

Thus the delta functions impose the conditions

$$\int d^4x\, Tr\left(F^A_{\mu\tau}Q_\tau\right) = 0$$

$$\int d^4x\, Tr\left(x_\sigma F^A_{\sigma\tau}Q_\tau\right) = 0. \tag{11.150}$$

But these conditions are exactly the conditions that the quantum fluctuation Q_ν be orthogonal to the zero modes corresponding to translations and scale transformations, respectively. Indeed, for translations we simply transform the classical solution A_τ by the broken symmetry, translation in the x_σ direction generated by a simple partial derivative in that direction. However, this is not gauge-invariant, hence we also perform a gauge transformation, $\delta_\sigma A_\tau = -D^A_\tau(A_\sigma)$, for an infinitesimal gauge transformation, to give

$$\psi^{tr.}_{\sigma\tau} = \partial_\sigma A_\tau - D^A_\tau(A_\sigma) = F^A_{\sigma\tau}. \tag{11.151}$$

The normalized zero mode is $\hat{\psi}^{tr.}_{\sigma\tau} = \frac{1}{\sqrt{N^{tr.}}}\psi^{tr.}_{\sigma\tau}$, with $N^{tr.}$ defined by

$$N^{tr.} = -\int d^4x\, Tr(\psi^{tr.}_{\sigma\tau})^2 = -\int d^4x \sum_{\substack{\tau \\ \sigma \text{ fixed}}} Tr\left(F^A_{\sigma\tau}F^A_{\sigma\tau}\right)$$

$$= -\frac{1}{4}\int d^4x \sum_{\sigma,\tau} Tr\left(F^A_{\sigma\tau}F^A_{\sigma\tau}\right) = g^2 S_0. \tag{11.152}$$

We wish to keep track of powers of g and hence we note that the normalization factor does not have any powers of g since $S_0 \sim 1/g^2$. For the scale transformation, the infinitesimal generator is $(1+x_\sigma\partial_\sigma)A_\tau$ and then the infinitesimal variation of the gauge field, made gauge-invariant, gives the zero mode

$$\psi^{sc.}_\tau = (1+x_\sigma\partial_\sigma)A_\tau - D^A_\tau(x_\sigma A_\sigma) = x_\sigma F^A_{\sigma\tau}. \tag{11.153}$$

A simple analysis also shows that the normalized zero mode, $\hat{\psi}^{sc.}_\tau = \psi^{sc.}_\tau/\sqrt{N^{sc.}}$, will not have have any net powers of g.

The delta functions impose that the integration over Q should be restricted to the function space that is orthogonal to these zero modes. But in writing Q in a sum over normal modes there is a subtlety involved. We should add a factor of g in the expansion

$$Q_\tau = C^{tr.}_\sigma g\frac{F^A_{\sigma\tau}}{\sqrt{N^{tr.}}} + C^{sc.}g\frac{x_\sigma F^A_{\sigma\tau}}{\sqrt{N^{sc.}}} + \sum_\xi C^\xi g\hat{\psi}^\xi_\tau, \tag{11.154}$$

where $\hat{\psi}^\xi_\tau$ are the non-zero modes. The integration measure is the usual infinite product

$$\mathcal{D}(Q) = \prod_\sigma \frac{dC^{tr.}_\sigma}{\sqrt{2\pi}} \frac{dC^{sc.}}{\sqrt{2\pi}} \prod_\xi \frac{dC^\xi}{\sqrt{2\pi}} \tag{11.155}$$

as long as the same conventions are followed in the numerator and the denominator for the functional integral. The reason for the extra factor of g comes from the exponent in the integrand. We examine the exponent more carefully,

$$\int d^4x \frac{1}{2} Q \cdot \frac{\delta^2}{\delta Q^2} \mathcal{L}(A_\mu^a(x)) \cdot Q = -\frac{1}{g^2} \frac{1}{2} \int d^4x Tr(Q_\mu (D_\nu^A D_\mu^A - D_\sigma^A D_\sigma^A \delta_{\mu\nu} - F_{\mu\nu}^A) Q_\nu).$$
(11.156)

The using the expansion in Equation (11.154) with its extra factor of g, and where $\hat{\psi}_\mu^\xi$ explicitly are the normalized non-zero eigenfunctions of the operator of the second-order variation $(D_\nu^A D_\mu^A - D_\sigma^A D_\sigma^A \delta_{\mu\nu} - F_{\mu\nu}^A)$ with eigenvalues ϵ_ξ, we get

$$\int d^4x \frac{1}{2} Q \cdot \frac{\delta^2}{\delta Q^2} \mathcal{L}(A_\mu^a(x)) \cdot Q = \frac{1}{2} \sum_\xi \epsilon_\xi \left(C^\xi\right)^2.$$
(11.157)

Notice that ϵ_ξ contains no powers of g, nor does the determinant that ensues after integrating over the C^ξ and the zero modes simply drop out in the exponent and the delta functions control the integration over their coefficients. Then the delta function for translations gives

$$\delta^4 \left(\frac{1}{g^2} \int d^4x F_{\sigma\tau}^A g \frac{F_{\sigma\tau}^A}{\sqrt{N^{tr.}}} C_\sigma^{tr.}\right) = \delta^4 \left(\frac{\sqrt{N^{tr.}}}{g} C_\sigma^{tr.}\right) = \left(\frac{g}{\sqrt{N^{tr.}}}\right)^4 \delta^4(C_\sigma^{tr.}).$$
(11.158)

The delta function for scale transformation also gives a factor of $\frac{g}{\sqrt{N^{sc.}}}$, giving an overall factor of g^5. Thus writing a prime as usual to indicate the restriction, we get

$$Z(T) = e^{-S_0} S_0^5 g^5 \frac{1}{(\sqrt{N^{tr.}})^4} \frac{1}{(\sqrt{N^{sc.}})} \int \frac{d(\lambda^2)}{\lambda^2} \int \frac{d^4R}{\lambda^4} \int$$
$$\mathcal{D}'(Q_\mu^a) e^{-\int d^4x \frac{1}{2} Q \cdot \frac{\delta^2}{\delta Q^2} \mathcal{L}(A_\mu^a(x)) \cdot Q}$$
(11.159)

with

$$S_0 = \frac{8\pi^2}{g^2}.$$
(11.160)

The overall power of g is then g^{-5} due to the five zero modes that we have treated. There are in fact more zero modes associated with the global gauge transformations and the rotation group [112, 98, 24, 12, 65]. As we have seen for $SU(2)$, the diagonal subgroup of these two is unbroken, but the anti-diagonal subgroup is broken with three broken generators. We will not analyse these explicitly, it will suffice to say that they give exactly another power of g^{-3}, for the gauge group $SU(2)$ giving the total, overall factor of g to be g^{-8}. In the case of interest, QCD, the gauge group is $SU(3)$ with eight generators. We imagine putting the instanton solution in an $SU(2)$ subgroup of $SU(3)$, but then one generator always commutes with the $SU(2)$ subgroup. For example, if we generate the subgroup with $\lambda_1, \lambda_2, \lambda_3$, then λ_8 commutes with these three matrices. Thus we find there are only seven broken generators. The $SU(2)$ subgroup mixes in an identical way with the rotation group as in the case when the entire group was $SU(2)$. Thus the upshot is there are seven additional zero modes, which give a factor of g^{-7} and an overall factor of g^{-12}.

We have computed the contribution of a single instanton to the transition between the two vacua $|0\rangle$ and $|1\rangle$. As usual, multi-instanton configurations are negligible except for those corresponding to well-separated single instantons.

$$
Z(|0\rangle \to |1\rangle) = \sum_{n,\bar{n}=-\infty}^{\infty} \delta_{n-\bar{n},1} \frac{1}{n!\bar{n}!} \left(\int d^4R \int_0^{\infty} \frac{d\lambda}{\lambda^5} g^{-12} e^{-\frac{8\pi^2}{g^2}} K \right)^{n+\bar{n}}
$$

$$
= \sum_{n,\bar{n}=-\infty}^{\infty} \delta_{n-\bar{n},1} \frac{1}{n!\bar{n}!} \left(VT \int_0^{\infty} \frac{d\lambda}{\lambda^5} g^{-12} e^{-\frac{8\pi^2}{g^2}} K \right)^{n+\bar{n}} \quad (11.161)
$$

where K is the determinantal factor including various other normalization factors and constants independent of g. This result directly generalizes to the amplitude

$$
Z(|m\rangle \to |\tilde{m}\rangle) = \sum_{n,\bar{n}=-\infty}^{\infty} \delta_{n-\bar{n},m-\tilde{m}} \frac{1}{n!\bar{n}!} \left(VT \int_0^{\infty} \frac{d\lambda}{\lambda^5} g^{-12} e^{-\frac{8\pi^2}{g^2}} K \right)^{n+\bar{n}}
$$

$$
(11.162)
$$

and finally to

$$
Z(\theta) = \langle \theta | e^{-HT} | \theta \rangle
$$

$$
= \sum_{m,\tilde{m}=-\infty}^{\infty} e^{i\theta(m-\tilde{m})} \langle \tilde{m} | e^{-HT} | m \rangle
$$

$$
= \sum_{m,\tilde{m}-\infty}^{\infty} e^{i\theta(m-\tilde{m})} Z(|m\rangle \to |\tilde{m}\rangle)
$$

$$
= \left(\sum_{\tilde{m}=-\infty}^{\infty} \right) \exp \left(2\cos\theta VT \int_0^{\infty} \frac{d\lambda}{\lambda^5} g^{-12} e^{-\frac{8\pi^2}{g^2}} K \right). \quad (11.163)
$$

The infinite, constant prefactor is simply a consequence of the plane wave normalization of the theta vacuum states.

Then from Equation (11.65) and Equation (11.66) we get the energy of the ground state

$$
E(\theta)/V = -\cos\theta \int_0^{\infty} \frac{d\lambda}{\lambda^5} g^{-12} e^{-\frac{8\pi^2}{g^2}} K. \quad (11.164)
$$

We have purposely left the g-dependent factors inside the integration over λ for a reason, and we have absorbed all constant factors into K. This is because evaluation of the determinants requires renormalization of the coupling constant, and renormalization inserts a scale dependence into g and K. The infinite product of eigenvalues of the operator corresponding to the second variation of the action in the presence of the instanton, Equation (11.156), is not rendered finite when divided by the same infinite product but in the absence of the instanton. We have to add counterterms with infinite coefficients so that the divergences are absorbed. Adding the counterterm proportional to $\sim F_{\mu\nu}F_{\mu\nu}$ exactly renormalizes the value of the coupling constant g. However, the renormalization inserts a dimensionful mass scale, M, into the theory, which fixes the physical, finite, observed value of the coupling constant at that scale. The coupling constant obeys the equation

$$\frac{1}{g^2(\lambda)} = \frac{1}{g^2(M)} - \frac{11}{8\pi^2} \ln(\lambda M) + o(g^2), \qquad (11.165)$$

where the λ dependence comes from the simple fact that g is a dimensionless coupling constant, since the only dimensionful parameter that exists, apart from the renormalization scale M, is the instantons scale λ, the two must come together in the dimensionless combination λM. The factor $-\frac{11}{8\pi^2}$ is the famous result of asymptotic freedom for the beta function of QCD, which is a long, hard calculation in perturbation theory [57, 102], which we will not describe here. Asymptotic freedom means that as the scale λ gets smaller $\lambda \ll 1/M$, and the instanton size goes to zero, the coupling constant $g^2(\lambda)$ becomes smaller as $\ln \lambda M$ is negative and the right-hand side becomes larger. Indeed, replacing the solution Equation (11.165) in the expression for the energy gives

$$E(\theta)/V$$

$$= -\cos\theta \int_0^\infty \frac{d\lambda}{\lambda^5} \left(\frac{1}{g^2(M)} - \frac{11}{8\pi^2}\ln(\lambda M)\right)^6 e^{-8\pi^2\left(\frac{1}{g^2(M)} - \frac{11}{8\pi^2}\ln(\lambda M)\right)} K(\lambda M)$$

$$= -\cos\theta \int_0^\infty \frac{d\lambda}{\lambda^5} g^{-12}(M) e^{-\frac{8\pi^2}{g^2(M)}} \left(1 + o\left(g^2(M)\ln(\lambda M)\right)\right) e^{11\ln(\lambda M)} K(\lambda M)$$

$$= -\cos\theta \int_0^\infty d\lambda \lambda^6 M^{11} g^{-12}(M) e^{-\frac{8\pi^2}{g^2(M)}} \left(1 + o\left(g^2(M)\ln(\lambda M)\right)\right) K(\lambda M).$$

$$(11.166)$$

Thus for small λ, in the ultraviolet, the integral is perfectly convergent; however, in the infrared, as $\lambda \to \infty$ the integral is obviously divergent. Thus the integral is well-behaved in the region where we trust our calculations, when $g \to 0$, but does not make sense in the regions where $g \gg 1$, where we do not trust our calculations. Indeed, we expect new, non-perturbative (not instanton effects, which are also non-perturbative but only valid for small g) strong coupling effects to kick in as g becomes large, effects which we have made no pretence to be able to compute. Thus we stop the calculation at this point, content with the expectation that large coupling, confinement-related effects cure the behaviour of this integral.

11.6 Quarks in QCD

We will next consider the question of quarks in QCD. The quarks come in the fundamental representation of $SU(3)$, which is generated exactly by the 3×3

Gell-Mann matrices of Equation (11.6) and in six flavours, up, down, strange, charm, top and bottom, which we will denote by a label a, and correspond to Dirac fields $\psi_a(x)$. The colour index is suppressed but takes on three values, 1,2 and 3, thus the Dirac field is a three-component column for each flavour index. The Lagrangian density in Minkowski spacetime is then given by

$$\mathcal{L} = \frac{1}{4g^2}Tr[F_{\mu\nu}(x)F^{\mu\nu}(x)] + \sum_a \left(\bar{\psi}^a(x)i\gamma^\mu(\partial_\mu + A_\mu)\psi^a(x) - m^a\bar{\psi}^a(x)\psi^a(x)\right),$$

(11.167)

where the gauge field is a 3×3 anti-hermitean matrix, $A_\mu = iA_\mu^a\lambda_a$ $F_{\mu\nu} = \partial_\mu A_\nu - \partial_\nu A_\mu + [A_\mu, A_\nu]$ is also anti-hermitean. The γ^μ are the usual Dirac matrices. The masses m^α are quite small for the up and down quarks, less than 10 MeV. Thus the massless limit is a reasonably good approximation when considering processes that largely imply only the up and down quarks. This limit has a higher symmetry, called chiral symmetry which is spontaneously broken, and can be treated with chiral perturbation theory. The strange quark mass is a little more, around 95 MeV, but still within the purview chiral symmetry and chiral perturbation theory. The charm mass is about 1.3 GeV, the bottom mass is 4.2 GeV, and the top mass is 173 GeV. Neglecting these masses is not a good approximation. In what follows, we will restrict our considerations to the up and down quarks and neglect their masses, which is a rather good approximation. Then the fermionic part of the Lagrangian density is

$$\mathcal{L} = \sum_{a=\text{up,down}} \bar{\psi}^a(x)i\gamma_\mu(\partial_\mu + A_\mu)\psi^a(x).$$

(11.168)

The Lagrangian in this case has a symmetry $SU_L(2) \times SU_R(2)$, called chiral symmetry. The subscripts L and R correspond to independent $SU(2)$ transformations on the left-handed and right-handed components of the Dirac spinor. If we write the spinor fields as

$$\psi = \begin{pmatrix} \psi_u \\ \psi_d \end{pmatrix}$$

(11.169)

then the chiral transformation corresponds to

$$\psi \to e^{i\left(\frac{1-\gamma_5}{2}\right)\vec{\alpha}_L\cdot\vec{\sigma}} e^{i\left(\frac{1+\gamma_5}{2}\right)\vec{\alpha}_R\cdot\vec{\sigma}}\psi,$$

(11.170)

where $\vec{\alpha}_{L,R}$ are independent parameters of the two $SU(2)$ transformations and the $\vec{\sigma}$ are the Pauli matrices. The chiral projection operators, $\frac{1\pm\gamma_5}{2}$, project onto the left-handed and right-handed components of the Dirac spinor

$$\psi = \psi_L + \psi_R = \left(\frac{1-\gamma_5}{2}\right)\psi + \left(\frac{1+\gamma_5}{2}\right)\psi.$$

(11.171)

In the chiral representation of the Dirac matrices,

$$\gamma_5 = \begin{pmatrix} 1 & 0 & 0 & 0 \\ 0 & 1 & 0 & 0 \\ 0 & 0 & -1 & 0 \\ 0 & 0 & 0 & -1 \end{pmatrix} \tag{11.172}$$

so that

$$\frac{1-\gamma_5}{2} = \begin{pmatrix} 0 & 0 & 0 & 0 \\ 0 & 0 & 0 & 0 \\ 0 & 0 & 1 & 0 \\ 0 & 0 & 0 & 1 \end{pmatrix}, \quad \frac{1+\gamma_5}{2} = \begin{pmatrix} 1 & 0 & 0 & 0 \\ 0 & 1 & 0 & 0 \\ 0 & 0 & 0 & 0 \\ 0 & 0 & 0 & 0 \end{pmatrix} \tag{11.173}$$

The Lagrangian Equation (11.169) is also invariant under two $U(1)$ symmetries, $U_V(1) \times U_A(1)$,

$$\psi \rightarrow e^{i\theta} e^{i\gamma_5 \theta'} \psi \tag{11.174}$$

for independent parameters θ and θ'. The corresponding conserved currents, via Noether's theorem, are denoted

$$j^\mu = \bar{\psi}\gamma^\mu\psi \quad \text{and} \quad j_5^\mu = \bar{\psi}\gamma^\mu\gamma_5\psi. \tag{11.175}$$

The chiral symmetry group $SU_L(2) \times SU_R(2)$ is spontaneously broken to the diagonal subgroup $SU_D(2)$, which is identified as the isospin group, in this case with just two flavours, up and down. Due to this spontaneous symmetry-breaking, the Goldstone theorem [56] implies the existence of three Goldstone bosons, massless scalar fields, which are then identified with the pions. The pions are not massless, however; the mass terms for the up and down quarks softly but explictly break the chiral symmetry. The consequence of this explict breaking is to give the putative Goldstone bosons, the pions, a small mass. This analysis is called chiral perturbation theory [117]. The $U_V(1)$ symmetry corresponds to the baryonic charge and is presumed to be conserved. The one question that remains is what happens to the $U_A(1)$, how does this symmetry manifest itself? If it is not broken, spontaneously or explicitly, then it should be associated with a conserved quantum number. We do not see any such additional conserved quantum number. If it is spontaneously broken, then we should see another corresponding massless Goldstone boson. It can be shown that this does not correspond to the η, Weinberg has shown [118] that the mass of such a putative Goldstone bosons must satisfy the inequality $m_{G.B.} \leq \sqrt{m_\pi}$. This lack of understanding of how the $U_A(1)$ symmetry manifests itself is called the $U(1)$ problem.

The $U(1)$ problem is related to instantons, the theta vacua and the chiral anomaly, which we will explain in this section and the next. The upshot is that the $U_A(1)$ symmetry is actually explicitly broken, due to a quantum effect, called the chiral anomaly. To understand the chiral anomaly it is easiest to work in a function integral formulation for the fermionic fields, the subject to which we will now turn.

11.6.1 Quantum Fermi Fields

Canonical quantization of fermionic fields demands that the fermions satisfy equal time anti-commutation relations

$$\{\psi(\vec{x},t), i\psi^\dagger(\vec{y},t)\} = -i\hbar\delta^3(\vec{x}-\vec{y}), \tag{11.176}$$

where $\{A,B\} = AB + BA$. Why does the anti-commutator arise? For a free field with equation of motion

$$(i\gamma^\mu\partial_\mu - m)\psi = 0 \tag{11.177}$$

we can construct the solution by simple Fourier transformation

$$\psi = \int \frac{d^3k}{(2\pi)^3} \frac{m}{(\vec{k}^2+m^2)^{1/2}} \sum_{\alpha=1,2} \left(b_\alpha(k)u^\alpha(k)e^{-ik\cdot x} + d_\alpha^\dagger(k)v^\alpha(k)e^{ik\cdot x}\right), \tag{11.178}$$

where $k \equiv ((\vec{k}^2+m^2)^{1/2}, \vec{k})$, $u^\alpha(k)$ and $v^\alpha(k)$ are specific, orthonormalized spinor solutions of the Dirac equation (11.177) of positive and negative energy, respectively, while $b_\alpha(k)$ and $d^\dagger(k)$ are arbitrary, operator valued coefficients. The expression for the Hamiltonian (energy) then becomes

$$\mathcal{H} = \int \frac{d^3k}{(2\pi)^3} (\vec{k}^2+m^2)^{1/2} \sum_\alpha \left(b_\alpha^\dagger(k)b_\alpha(k) - d_\alpha(k)d_\alpha^\dagger(k)\right), \tag{11.179}$$

where we have not changed the order of the operators in the expression for the Hamiltonian. The order of the d_α and the $d_\alpha^\dagger$ occurs because we expanded ψ with $d_\alpha^\dagger$ rather than d_α but the minus sign occurs because the v^αs correspond to negative energy solutions of the Dirac equation. If we had used d_α in the expansion of ψ we would have arrived at the expression in Equation (11.179) with the d_α and the $d_\alpha^\dagger$ interchanged; however, the minus sign would still be there. Now if we want to have $\mathcal{H} \geq 0$, up to a constant, we need

$$d_\alpha(k)d_\alpha^\dagger(k) = -d_\alpha^\dagger(k)d_\alpha(k) + 1, \tag{11.180}$$

where we have chosen the constant to be 1 for the case of discrete k. For a continuum of k's we get

$$\{d_\alpha(k), d_\alpha^\dagger(k')\} = \delta^4(k-k') \tag{11.181}$$

and the Hamiltonian, up to a constant (which can very well be an infinite constant!) is

$$\mathcal{H} = \int \frac{d^3k}{(2\pi)^3} (\vec{k}^2+m^2)^{1/2} \sum_\alpha \left(b_\alpha^\dagger(k)b_\alpha(k) + d_\alpha^\dagger(k)d_\alpha(k)\right), \tag{11.182}$$

a positive semi-definite form. As the b_αs and the d_αs are equivalent, we must choose anti-commutation relations for both.

11.6.2 Fermionic Functional Integral

The limit $\hbar \to 0$ in the canonical anti-commutation relations Equation (11.176) does not yield ordinary, commuting c-number fields in the classical limit. The fields become anti-commuting fields, so-called Grassmann number-valued fields whose anti-commutator, rather than commutator, vanishes

$$\{\psi(\vec{x},t), \psi^{\dagger}(\vec{y},t)\} = 0. \tag{11.183}$$

Thus the classical limit gives fields that are elements of an infinite dimensional Grassmann algebra from an infinite dimensional Clifford algebra in the quantum domain. Then, if there is to be a Feynman path-integral description of fermions, the integral should be defined over the classical space of fields, fields that are Grassmann algebra-valued. Such an integral can be formally defined. For free theories, perhaps all such formalism is rather unnecessary. However, for interacting theories of fermions, the functional integral description must at least be able to generate the perturbative expansion. In fact, we can almost think that the fermionic functional integral representation for the amplitudes of a quantum field theory with fermions is simply a very compact and efficient notation that can and does serve as a means of generating the perturbative expansion.

Abstractly, an integral is a linear map that takes a space of functions to the real numbers. We will define the functional integral over a Grassmann number in this way, first for a finite set of Grassmann numbers, and then generalize to the infinite limit. A Grassmann number θ satisfies

$$\{\theta, \theta\} = \theta\theta + \theta\theta = 2\theta\theta = 2\theta^2 = 0. \tag{11.184}$$

We define a differential operator $\frac{d}{d\theta}$ by the very reasonable rules for any other anti-commuting number β and for a c-number a,

$$\frac{d}{d\theta}\theta = 1, \quad \frac{d}{d\theta}\beta = \frac{d}{d\theta}a = 0. \tag{11.185}$$

The derivative operator should be thought of as a Grassmann-valued operator; it should anti-commute with other Grassmann numbers. A general function of $f(\theta)$, *i.e.* a commutative function, can be expanded in two terms

$$f(\theta) = a + \beta\theta, \tag{11.186}$$

where a is real while β is Grassmannian. Then

$$\frac{d}{d\theta}f(\theta) = \frac{d}{d\theta}a + \frac{d}{d\theta}\beta\theta = -\frac{d}{d\theta}\theta\beta = -\beta. \tag{11.187}$$

The idea that f is a commutative function means that it is composed of an even number of Grassmann numbers, 0 and 2 in this case. β is a Grassmann number, hence

$$\beta^2 = 0, \quad \{\beta, \theta\} = 0 \tag{11.188}$$

then it is easy to verify

$$[f, \beta] = [f, \theta] = 0. \tag{11.189}$$

Then clearly

$$\frac{d}{d\theta}\frac{d}{d\theta} = \frac{d^2}{d\theta^2} = 0. \tag{11.190}$$

This means that the integral can in no way be the inverse of differentiation, the derivative is a nilpotent operator. However, we will define it, following Berezin [13] to be a linear map from the space of Grassmann numbers to the real numbers, we define

$$\int d\theta\, 1 = 0 \quad \int d\theta\, \theta = 1, \tag{11.191}$$

which implies

$$\int d\theta\, f(\theta) = \int d\theta\, (a + \beta\theta) = 0 + \int d\theta\, \beta\theta = -\int d\theta\, \theta\beta = -\beta. \tag{11.192}$$

For N Grassmann numbers we have the algebra

$$\{\theta_i, \theta_j\} = 0$$

$$\left\{\frac{d}{d\theta_i}, \theta_j\right\} = \delta_{ij}$$

$$\left\{\frac{d}{d\theta_i}, \frac{d}{d\theta_j}\right\} = 0, \tag{11.193}$$

for $i, j = 1, \cdots N$. Then a general, commutative function is expanded as

$$f(\theta_i) = a + c_i\theta_i + c_{ij}\theta_i\theta_j + \cdots + c\theta_1\theta_2\cdots\theta_N \tag{11.194}$$

and we notice it has a finite number of terms. $c_{ijkl\cdots}$ is Grassmannian if the number of indices is odd but a real number if the number of indices is even. The integration rules generalize as

$$\int d\theta_i\, 1 = 0, \quad \int d\theta_i\, \theta_j = \delta_{ij} \tag{11.195}$$

and by convention and consistency

$$\int d\theta_1 d\theta_2\, \theta_1\theta_2 \equiv \int d\theta_1 \left(\int d\theta_2(-\theta_2)\right)\theta_1 = -\int d\theta_1\, \theta_1 = -1. \tag{11.196}$$

Then it is easy to see for an anti-symmetric matrix M (clearly any symmetric part of M will not contribute)

$$I_N(M) = \int d\theta_1 \cdots d\theta_N e^{-\sum_{ij}\theta_i M_{ij}\theta_j} = \begin{cases} 2^{N/2}\sqrt{\det(M)}, & \text{for } N \text{ even} \\ 0 & \text{for } N \text{ odd} \end{cases}. \tag{11.197}$$

For an invertible, anti-symmetric M and a set of Grassmann parameters χ_i, $i = 1 \cdots N$ and $\{\chi_i, \chi_j\} = \{\chi_i, \theta_j\} = 0$, we can compute

$$\mathcal{I}_N(M; \chi) = \int d\theta_1 \cdots d\theta_N e^{-\sum_{ij}\theta_i M_{ij}\theta_j + \sum_j \chi_j\theta_j} \tag{11.198}$$

as follows. Translating the integration variable $\theta_i = \theta_i' - \frac{1}{2}M_{ij}^{-1}\chi_j$, we get, using matrix notation

$$
\theta^T M\theta - \chi^T\theta = \left(\theta' - \frac{1}{2}M^{-1}\chi\right)^T M\left(\theta' - \frac{1}{2}M^{-1}\chi\right) - \chi^T\left(\theta' - \frac{1}{2}M^{-1}\chi\right)
$$
$$
= \theta'^T M\theta' - \frac{1}{2}(\chi^T M^{-1^T}M\theta' + \theta'^T\chi) + \frac{1}{4}\chi^T M^{-1^T}\chi
$$
$$
- \chi^T\theta' + \frac{1}{2}\chi^T M^{-1}\chi
$$
$$
= \theta'^T M\theta' - \frac{1}{2}(-\chi^T\theta' - \chi^T\theta') - \frac{1}{4}\chi^T M^{-1}\chi
$$
$$
- \chi^T\theta' + \frac{1}{2}\chi^T M^{-1}\chi
$$
$$
= \theta'^T M\theta' + \frac{1}{4}\chi^T M^{-1}\chi \tag{11.199}
$$

using $M^{-1^T}M = (M^T M^{-1})^T = (-MM^{-1})^T = -\mathbb{I}^T = -\mathbb{I}$ since M is antisymmetric and the fact that the χ is also anti-commuting. Then

$$
\mathcal{I}_N(M;\chi) = \int d\theta_1' \cdots d\theta_N' e^{-\Sigma_{ij}\left(\theta_i' M_{ij}\theta_j' - \frac{1}{4}\chi_i M^{-1}\chi_j\right)}
$$
$$
= \begin{cases} 2^{N/2}\sqrt{\det M}\, e^{-\Sigma_{ij}\frac{1}{4}\chi_i M^{-1}\chi_j}, & \text{for } N \text{ even} \\ 0 & \text{for } N \text{ odd} \end{cases}. \tag{11.200}
$$

For complex fields, we have the equivalent of complex Grassmann numbers

$$
\eta = \frac{\theta_1 + i\theta_2}{\sqrt{2}}, \quad \eta^* = \frac{\theta_1 - i\theta_2}{\sqrt{2}}. \tag{11.201}
$$

Considering the 2×2 case, we impose $-\sum_{i,j}\theta_i M_{ij}\theta_j = i\eta^*\tilde{M}\eta$ which gives $\tilde{M} = 2M_{12}$ and we have

$$
\int d\eta^* d\eta\, e^{i\eta^*\tilde{M}\eta} = \det(\tilde{M}), \tag{11.202}
$$

where the integration is done by treating η and η^* as completely independent Grassmann variables. Dropping the tilde, the integration formula generalizes as

$$
\int \prod_{i,j} d\eta_i^* d\eta_j\, e^{i\Sigma_{i,j}\eta_i^* M_{ij}\eta_j} = \det(M) \tag{11.203}
$$

and with sources, suppressing the indices and summation signs,

$$
\int \prod d\eta^* d\eta\, e^{i\eta^* M\eta + i\xi^*\eta + i\eta^*\xi} = \det(M)e^{-i\xi^* M^{-1}\xi}. \tag{11.204}
$$

Then boldly generalizing to infinite dimensional integrals we get for the fermionic field

$$
\int \mathcal{D}(\psi,\bar{\psi})e^{i\int d^4 x\bar{\psi}(i\gamma^\mu\partial_\mu - m)\psi} = \det(i\gamma^\mu\partial_\mu - m) \tag{11.205}
$$

and including sources

$$\int \mathcal{D}(\psi,\bar{\psi})e^{i\int d^4x\bar{\psi}(i\gamma^\mu\partial_\mu-m)\psi+\bar{\psi}\zeta+\bar{\zeta}\psi} = det(i\gamma^\mu\partial_\mu-m)e^{-i\int d^4x\bar{\zeta}(i\gamma^\mu\partial_\mu-m)^{-1}\zeta}$$

$$= \mathcal{N}'e^{-i\int d^4p\bar{\zeta}(p)\frac{1}{\not{p}-m}\zeta(p)}. \qquad (11.206)$$

Then for a general gauge interaction the usual perturbative expansion ensues from the coupling (where we have expressly put the coupling constant e)

$$\mathcal{L}' = ie\bar{\psi}\gamma^\mu A_\mu\psi \qquad (11.207)$$

then

$$Z(\zeta,A) = \int \mathcal{D}(\psi,\bar{\psi})e^{i\int d^4x\mathcal{L}+\bar{\zeta}\psi+\bar{\psi}\zeta+ie\bar{\psi}\gamma^\mu A_\mu\psi}$$

$$= \mathcal{N}'e^{i\int d^4x\left(\frac{\delta}{\delta\bar{\zeta}(x)}ie\gamma^\mu A_\mu\frac{\delta}{\delta\zeta(x)}\right)}e^{-i\int d^4p\bar{\zeta}(p)\frac{1}{\not{p}-m}\zeta(p)}. \qquad (11.208)$$

The derivatives with respect to $\zeta(x)$ and $\bar{\zeta}(x)$ can be trivially converted into derivatives with respect to $\zeta(p)$ and $\bar{\zeta}(p)$ by Fourier transformation. This gives rise to the usual perturbation expansion expressed in Feynman diagrams [101].

Reverting back to Euclidean space, the action is

$$S_E = -\int d^4x\bar{\psi}(i\gamma_\mu\partial_\mu-im)\psi, \qquad (11.209)$$

where the γ_μ matrices satisfy the Clifford algebra

$$\{\gamma_\mu,\gamma_\nu\} = 2\delta_{\mu\nu}. \qquad (11.210)$$

The fields $\bar{\psi}$ and ψ are no longer related to each other, but are in fact completely independent Grassmann-valued fields. We can infer this from many points of view. First, and most importantly, if the formula Equation (11.203) is to work, the integration variables η and η^* are completely independent. First of all, η and η^* satisfy

$$\{\eta_i,\eta_j\} = \{\eta_i^*,\eta_j^*\} = \{\eta_i^*,\eta_j\} = 0. \qquad (11.211)$$

Then if η^* were the adjoint of η, i.e. $\eta^* = \eta^\dagger C$, where C is a fixed matrix akin to a charge conjugation matrix, then the last relation would imply (multiplying by C^{-1}) and contracting together

$$\sum_i(\eta_i\eta_i^\dagger + \eta_i^\dagger\eta_i) = 0. \qquad (11.212)$$

This says that the sum of two positive operators vanishes, requiring the operators to be zero. Additionally, the Euclidean Dirac fields transform according to the $(\frac{1}{2},0)\oplus(0,\frac{1}{2})$ representation of the four-dimensional Euclidean rotation group $SO(4) = SU(2)\times SU(2)$. The two $SU(2)$ subgroups are totally independent of one another, hermitean conjugation does not take one into the other, as is the case in Minkowski spacetime.

We decompose ψ and $\bar{\psi}$ with a complete set of orthonormal spinor solutions of the Dirac equation

$$\psi = \sum_r a_r \psi_r \quad \bar{\psi} = \sum_r \bar{a}_r \bar{\psi}_r, \tag{11.213}$$

where the coefficients a_r and $\bar{a}_r$ are independent Grassmann numbers and

$$\int d^4 x \psi_r^\dagger \psi_s = \int d^4 x \bar{\psi}_r \bar{\psi}_s^\dagger = \delta_{rs}. \tag{11.214}$$

Then we define the functional integration measure as

$$\mathcal{D}(\psi, \bar{\psi}) = \prod_r da_r d\bar{a}_r. \tag{11.215}$$

Then

$$S_E = -\int d^4 x \bar{\psi}(i\gamma_\mu \partial_\mu)\psi = -\sum_r \lambda_r \bar{a}_r a_r, \tag{11.216}$$

where

$$(i\gamma_\mu \partial_\mu)\psi_r = \lambda_r \psi_r. \tag{11.217}$$

Then the integral

$$\int \mathcal{D}(\psi, \bar{\psi})e^{-S_E} = \int \prod_s da_s d\bar{a}_s e^{\sum_r \lambda_r \bar{a}_r a_r} = \prod_r \lambda_r = det(i\gamma_\mu \partial_\mu). \tag{11.218}$$

In the massless limit, $m \to 0$, the action Equation (11.209) is invariant under global chiral transformations, decomposed as vector and axial transformations

$$\psi \to e^{i(\alpha + \beta\gamma_5)}\psi \quad \bar{\psi} \to \bar{\psi}e^{i(\alpha + \beta\gamma_5)}. \tag{11.219}$$

The chiral anomaly corresponds to the fact that it is impossible to define the functional integral while simultaneously keeping the axial gauge symmetry and the vector gauge symmetry. The full chiral symmetry of two-flavour QCD is

$$SU_V(2) \times SU_A(2) \times SU_V(1) \times U_A(1), \tag{11.220}$$

but the $SU_A(2)$ is spontaneously broken, giving rise to three massless Goldstone bosons, the pions,

$$SU_V(2) \times SU_A(2) \times U_V(1) \times U_A(1) \to SU_V(2) \times U_V(1) \times U_A(1). \tag{11.221}$$

The anomaly results from the impossibility to preserve the remaining chiral symmetry in the quantum theory. Fundamentally, the anomaly results because of divergences in the naive, original theory. Then to make sense of the theory these divergences must be removed; this is done in a rather brutal fashion and is called regularization. The brutality of the regularization means that it seems necessary to explicitly break at least some of the chiral symmetry of the original Lagrangian. Indeed, there is no known regularization that can preserve all of the chiral symmetry and it is understood that no such regularization exists. The hope and expectation was that once the regularization is removed, the full chiral symmetry of the theory would return. The anomaly corresponds to the fact that this is not the case. In fact, upon removing the regularization it is not possible to preserve both the $U_V(1)$ and the $U_A(1)$ symmetries.

11.6.3 The Axial Anomaly

Conceptually, the clearest method for seeing this was discovered by Fujikawa [50][3]. He considered the fermionic functional integral

$$\mathcal{I} = \int \mathcal{D}(\psi, \bar\psi) e^{\int d^4x \bar\psi(i\gamma_\mu D_\mu)\psi} \qquad (11.222)$$

and realized that the anomaly comes from the inability to define the functional integration measure in a manner that is invariant under all the chiral transformations. Here we generalize further, allowing the covariant derivative to include gauge fields, in principle, for all the global symmetries. However, we will find that some global symmetries are not preserved in the quantum theory, and then adding the gauge fields corresponding to those symmetries is inopportune. Their quantization makes no sense as renormalizability requires gauge-invariance. Thus we imagine adding gauge fields for all symmetries that can be preserved at the quantum level. For the case of QCD, this corresponds to gauge fields for the colour gauge symmetry $SU_c(3)$ and the $U_V(1)$ symmetry. Gauging the chiral $SU_V(2) \times SU_A(2)$ actually corresponds to part of the gauge group of the weak interactions, but we shall not develop this theory here. We will expand the fields slightly differently from Equation (11.213) as

$$\psi = \sum_r a_r \varphi_r \qquad \bar\psi = \sum_r \varphi_r^\dagger \bar a_r \qquad (11.223)$$

with

$$i\slashed{D}\varphi_r = \lambda_r \varphi_r \qquad \int d^4x \varphi_r^\dagger \varphi_s = \delta_{rs} \qquad (11.224)$$

and

$$\mathcal{D}(\psi, \bar\psi) = \prod_r d\bar a_r da_r. \qquad (11.225)$$

For a local axial transformation $\psi(x) \to e^{i\beta(x)\gamma_5}\psi(x) \approx \psi(x) + i\beta(x)\gamma_5\psi(x)$ and $\bar\psi(x) \to \bar\psi(x)e^{i\beta(x)\gamma_5} \approx \bar\psi(x) + i\beta(x)\bar\psi(x)\gamma_5$ the Lagrangian is not invariant

$$\mathcal{L}(x) \to \mathcal{L}(x) - (\partial_\mu\beta(x))\bar\psi(x)\gamma_\mu\gamma_5\psi(x). \qquad (11.226)$$

However,

$$\psi(x) \to \psi'(x) = \sum_r a'_r \varphi_r(x) = \sum_r a_r e^{i\beta(x)\gamma_5}\varphi_r(x). \qquad (11.227)$$

---

[3] We note that Fujikawa used anti-hermitean Euclidean Dirac matrices. We stick with hermitean Euclidean Dirac matrices.

Then

$$a'_r = \sum_s \int d^4x \varphi_r^\dagger e^{i\beta(x)\gamma_5}\varphi_s(x)a_s \equiv \sum_s C_{rs}a_s \qquad (11.228)$$

and

$$\prod_r da'_r = (detC)^{-1}\prod_s da_s. \qquad (11.229)$$

Interestingly, the power of the determinant is -1. This is because the Grassmann integration actually behaves a lot like differentiation. Indeed,

$$1 = \int d(\lambda a)\,(\lambda a) = \int da\, J(\lambda a) = J\lambda \int da\, a = J\lambda, \qquad (11.230)$$

thus $J = 1/\lambda$ and $d(\lambda a) = da/\lambda$. The determinant of the matrix C_{rs} for an infinitesimal transformation is

$$
\begin{aligned}
&= det(C_{rs})^{-1} \, det\left(\delta_{rs} + i\int d^4x\beta(x)\varphi_r^\dagger(x)\gamma_5\varphi_s(x)\right)^{-1}\\
&= \exp\left(-Tr\ln\left(\delta_{rs} + \int d^4x\,\beta\varphi_r^\dagger\gamma_5\varphi_s\right)\right)\\
&= \exp\left(-i\int d^4x\beta(x)\sum_r \varphi_r^\dagger\gamma_5\varphi_r\right)
\end{aligned}
\qquad (11.231)
$$

using the expansion of $\ln(1+\epsilon) \approx \epsilon$. We must not forget that an equal contribution will come from the variation of $\prod_r d\bar{a}_r$. We wish to evaluate $A(x) = \sum_r \varphi_r^\dagger(x)\gamma_5\varphi_r(x)$; however, the sum is surely hopelessly divergent. We regularize it with the eigenvalues of the Dirac operator, taking

$$A(x) \equiv \lim_{M\to\infty}\sum_r \varphi_r^\dagger(x)\gamma_5 e^{-(\lambda_r/M)^2}\varphi_r(x). \qquad (11.232)$$

It is this choice of regulator that puts the anomaly in the axial symmetry, preserving the vector symmetry in the quantum theory. Another choice can preserve the axial symmetry but not the vector. We can choose which symmetry we wish to preserve; however, we cannot preserve both. Writing $\varphi_r(x) = \langle x|r\rangle$, (note the ket $|r\rangle$ must span the matrix indices of the coordinate wave function $\varphi(x)$) we have

$$
\begin{aligned}
A(x) &= \lim_{M\to\infty}\sum_r \langle r|x\rangle\gamma_5 e^{-(\lambda_r/M)^2}\langle x|r\rangle = \lim_{M\to\infty}\sum_r Tr\left(\gamma_5\langle x|e^{-(\lambda_r/M)^2}|r\rangle\langle r|x\rangle\right)\\
&= \lim_{M\to\infty}\sum_r Tr\left(\gamma_5\langle x|e^{-(i\slashed{D}/M)^2}|r\rangle\langle r|x\rangle\right) = \lim_{M\to\infty}\lim_{x\to y}Tr\left(\gamma_5\langle x|e^{-(i\slashed{D}/M)^2}|y\rangle\right)\\
&= \lim_{M\to\infty}\lim_{x\to y}Tr\left(\gamma_5\langle x|e^{-(i\slashed{D}/M)^2}\int\frac{d^4k}{(2\pi)^4}|k\rangle\langle k|y\rangle\right)\\
&= \lim_{M\to\infty}\lim_{x\to y}Tr\left(\gamma_5 e^{-(i\slashed{D}(x)/M)^2}\langle x|\int\frac{d^4k}{(2\pi)^4}|k\rangle\langle k|y\rangle\right)\\
&= \lim_{M\to\infty}\lim_{x\to y}Tr\left(\gamma_5\int\frac{d^4k}{(2\pi)^4}e^{-(i\slashed{D}(x)/M)^2}e^{ik\cdot x}e^{-ik\cdot y}\right) \qquad (11.233)
\end{aligned}
$$

and we should be aware that the Tr is over Dirac and internal indices. Now $i\slashed{D}(x)e^{ik\cdot x} = e^{ik\cdot x}(-\slashed{k}+i\slashed{D}(x))$, thus

$$
\begin{aligned}
A(x) &= \lim_{M\to\infty} \lim_{x\to y} Tr\left(\gamma_5 \int \frac{d^4k}{(2\pi)^4} e^{-ik\cdot y} e^{ik\cdot x} e^{-((-\slashed{k}+i\slashed{D}(x))/M)^2}\right)\\
&= \lim_{M\to\infty} Tr\left(\gamma_5 \int \frac{d^4k}{(2\pi)^4} e^{-((-\slashed{k}+i\slashed{D}(x))/M)^2}\right)\\
&= \lim_{M\to\infty} Tr\left(\gamma_5 \int \frac{d^4k}{(2\pi)^4} e^{-(\frac{\{\gamma_\mu,\gamma_\nu\}}{2}+\frac{[\gamma_\mu,\gamma_\nu]}{2})(-k_\mu+iD_\mu(x))(-k_\nu+iD_\nu(x))/M^2}\right)\\
&= \lim_{M\to\infty} Tr\left(\gamma_5 \int \frac{d^4k}{(2\pi)^4} e^{-(\delta_{\mu\nu}+\frac{[\gamma_\mu,\gamma_\nu]}{2})(-k_\mu+iD_\mu(x))(-k_\nu+iD_\nu(x))/M^2}\right)\\
&= \lim_{M\to\infty} Tr\left(\gamma_5 \int \frac{d^4k}{(2\pi)^4} e^{-\left((-k+iD(x))^2-\frac{[\gamma_\mu,\gamma_\nu]}{2}\frac{F_{\mu\nu}}{2}\right)/M^2}\right)\\
&= \lim_{M\to\infty} Tr\left(\gamma_5 \int \frac{d^4k}{(2\pi)^4} e^{-k^2/M^2} e^{\left(2ik\cdot D(x)+D^2(x)+\frac{1}{4}[\gamma_\mu,\gamma_\nu]F_{\mu\nu}\right)/M^2}\right)\\
&= \lim_{M\to\infty} Tr\left(\gamma_5 \int \frac{d^4k}{(2\pi)^4} e^{-k^2/M^2}\left(1+\cdots+\frac{1}{2}\left(\frac{1}{4}[\gamma_\mu,\gamma_\nu]F_{\mu\nu}\right)^2/M^4+\cdots\right)\right).
\end{aligned}
$$

$$(11.234)$$

The first term in the expansion of the exponential that survives the Dirac trace is shown and, although there will be other terms in the higher orders that survive this trace, they will have higher powers of M in the denominator. The Gaussian integral only gives a factor of M^4, hence in the limit $M\to\infty$ this is the only term that survives. Thus we get

$$
A(x) = \lim_{M\to\infty} \int \frac{d^4k}{(2\pi)^4} Tr\left(\frac{1}{2}\frac{1}{M^4}\left(\gamma_5\frac{1}{4}[\gamma_\mu,\gamma_\nu]\frac{1}{4}[\gamma_\sigma,\gamma_\tau]\right)F_{\mu\nu}F_{\sigma\tau}\right) e^{-k^2/M^2}.
$$

$$(11.235)$$

Using that $Tr(\gamma_5\gamma_\mu\gamma_\nu\gamma_\sigma\gamma_\tau) = 4\epsilon_{\mu\nu\sigma\tau}$ and that the Gaussian integral is

$$
\int \frac{d^4k}{(2\pi)^4} e^{-k^2/M^2} = \frac{M^4}{16\pi^2}
$$

$$(11.236)$$

gives

$$
A(x) = \frac{1}{32\pi^2}\epsilon_{\mu\nu\sigma\tau} Tr\left(F_{\mu\nu}F_{\sigma\tau}\right).
$$

$$(11.237)$$

Therefore, the fermionic functional integration measure is not invariant under axial transformations and transforms as

$$
\mathcal{D}(\psi,\bar\psi) \to \mathcal{D}(\psi,\bar\psi)e^{-i\frac{1}{16\pi^2}\int d^4x\,\beta(x)\epsilon_{\mu\nu\sigma\tau} Tr\left(F_{\mu\nu}F_{\sigma\tau}\right)},
$$

$$(11.238)$$

where we get twice the variation since both ψ and $\bar\psi$ contribute to the measure.

11.6.4 The U(1) *Problem*

With the chiral anomaly, we understand that, at the quantum level, we cannot preserve all of the classical symmetry of the action. Only the subgroup $U_A(1)$ is explicitly broken by the chiral anomaly and the remaining subgroup is spontaneously broken $SU_V(2) \times SU_A(2) \times U_V(1) \times SU_c(3) \rightarrow SU_V(2) \times U_V(1) \times SU_c(3)$. Under a local axial $U_A(1)$ transformation, $-S_E$, minus the action (that appears in the exponent) transforms as, keeping only terms to first order,

$$\int d^4x \bar{\psi} \left(i\gamma_\mu D_\mu\right)\psi \rightarrow \int d^4x \bar{\psi}(1 + i\beta(x)\gamma_5)\left(i\gamma_\mu D_\mu\right)(1 + i\beta(x)\gamma_5)\psi$$

$$= \int d^4x \bar{\psi}\left(i\gamma_\mu D_\mu\right)\psi - (\partial_\mu\beta(x))\bar{\psi}\gamma_\mu\gamma_5\psi. \qquad (11.239)$$

The functional integral under a change of variables must be invariant. If we transform to the field $\psi' = (1 + i\beta(x)\gamma_5)\psi$, we get

$$\mathcal{I} = \int \mathcal{D}(\psi, \bar{\psi}) e^{\int d^4x \bar{\psi} i \slashed{D}\psi} = \int \mathcal{D}(\psi', \bar{\psi}') e^{\int d^4x \bar{\psi} i \slashed{D}\psi - (\partial_\mu\beta(x))\bar{\psi}\gamma_\mu\gamma_5\psi}$$

$$= \int \mathcal{D}(\psi, \bar{\psi}) e^{-i\int d^4x \beta(x)\left(\frac{1}{16\pi^2}\epsilon_{\mu\nu\sigma\tau}Tr\left(F_{\mu\nu}F_{\sigma\tau}\right) + i\partial_\mu\bar{\psi}\gamma_\mu\gamma_5\psi\right)} e^{\int d^4x \bar{\psi} i \slashed{D}\psi}.$$

$$(11.240)$$

Then invariance requires

$$\partial_\mu \langle \bar{\psi}\gamma_\mu\gamma_5\psi \rangle^A = i\frac{1}{16\pi^2}\epsilon_{\mu\nu\sigma\tau}Tr\left(F_{\mu\nu}F_{\sigma\tau}\right) = i\frac{C}{8\pi^2}\langle F_{\mu\nu}\tilde{F}_{\mu\nu}\rangle, \qquad (11.241)$$

where the matrix element on the left-hand side signifies the fermionic expectation value of the axial current operator $\bar{\psi}\gamma_\mu\gamma_5\psi$ in the presence of the background gauge fields. The latter equality is easily obtained for an arbitrary multiplet of fermions in a representation with hermitean generators T^a of $SU(n)$, and then C is the constant in $Tr\left(T^aT^b\right) = C\delta^{ab}$. The i on the right-hand side is expected and disappears upon Wick rotation back to Minkowski space.

We can demonstrate the so-called chiral Ward–Takahashi identities, which have to do with symmetries, and will be useful in our analysis later. Consider the m point function

$$\langle \phi^1(x_1)\cdots\phi^m(x_m) \rangle^A \equiv \frac{\int \mathcal{D}(\psi, \bar{\psi}) e^{-S_E(A)}\phi^1(x_1)\cdots\phi^m(x_m)}{\int \mathcal{D}(\psi, \bar{\psi}) e^{-S_E(A)}}, \qquad (11.242)$$

where the $\phi^i(x_i)$ are local, multi-linear functions of the fermionic fields where S_E is given in Equation (11.209). With the variations (taken in the opposite sense to Fujikawa as in Equation (11.239), to stay with the conventions of Coleman)

$$\delta\psi = -i\gamma_5\psi\delta\alpha(x) \quad \delta\bar{\psi} = -i\bar{\psi}\gamma_5\delta\alpha(x) \qquad (11.243)$$

we have

$$\delta\phi^i = \frac{\partial\phi^i}{\partial\alpha}\delta\alpha(x). \qquad (11.244)$$

For example,

$$\delta(\bar{\psi}\psi) = -2i\bar{\psi}\gamma_5\psi\delta\alpha(x). \tag{11.245}$$

But changing the variables in the functional integral must not make a difference, it must be invariant. Thus we get

$$0 = \delta\langle\phi^1(x_1)\cdots\phi^m(x_m)\rangle^A = \langle-\delta S_E\phi^1(x_1)\cdots\phi^m(x_m)\rangle$$
$$+ \left\langle\sum_r \phi^1(x_1)\cdots\delta\phi^r(x_r)\cdots\phi^m(x_m)\right\rangle. \tag{11.246}$$

Then, since

$$-\delta S_E = \int d^4x\,\bar{\psi}i\gamma_\mu(-i\partial_\mu\delta\alpha)\gamma_5\psi = \int d^4x\,(\partial_\mu\delta\alpha)j_{5\mu} = -\int d^4x\,\delta\alpha\partial_\mu j_{5\mu} \tag{11.247}$$

we get

$$0 = \langle(-\partial_\mu j_{5\mu(x)})\phi^1(x_1)\cdots\phi^m(x_m)\rangle + \sum_r \delta(x-x_r)\langle\phi^1(x_1)\cdots\frac{\partial\delta\phi^r(x_r)}{\partial\alpha}\cdots\phi^m(x_m)\rangle$$
$$-2M\langle\bar{\psi}(x)\gamma_5\psi(x)\phi^1(x_1)\cdots\phi^m(x_m)\rangle, \tag{11.248}$$

where the last term is there if we add a mass term that breaks the chiral symmetry explicitly and we will write $j_5(x) = \bar{\psi}(x)\gamma_5\psi(x)$. Then using Equation (11.241) and integrating over x, we get

$$2M\left\langle\int d^4x\,j_5(x)\phi^1(x_1)\cdots\phi^m(x_m)\right\rangle^A$$
$$= \frac{\partial}{\partial\alpha}\langle\phi^1(x_1)\cdots\phi^m(x_m)\rangle^A$$
$$-i\frac{C}{8\pi^2}\int d^4x\,\langle F_{\mu\nu}\tilde{F}_{\mu\nu}\rangle\langle\phi^1(x_1)\cdots\phi^m(x_m)\rangle^A$$
$$= \frac{\partial}{\partial\alpha}\langle\phi^1(x_1)\cdots\phi^m(x_m)\rangle^A - 4iC\nu\langle\phi^1(x_1)\cdots\phi^m(x_m)\rangle^A. \tag{11.249}$$

But what is the effect of the fermions on the instanton? The instantons must still be solutions of the equations of motion

$$D_\mu F_{\mu\nu} = j_\nu \quad \text{with} \quad D_\nu j_\nu = 0 \quad \text{and} \quad i\gamma_\mu D_\mu\psi = 0. \tag{11.250}$$

These equations have a perfectly good solution, $\psi = 0$ and $D_\mu F_{\mu\nu} = 0$. The latter equation is satisfied by the instantons' and hence the instatons' configuration is unchanged by the fermions. All the previously found formulae must still be valid,

$$E(\theta)/V = -2K\cos\theta\,e^{-S_0} \tag{11.251}$$

and

$$\langle\theta|F\tilde{F}|\theta\rangle = -64\pi^2 iK\,e^{-S_0}\sin\theta. \tag{11.252}$$

The only change that occurs is that the Gaussian integral about the instanton configuration over the gauge fields is appended with a functional integral over the fermion fields (which is in a sense also Gaussian as the fermion fields only enter quadratically) in the presence of the instanton background

$$K \to K \frac{det(i\gamma_\mu (\partial_\mu + A_\mu))}{det(i\gamma_\mu (\partial_\mu))}. \qquad (11.253)$$

The consequences of this change are profound. The fermionic determinant in the presence of the instanton vanishes exactly, giving

$$E(\theta)/V = 0 \quad \langle \theta | F\tilde{F} | \theta \rangle = 0. \qquad (11.254)$$

This means that all the theta vacua become degenerate in energy, that the $U_A(1)$ symmetry is spontaneously broken. The $U(1)$ problem corresponds to the question, "Why is there no corresponding massless Goldstone boson?" What we will find is that the massless boson never contributes to gauge-invariant matrix elements and therefore is not physically manifested.

Why does the fermion determinant vanish? It is because in the presence of an instanton, there is necessarily a zero energy mode to the Dirac equation. Evidently

$$\int d\theta d\bar{\theta}\, e^{0 \times \bar{\theta}\theta} = \int d\theta d\bar{\theta}\, 1 = 0, \qquad (11.255)$$

which is quite unlike the bosonic case

$$\int d\varphi\, e^{-0 \times \varphi^2} = \int d\varphi\, 1 = \infty (= \frac{1}{0}). \qquad (11.256)$$

The zero mode follows from a deep theorem, the Atiyah–Singer index theorem [8]. However, we can quite easily establish the existence of the zero mode directly using the simplest chiral Ward–Takahashi relation. We will work with an $SU(2)$ gauge group with one doublet of fermions for simplicity. The Dirac equation for eigenmode of energy λ_r is

$$i\slashed{D}\psi_r = \lambda_r \psi_r \qquad (11.257)$$

but then

$$i\slashed{D}\gamma_5 \psi_r = -\gamma_5 i\slashed{D}\psi_r = -\lambda_r \gamma_5 \psi_r. \qquad (11.258)$$

Thus for each mode ψ_r of energy λ_r there is a matching eigenmode $\gamma_5 \psi_r$ of energy $-\lambda_r$. But what happens if $\lambda_r = 0$? Let ψ_r^0 be a zero mode, $i\slashed{D}\psi_r^0 = 0$, but then obviously $i\slashed{D}\gamma_5 \psi_r^0 = 0$. We can choose the zero mode to be an eigenmode of γ_5: with $\psi_r^{0\pm} = \frac{1 \pm \gamma_5}{2}\psi_r^0$ we have $\gamma_5 \psi_r^{0\pm} = \pm\psi_r^{0\pm}$ with $i\slashed{D}\psi_r^{0\pm} = 0$ and $\frac{1\pm\gamma_5}{2}\psi_r^{0\mp} = 0$. The eigenvalue of γ_5 of the zero mode is called its chirality, which we will call χ_r for zero mode ψ_r^0. We do not know if $\psi_r^{0+} = 0$ or perhaps $\psi_r^{0-} = 0$, or possibly neither vanishes (in which case there are two zero modes, of chirality ± 1, respectively); however, both cannot vanish if $\psi_r^{0+} + \psi_r^{0-} = \psi_r^0 \neq 0$.

Let n_+ be the number of zero modes with positive chirality and n_- be the number of zero modes with negative chirality. The Atiyah–Singer index theorem states that $n_+ - n_- = \nu$. We can prove this theorem using the chiral Ward–Takahashi identities. Consider the simplest identity, without any fields ϕ^r, for a single doublet of fermions in $SU(2)$, $C = 1/2$, we have,

$$-2i\nu = 2M \int d^4x\, \langle \bar\psi \gamma_5 \psi \rangle^A = 2M \frac{\int \mathcal{D}(\psi,\bar\psi)\, e^{-S_E} \int d^4x\, \bar\psi \gamma_5 \psi}{\int \mathcal{D}(\psi,\bar\psi)\, e^{-S_E}} \tag{11.259}$$

with $S_E = \int d^4x\, \bar\psi \left(i\slashed{D} - iM \right) \psi$ and the solutions of the Dirac equation are unchanged from the massless case, only the energy eigenvalues are shifted

$$i(\slashed{D} - M)\psi_r = (\lambda_r - iM)\psi_r. \tag{11.260}$$

Clearly,

$$i(\slashed{D} - M)\gamma_5\psi_r = (-\lambda_r - iM)\gamma_5\psi_r, \tag{11.261}$$

but the eigenmodes ψ_r and $\gamma_5\psi_r$ must be orthogonal if $\lambda_r \neq 0$ as they are actually eigenmodes of the hermitean operator $i\slashed{D}$. We observe

$$\int d^4x\, \bar\psi_s \gamma_5 \psi_s = 0 \quad \text{if} \quad \lambda_s \neq 0, \tag{11.262}$$

but for the zero modes

$$\int d^4x\, \bar\psi_s^0 \gamma_5 \psi_s^0 = \chi_s, \tag{11.263}$$

thus

$$\int d^4x\, \bar\psi \gamma_5 \psi = \sum_{s,\lambda_s=0} \chi_s \bar b_s a_s. \tag{11.264}$$

The ψ_r are a complete and orthonormal basis of the space of fermion fields, hence we can write

$$\psi = \sum_r a_r \psi_r \quad \bar\psi = \sum_r \psi_r^\dagger \bar b_r \tag{11.265}$$

with Grassmann coefficients a_r and $\bar b_r$. Then the functional integral is given by

$$-2i\nu = 2M \frac{\int \prod_r da_r d\bar b_r\, e^{\sum_r (\lambda_r - iM)\bar b_r a_r} \int d^4x\, \bar\psi \gamma_5 \psi}{\prod_r (\lambda_r - iM)}$$
$$= 2M \frac{\int \sum_{s,\lambda_s=0} \prod_{r\neq s} (\lambda_r - iM)\chi_s}{\prod_r (\lambda_r - iM)} \tag{11.266}$$

as the fermionic integral gives $(\lambda_r - iM)$ for all the non-zero modes but a factor of 1 for the zero mode in the sum $\sum_{s,\lambda_s=0} \chi_s \bar b_s a_s$. The infinite product cancels between numerator and denominator for all the non-zero modes, and therefore the chiral Ward identity gives

$$-2i\nu = 2i \sum_s \chi_s = 2i(n_+ - n_-). \tag{11.267}$$

Thus $\nu = n_- - n_+$, which cannot be satisfied unless there are at least ν zero modes.

For the case $\nu = 1$ we can easily show that there is one zero mode with negative chirality and no zero modes of positive chirality. We note that the instanton configuration is self dual, $F_{\mu\nu} = \tilde{F}_{\mu\nu}$. We assume that there is a positive chirality zero mode $\slashed{D}\psi^{0+} = 0$ and $\gamma_5\psi^{0+} = \psi^{0+}$. Then

$$0 = (\slashed{D})^2\psi^{0+} = \left(\frac{1}{2}\{\gamma_\mu,\gamma_\nu\} + \frac{1}{2}[\gamma_\mu,\gamma_\nu]\right) D_\mu D_\nu \psi^{0+}$$

$$= D_\mu D_\mu \psi^{0+} + \frac{1}{4}[\gamma_\mu,\gamma_\nu]F_{\mu\nu}\psi^{0+}$$

$$= D^2\psi^{0+} + \frac{1}{4}[\gamma_\mu,\gamma_\nu]F_{\mu\nu}\psi^{0+}. \tag{11.268}$$

However,

$$F_{\mu\nu}\frac{1}{2}[\gamma_\mu,\gamma_\nu]\psi^{0+} F_{\mu\nu}\frac{1}{2}[\gamma_\mu,\gamma_\nu]\gamma_5\psi^{0+} = F_{\mu\nu}\left(-\frac{1}{2}\epsilon_{\mu\nu\sigma\tau}\gamma_\sigma\gamma_\tau\right)\psi^{0+}$$

$$= -\tilde{F}_{\mu\nu}\gamma_\mu\gamma_\nu\psi^{0+} = -F_{\mu\nu}\frac{1}{2}[\gamma_\mu,\gamma_\nu]\psi^{0+}. \tag{11.269}$$

Therefore, $F_{\mu\nu}\frac{1}{2}[\gamma_\mu,\gamma_\nu]\psi^{0+} = 0$ and consequently $D^2\psi^{0+} = 0$. Then

$$0 = \int d^4x (\psi^{0+})^\dagger (-D^2\psi^{0+}) = \int d^4x (D_\mu\psi^{0+})^\dagger (D_\mu\psi^{0+}), \tag{11.270}$$

which is positive unless $D_\mu\psi^{0+} = 0$ identically. Then, in the gauge $A_3 = 0$, this requires $\partial_3\psi^{0+} = 0$. However, this is inconsistent for a normalizable wave function except if $\psi^{0+} = 0$. Therefore, in fact no positive chirality zero mode can exist. Of course the analysis fails for a negative chirality solution, we cannot conclude $F_{\mu\nu}\frac{1}{2}[\gamma_\mu,\gamma_\nu]\psi^{0+} = 0$ for a negative chirality zero mode, and there has to be exactly one negative chirality zero mode so that $\nu = n_- - n_+$ is satisfied.

Therefore, the fermionic functional integral simply makes the contribution from all non-zero instanton sectors vanish. Thus the theta vacua are all degenerate in energy, and the chiral symmetry is certainly spontaneously broken.

11.6.5 Why is there no Goldstone Boson?

To see the non-existence of a Goldstone boson we must modify our chiral Ward identities. The following matrix element no longer makes sense in the non-zero instanton sector as the denominator vanishes,

$$\langle \phi^1(x_1)\cdots\phi^m(x_m)\rangle^A = \frac{\int \mathcal{D}(\psi,\bar{\psi})e^{-S_E(\psi,\bar{\psi})}\phi^1(x_1)\cdots\phi^m(x_m)}{\int \mathcal{D}(\psi,\bar{\psi})e^{-S_E(\psi,\bar{\psi})}}; \tag{11.271}$$

however, if we consider just the numerator

$$\langle\langle \phi^1(x_1)\cdots\phi^m(x_m)\rangle\rangle^A \equiv \int \mathcal{D}(\psi,\bar{\psi})e^{-S_E(\psi,\bar{\psi})}\phi^1(x_1)\cdots\phi^m(x_m) \tag{11.272}$$

then formally the symmetry properties are identical, and we find

$$\left(\frac{\partial}{\partial\alpha} - 2i\nu\right)\langle\langle\phi^1(x_1)\cdots\phi^m(x_m)\rangle\rangle^A = 0. \tag{11.273}$$

Now the matrix element in a theta vacuum is given by

$$\langle\theta|\phi^1(x_1)\cdots\phi^m(x_m)|\theta\rangle^A$$
$$= \frac{\int\mathcal{D}(A)e^{-S_E(A)}e^{i\nu\theta}\int\mathcal{D}(\psi,\bar\psi)e^{-S_E(\psi,\bar\psi)}\phi^1(x_1)\cdots\phi^m(x_m)}{\int\mathcal{D}(A)e^{-S_E(A)}e^{i\nu\theta}\int\mathcal{D}(\psi,\bar\psi)e^{-S_E(\psi,\bar\psi)}}$$
$$= \frac{\int\mathcal{D}(A)e^{-S_E(A)}e^{i\nu\theta}\langle\langle\phi^1(x_1)\cdots\phi^m(x_m)\rangle\rangle^A}{\int\mathcal{D}(A)e^{-S_E(A)}e^{i\nu\theta}\langle\langle 1\rangle\rangle^A} \tag{11.274}$$

and now the denominator does not vanish for $\int\mathcal{D}(\psi,\bar\psi)e^{-S_E(\psi,\bar\psi)}1 \neq 0$ for the sector $\nu = 0$. Thus clearly

$$\left(\frac{\partial}{\partial\alpha} - 2\frac{\partial}{\partial\theta}\right)\langle\theta|\phi^1(x_1)\cdots\phi^m(x_m)|\theta\rangle = 0. \tag{11.275}$$

This is quite interesting. It means that the $U_A(1)$ transformation corresponds equivalently to a change in θ, *i.e.* a $U_A(1)$ transformation changes one theta vacuum into another. The chiral symmetry is therefore spontaneously broken, and the degenerate set of vacua are exactly the theta vacua.

In summary, we have first found the degenerate, classical vacua and their quantum counterparts, $|n\rangle$. Then instantons have the effect of breaking the degeneracy obtained by quantum tunnelling between the different $|n\rangle$ vacua, and the new combinations $|\theta\rangle = \sum_n e^{in\theta}|n\rangle$ are the new energy eigenstates with spectrum

$$E(\theta)/V = -2K\cos\theta e^{-S_0}, \tag{11.276}$$

where S_0 is the classical Euclidean action of one instanton. The parameter θ has nothing to do with chiral symmetry; indeed, there are no fermions yet. But once massless fermions are added to the theory, all the effects of the instantons disappear, due to the appearance of a fermionic zero mode. The $|\theta\rangle$ states suddenly become degenerate, and a chiral transformation corresponds exactly to a transformation of θ. The chiral symmetry is spontaneously broken as there exist infinitely many vacua which are transformed into each other by the action of a chiral transformation. There is one possible way that the system could escape these conclusions, if $\partial/\partial\alpha\langle\theta|\phi^1(x_1)\cdots\phi^m(x_m)|\theta\rangle = \partial/\partial\theta\langle\theta|\phi^1(x_1)\cdots\phi^m(x_m)|\theta\rangle = 0$, *i.e.* nothing depends on α or θ. This would mean that chiral symmetry is manifest and not spontaneously broken, and the vacua $|\theta\rangle$ are just copies of a single, unique vacuum state. It is easy to dispose of this possibility. If we calculate

$$\langle\theta|\sigma_\pm|\theta\rangle, \tag{11.277}$$

where $\sigma_\pm = \bar{\psi}\left(\frac{1 \pm \gamma_5}{2}\right)\psi$ then

$$\frac{\partial}{\partial \alpha}\sigma_\pm = \pm 2i\sigma_\pm. \tag{11.278}$$

Then if $\langle\theta|\sigma_\pm|\theta\rangle \neq 0$ we have $\frac{\partial}{\partial\alpha}\sigma_\pm \neq 0$, which then requires that the symmetry is spontaneously broken. We will calculate $\langle\theta|\sigma_\pm|\theta\rangle$ next and show that it cannot vanish.

We have already done the Gaussian functional integral about the classical critical point, the instanton solution, up to a final integral over the scale size

$$K = \frac{2}{g^8}\int_o^\infty \frac{d\lambda}{\lambda^5}f(\lambda M), \tag{11.279}$$

where M is a renormalization point scale (represented by a mass), but now we should append this result with a fermionic functional integral. For a fermion in a background of a configuration of n well-separated instantons and anti-instantons, there are n fermionic zero modes to the operator $i\not{D}$. Then the corresponding fermionic functional integral over the corresponding Grassmann coefficients vanishes,

$$\int d\bar{a}_r da_r e^{0 \times \bar{a}_r a_r}\phi(x) = 0, \tag{11.280}$$

unless ϕ contains exactly the bilinear $\bar{a}_r a_r$, for each zero mode. This requires $2n$ fermionic fields. We are interested in the bilinear $\sigma_\pm$, which contains two fermionic fields. Hence the fermionic functional integral vanishes in all sectors of the gauge field except for the sector with $n = 1$. Indeed, we must have only exactly one instanton or one anti-instanton so that there is exactly one zero mode. We cannot have a configuration of n_+ instantons and n_- anti-instantons with $n_+ - n_- = \pm 1$, since this configuration will have $n = n_+ + n_- > 1$ fermionic zero modes and the fermionic functional integral will vanish.

Then in the sector with just one anti-instanton with $n_- = 1$, $n_+ = 0$ and self dual instanton fields, $F_{\mu\nu} = \tilde{F}_{\mu\nu}$ the fermionic functional integral will have just one term

$$\int d\bar{a}_r da_r e^{\sum_{r,\lambda_r \neq 0}\lambda_r \bar{a}_r a_r}(\psi^{0-})^\dagger\left(\frac{1-\gamma_5}{2}\right)\psi^{0-} = (\psi^{0-})^\dagger\left(\frac{1-\gamma_5}{2}\right)\psi^{0-}\prod_{\lambda_r \neq 0}\lambda_r$$

$$= (\psi^{0-})^\dagger\psi^{0-}\left(det'i\not{D}\right) \tag{11.281}$$

as the zero mode is a chirality -1, $\left(\frac{1-\gamma_5}{2}\right)\psi^{0-} = \psi^{0-}$. The fermionic zero modes satisfy $i\not{D}\psi^{0-}(x) = i\gamma_\mu(\partial_\mu + A_\mu(x))\psi^{0-}(x) = 0$. If we move the position of the anti-instanton, we change $A_\mu(x) \to A_\mu(x+X)$, then evidently $\psi^{0-}(x) \to \psi^{0-}(x+X)$ and $i\gamma_\mu(\partial_\mu + A_\mu(x+X))\psi^{0-}(x+X) = 0$. For the case of an instanton, $\nu = 1$ with $n_+ = 1, n_- = 0$, which can in principle also contribute, we immediately get a vanishing contribution since $(\psi^{0+})^\dagger(x)\left(\frac{1-\gamma_5}{2}\right)\psi^{0+}(x) = 0$ as the zero mode has chirality $+1$ as $\left(\frac{1-\gamma_5}{2}\right)\psi^{0+}(x) = 0$. In the denominator only $\nu = 0$ can contribute, and the Gaussian integral is done around the configuration $A_\mu = 0$.

We perform the fermionic functional integral first as a functional of the gauge fields. In the sector $\nu = 1$ we must integrate over the position of the anti-instanton since nothing depends on the position of the anti-instanton, which gives a factor of TV. For σ_+ only the sector $\nu = -1$ contributes. The action remains the same and $e^{\nu\theta} = e^{\pm\theta}$ for the two sectors $\nu = \pm 1$. Hence finally we get

$$\langle\theta|\sigma_\pm|\theta\rangle = \int_0^\infty \frac{d\lambda}{\lambda^5} e^{\frac{-8\pi^2}{g^2}} e^{\mp i\theta} g^{-8} f(\lambda M) \frac{det'(i\slashed{D})}{det(i\slashed{\partial})}. \tag{11.282}$$

Dimensional analysis gives

$$\frac{det'(i\slashed{D})}{det(i\slashed{\partial})} = \lambda h(\lambda M) \tag{11.283}$$

for some dimensionless function $h(\lambda M)$, thus

$$\langle\theta|\sigma_\pm|\theta\rangle = \int_0^\infty \frac{d\lambda}{\lambda^4} e^{\frac{-8\pi^2}{g^2}} e^{\mp i\theta} g^{-8} f(\lambda M) h(\lambda M) \neq 0. \tag{11.284}$$

This amplitude also satisfies the chiral Ward identity

$$\frac{\partial}{\partial\alpha}\langle\theta|\sigma_\pm|\theta\rangle = \langle\theta|\pm 2i\sigma_\pm|\theta\rangle = -2\frac{\partial}{\partial\theta}e^{\mp i\theta}\int_0^\infty \frac{d\lambda}{\lambda^4} e^{\frac{-8\pi^2}{g^2}} e^{\mp i\theta} g^{-8} f(\lambda M) h(\lambda M) \tag{11.285}$$

$$\left(\frac{\partial}{\partial\alpha} + 2\frac{\partial}{\partial\theta}\right)\langle\theta|\sigma_\pm|\theta\rangle = 0. \tag{11.286}$$

Thus the symmetry transformation, corresponding to a change in α,

$$\frac{\partial}{\partial\alpha}\langle\theta|\sigma_\pm|\theta\rangle \neq 0 \tag{11.287}$$

and the symmetry is spontaneously broken. But there is no Goldstone boson. Such a boson must be a pseudoscalar and should give a pole at $p^2 = 0$ in any matrix element (now continued back to Minkowski space) such as

$$\langle\theta|\sigma_+(x)\sigma_-(x)|\theta\rangle = \sum_n \langle\theta|\sigma_+(x)|n\rangle\langle n|\sigma_-(0)|\theta\rangle$$

$$= \int \frac{d^3p}{(2\pi)^3}\langle\theta|\sigma_+(x)|GB\,\vec{p}\rangle\langle GB\,\vec{p}|\sigma_-(0)|\theta\rangle\cdots$$

$$= \int \frac{d^3p}{(2\pi)^3}\langle\theta|e^{i\hat{P}_\mu\cdot x^\mu}\sigma_+(0)e^{-i\hat{P}_\mu\cdot x^\mu}|GB\vec{p}\rangle\langle GB\,\vec{p}|\sigma_-(0)|\theta\rangle\cdots$$

$$= \int \frac{d^3p}{(2\pi)^3}e^{-ip_\mu\cdot x^\mu}\langle\theta|\sigma_+(0)|GB\,\vec{p}\rangle\langle GB\,\vec{p}|\sigma_-(0)|\theta\rangle\cdots$$

$$= \int \frac{d^4p}{(2\pi)^4}e^{-ip_\mu\cdot x^\mu}\delta(p^2)\langle\theta|\sigma_+(0)|GB\,\vec{p}\rangle\langle GB\,\vec{p}|\sigma_-(0)|\theta\rangle\cdots \tag{11.288}$$

where $p_0 = |\vec{p}|$ and $|GB\,\vec{p}\rangle$ is a state with one Goldstone boson of momentum p_μ. Then using (note $p^\mu p_\mu \equiv p^2$)

$$\delta(p^\mu p_\mu) = \frac{1}{\pi} \Im m \left(\frac{1}{p^2 + i\epsilon} \right) \tag{11.289}$$

we get

$$\langle \theta | \sigma_+(x) \sigma_-(x) | \theta \rangle = \Im m \int \frac{d^4 p}{(2\pi)^4} e^{-i p_\mu \cdot x^\mu} \left(\frac{1}{p^2 + i\epsilon} \right)$$
$$\langle \theta | \sigma_+(0) | GB\,\vec{p} \rangle \langle GB\,\vec{p} | \sigma_-(0) | \theta \rangle \cdots \tag{11.290}$$

and the singularity appears because of the existence of a massless particle. However, in the calculation, there is only a contribution from the sector $n_+ + n_- = 2$, which can be from $\nu = 0, \nu = \pm 2$. Because there is one σ_+ operator and one σ_-, the sectors with $\nu = \pm 2$ corresponding to two instantons or to two anti-instantons simply vanish. Only the sector $\nu = 0$ remains. Here there are two possible contributions, one is the normal, perturbative contribution without any instantons. It is straightforward to verify that there is no massless pole in the perturbative calculation. The only non-trivial configurations come in the sector $\nu = 0$ that correspond to a well-separated pair of one instanton and one anti-instanton. This contribution will simply be a constant

$$\langle \theta | \sigma_+(x) \sigma_-(0) | \theta \rangle = \langle \theta | \sigma_+(x) | \theta \rangle \langle \theta | \sigma_-(0) | \theta \rangle$$
$$= \left(\int_0^\infty \frac{d\lambda}{\lambda^4} e^{\frac{-8\pi^2}{g^2}} e^{\mp i\theta} g^{-8} f(\lambda M) h(\lambda M) \right)^2 \tag{11.291}$$

and certainly will not contain a massless pole. Indeed, if we check any matrix element of a set of gauge-invariant operators, we will find no massless pole.

However, if we consider a gauge-variant operator, for example

$$G_\mu = 4\epsilon_{\mu\nu\lambda\sigma} Tr \left(A_\nu \partial_\lambda A_\sigma + \frac{2}{3} A_\nu A_\lambda A_\sigma \right) \tag{11.292}$$

then matrix elements with G_μ inserted will contain a massless pole. This is because $\partial_\mu G_\mu = \epsilon_{\mu\nu\lambda\sigma} Tr\,(F_{\mu\nu} F_{\lambda\sigma})$. Hence any matrix element with G_μ inserted must have no pole when contracted with p_μ. This implies that the original gauge-variant matrix element must have a structure of the form $\dfrac{p_\mu}{p^2 + i\epsilon}$, *i.e.* exactly a massless pole. For example, consider

$$\langle \theta | G_\mu(x) \sigma_-(0) | \theta \rangle = \int d^4 p\, e^{i p_\mu x^\mu} p_\mu \Sigma_-(p) \tag{11.293}$$

from Lorentz invariance. Then the divergence

$$\int d^4 x \langle \theta | \partial_\mu G_\mu(x) \sigma_-(0) | \theta \rangle = \int d^4 x\, d^4 p\, e^{i p_\mu x^\mu} i p^2 \Sigma_-(p) \tag{11.294}$$

must have a pole at $p^2 = 0$ if it does not vanish. This would then require

$$\Sigma_-(p) = \frac{C}{p^2 - i\epsilon} + \cdots .\tag{11.295}$$

However, $\int d^4x \, \partial_\mu G_\mu = 32\pi^2 \nu$, thus

$$\int d^4x \langle \theta | \partial_\mu G_\mu(x) \sigma_-(0) | \theta \rangle = 32\pi^2 \langle \theta | \sigma_-(0) | \theta \rangle \neq 0, \tag{11.296}$$

where the contribution to the matrix element of σ_- is only from the sector with $\nu = 1$. Thus $\Sigma_-(p)$ must have a pole at zero momentum.

12

Instantons, Supersymmetry and Morse Theory

As a final application of instanton methods, we will present an exposition of instantons in supersymmetric theories. The great simplification that occurs because of supersymmetry is the exact pairing of fermionic states with bosonic states, which makes the calculation of the fluctuation determinant very simple. The fermionic determinant exactly cancels the bosonic one.

Morse theory and the Morse inequalities concern the critical points of a function defined on a compact, Riemannian manifold, and the global topological aspects of the manifold. It was the genius of Witten [125, 123, 124] to point out that there is a deep connection between Morse theory and supersymmetric quantum mechanics defined on a manifold. This is what we hope to recount in this chapter.

We will require some familiarity with certain concepts from differential geometry which we will review here, but the reader should refer to more detailed texts [42, 60, 26] for a more complete picture.

12.1 A Little Differential Geometry

12.1.1 Riemannian Manifolds

We consider a compact, n-dimensional Riemannian manifold. A manifold is a point set with a topology (the definition of the open sets in $\mathcal{M}$) that is locally homeomorphic to $\mathbf{R}^n$. This means that each point in the manifold is contained in an open subset U_i of the manifold which can be mapped to $\mathbf{R}^n$ by a homeomorphism $\phi_i \ni \phi_i(U_i) \subseteq \mathbf{R}^n$. Homeomorphism means the mapping takes open sets in U_i to open sets in $\mathbf{R}^n$. The set of the U_is cover the manifold, i.e. $\cup_i U_i = \mathcal{M}$. Any such set of U_is is called an atlas and each individual U_i provides a coordinate chart. If two different coordinate charts U_i and U_j have a non-empty intersection, $U_i \cap U_j \neq \varnothing$, then the function $\phi_i \circ \phi_j^{-1}$ which maps points in $\phi_j(U_i \cap U_j) \to \phi_i(U_i \cap U_j)$, *i.e.* defines a function from $\mathbf{R}^n \to \mathbf{R}^n$ must be k times differentiable. This defines a $\mathbb{C}^k$ manifold. We will always simply take $\mathbb{C}^\infty$ manifolds. The (local) coordinates of each point in a given U_i are just the coordinates of the point to which it is mapped in $\mathbf{R}^n$.

12.1.2 The Tangent Space, Cotangent Space and Tensors

The manifold has a tangent space at each point P, $T_P(\mathcal{M})$, which is defined as the space of linear mappings of real-valued functions defined on the manifold to the real numbers which satisfy the Liebniz rule, $\vec{v}(fg) = (\vec{v}f)g + f(\vec{v}g)$. The dimension of the tangent space is also n. The elements of the tangent space are called vectors. A basis of the tangent space can be trivially given in terms of a system of local coordinates. If x^i are a set of coordinates at a point P of the manifold, then any linear mapping that satisfies the Liebniz rule, on the space of functions defined on the manifold at the point P can be defined by

$$\vec{v} : f(x) \to \mathbf{R} \ni \vec{v}(f) = v^i \partial_i f(x)|_P. \tag{12.1}$$

Thus a vector is equivalent to a set of n components $\vec{v} \equiv (v^1, v^2, \cdots, v^n)$. If the components of the vector are smoothly varying functions of the coordinates $v^i(x)$, then we define a vector field. The cotangent space $T_P^*(\mathcal{M})$ at the point P is simply defined as the dual vector space of the tangent space at the point P. The dual vector space of a given vector space is simply the space of linear mappings of the vector space to the real numbers, thus $T_P^*(\mathcal{M}) : T_P(\mathcal{M}) \to \mathbf{R}$. The dimensionality of $T_P^*(\mathcal{M})$ is also n. If we have an arbitrary basis E_i of $T_p(\mathcal{M})$, then the dual basis of $T_P^*(\mathcal{M})$ is defined by the condition

$$\langle E_i, e^j \rangle = \delta_i^j. \tag{12.2}$$

We name the dual basis to the coordinate basis ∂_i using the notation dx^j so that

$$\langle \partial_i, dx^j \rangle = \delta_i^j. \tag{12.3}$$

A general dual vector or "co-vector" can be written as $\vec{u}^* = u_j dx^j$ and then for a general vector $\vec{v} = v^i \partial_i$ we have

$$\langle \vec{v}, \vec{u}^* \rangle = v^i u_j \langle \partial_i, dx^j \rangle = v^i u_j \delta_i^j = v^i u_i. \tag{12.4}$$

If we change our system of coordinates of the coordinate chart at the point p, $x^i \to x'^j$, then the coordinate basis vectors of the tangent space transform simply as $\partial_i = \frac{\partial x'^j}{\partial x^i} \partial_j'$ or equivalently $\partial_j' = \frac{\partial x^i}{\partial x'^j} \partial_i$. But then the new dual basis vectors must be given by $dx'^j = \frac{\partial x'^j}{\partial x^i} dx^i$ or equivalently $dx^i = \frac{\partial x^i}{\partial x'^j} dx'^j$ so that the inner product between ∂_i and dx^j is preserved, *i.e.*

$$\begin{aligned}
\langle \partial_i, dx^j \rangle &= \left\langle \frac{\partial x'^k}{\partial x^i} \partial_k', \frac{\partial x^j}{\partial x'^l} dx'^l \right\rangle = \frac{\partial x'^k}{\partial x^i} \frac{\partial x^j}{\partial x'^l} \langle \partial_k', dx'^l \rangle \\
&= \frac{\partial x'^k}{\partial x^i} \frac{\partial x^j}{\partial x'^l} \delta_k^l = \frac{\partial x'^k}{\partial x^i} \frac{\partial x^j}{\partial x'^k} = \delta_i^j.
\end{aligned} \tag{12.5}$$

This then gives the transformation properties of the covariant and contravariant components of of vectors and co-vectors, indeed, $\vec{v} = v'^j \partial'_j = v'^j \frac{\partial x^i}{\partial x'^j} \partial_i = v^i \partial_i$ and $\vec{u}* = u'_j dx'^j = u'_j \frac{\partial x'^j}{\partial x^i} dx^i = u_i dx^i$. Hence $v^i = v'^j \frac{\partial x^i}{\partial x'^j}$ and $u_i = u'_j \frac{\partial x'^j}{\partial x^i}$ or equivalently $v'^i = v^j \frac{\partial x'^i}{\partial x^j}$ and $u'_i = u_j \frac{\partial x^j}{\partial x'^i}$. Then the inner product between arbitrary vectors and co-vectors is invariant

$$\langle \vec{v}, \vec{u}^* \rangle = v^i u_i = v'^i u'_i. \tag{12.6}$$

We note the possibly confusing nomenclature: the components of vectors are said to transform contravariantly while the components of co-vectors are said to transform covariantly.

We can also take tensor products of the tangent space k times and the cotangent space l times,

$$\underbrace{T_P \otimes \cdots \otimes T_P}_{k} \otimes \underbrace{T_P^* \otimes \cdots \otimes T_P^*}_{l} \tag{12.7}$$

to define tensors (and tensor fields)

$$t = t^{i_1 \cdots i_k}_{j_1 \cdots j_l} \partial_{i_1} \otimes \cdots \otimes \partial_{i_k} \otimes dx^{j_1} \otimes \cdots \otimes dx^{j_l}. \tag{12.8}$$

We should stress that at this point there is no relationship between the tangent spaces, the cotangent spaces and their tensor products over distinct points. The construction is independently done over each point. To use a leading terminology, there is, at the moment, no connection between tangent spaces at neighbouring points. The ensemble of the tangent spaces over all the points in the manifold defines a larger manifold called the tangent bundle, a fibre bundle over the manifold $\mathcal{M}$. The base manifold is $\mathcal{M}$ and the fibre is T_P over the point P in $\mathcal{M}$. There is also the corresponding cotangent bundle constructed with the cotangent space. The complete spaces are fibre bundles, spaces that locally permit a decomposition into a Cartesian product of a patch of the base manifold $\mathcal{M}$ cross the fibre, which would be the tangent space in the case of the tangent bundle, etc.

12.2 The de Rham Cohomology

12.2.1 The Exterior Algebra

The de Rham cohomology concerns the ensemble of the set of spaces of the completely anti-symmetric tensor products of the dual tangent space. We start with the cotangent space, T_P^*. Any two basis elements dx and dy can form an anti-symmetric two-co-tensor defined as

$$dx \wedge dy = \frac{1}{2}(dx \otimes dy - dy \otimes dx). \tag{12.9}$$

The product $\wedge$ is called the Cartan wedge product or the exterior product. Then an arbitrary anti-symmetric two-co-tensor is given by

$$t = t_{ij} dx^i \wedge dx^j. \tag{12.10}$$

This construction obviously generalizes to the notion of anti-symmetric p co-tensors constructed over each point x of the manifold. The set of anti-symmetric p co-tensors forms a sub-space of the p-fold tensor product of the co-tangent space which we will call $\Lambda^p(x)$. The set of $\Lambda^p(x)$s for all the points of the manifold forms a fibre bundle over $\mathcal{M}$. The elements of $\Lambda^p(x)$ are called differential forms, or more precisely p-forms. The dimensionality of $\Lambda^p(x)$ is obviously $\binom{n}{p}$ the number of ways of choosing p basis vectors from the total set of n basis vectors. We add in $\Lambda^0(x) = \mathbf{R}$, simply the real line, and then we have $n+1$ spaces of differential forms, $\Lambda^0(x)$ to $\Lambda^n(x)$, since for $\Lambda^{n+1}(x)$ or higher, it is no longer possible to anti-symmetrize $n+1$ or more co-vectors and these spaces are just empty. The space of smooth p-forms corresponds to the choice of the anti-symmetric tensor component fields $f_{i_1 \cdots i_p}(x)$, the corresponding tensor field being $f_{i_1 \cdots i_p}(x) dx^{i_1} \wedge \cdots \wedge dx^{i_p}$, which we write as $C^\infty(\Lambda^p)$ which is a space of dimension $\binom{n}{p}$. It is obvious that $C^\infty(\Lambda^p)$ and $C^\infty(\Lambda^{n-p})$ have the same dimensionality. The wedge product serves as a product on the full space of the direct sum of all possible anti-symmetric tensor fields

$$\Lambda^* = \Lambda^0 \oplus \Lambda^1 \oplus \cdots \oplus \Lambda^n, \tag{12.11}$$

which then defines an algebra called Cartan's exterior algebra.

12.2.2 Exterior Derivative

We can define the exterior derivative of a p-form, an operation d, which takes p forms to $p+1$ forms

$$d : C^\infty(\Lambda^p) \to C^\infty(\Lambda^{p+1}) \quad \ni$$
$$d(f_{i_1 \cdots i_p}(x) dx^{i_1} \wedge \cdots \wedge dx^{i_p}) = \left(\frac{\partial}{\partial x^j} f_{i_1 \cdots i_p}(x) \right) dx^j \wedge dx^{i_1} \wedge \cdots \wedge dx^{i_p}. \tag{12.12}$$

Note the placement of the additional dx^j by convention to the left of all the other differential forms. Obviously

$$dd\omega_p = 0 \tag{12.13}$$

for any p-form ω_p. The chain rule also simply follows, for ω_p a p-form and χ_q a q-form

$$d(\omega_p \wedge \chi_q) = (d\omega_p) \wedge \chi_q + (-1)^p \omega_p \wedge (d\chi_q). \tag{12.14}$$

As $C^\infty(\Lambda^p)$ and $C^\infty(\Lambda^{n-p})$ have the same dimensionality, we can define a duality mapping between these spaces, called the Hodge $*$ duality transformation. We define

$$* : C^\infty(\Lambda^p) \to C^\infty(\Lambda^{n-p}) \quad \ni$$

$$*(dx^{i_1} \wedge \cdots \wedge dx^{i_p}) = \frac{1}{(n-p)!} \epsilon^{i_1 \cdots i_p}_{\quad i_{p+1} \cdots i_n} dx^{i_{p+1}} \wedge \cdots \wedge dx^{i_n}$$

$$(12.15)$$

where $\epsilon^{i_1 \cdots i_p}_{\quad i_{p+1} \cdots i_n}$ is the completely anti-symmetric tensor in n dimensions. We have been careful about keeping indices up or down; however, it is important to point out that nothing we are doing requires the definition of a metric on the manifold. The exterior algebra and exterior differentiation does not depend on a metric. We note that

$$* * \omega_p = (-1)^{p(n-p)} \omega_p.$$

$$(12.16)$$

12.2.3 Integration

The space $C^\infty(\Lambda^n)$ is one-dimensional, there is only one n-form, $dx^1 \wedge \cdots \wedge dx^n$, thus it is easy to see that $dx^{i_1} \wedge \cdots \wedge dx^{i_n} = \epsilon^{i_1 \cdots i_n} dx^1 \wedge \cdots \wedge dx^n$. This form can be identified with the volume form on the manifold and we can define the integration over the manifold with this volume form; one simply integrates in $\mathbf{R}^n$ in the charts of any given atlas, making sure not to double count the contributions from regions where the charts intersect. The integration is independent of the coordinate system, since the volume form transforms exactly by the Jacobian of the coordinate transformation, $dx^1 \wedge \cdots \wedge dx^n = det\left(\frac{\partial x^i}{\partial x'^j}\right) dx'^1 \wedge \cdots \wedge dx'^n$. This integration generalizes trivially to integration over sub-manifolds of $\mathcal{M}$ of a given dimensionality p of a p-form defined over the sub-manifold. With the notion of integration, we can define an inner product on the space of p-forms

$$(\omega_p, \chi_p) = \int_{\mathcal{M}} \omega_p \wedge *\chi_p.$$

$$(12.17)$$

In terms of the coefficients, $\omega_p = \omega_{i_1 \cdots i_p} dx^{i_1} \wedge \cdots \wedge dx^{i_p}$ and $\chi_p = \chi_{j_1 \cdots j_p} dx^{j_1} \wedge \cdots \wedge dx^{j_p}$ then

$$(\omega_p, \chi_p) = \int_{\mathcal{M}} \omega_{i_1 \cdots i_p} dx^{i_1} \wedge \cdots \wedge dx^{i_p} \wedge *(\chi_{j_1 \cdots j_p} dx^{j_1} \wedge \cdots \wedge dx^{j_p})$$

$$= \int_{\mathcal{M}} \omega_{i_1 \cdots i_p} \chi_{j_1 \cdots j_p} dx^{i_1} \wedge \cdots \wedge dx^{i_p} \wedge \frac{1}{(n-p)!} \epsilon^{j_1 \cdots j_p}_{\quad j_{p+1} \cdots j_n} dx^{j_{p+1}} \wedge \cdots \wedge dx^{j_n}$$

$$= \int_{\mathcal{M}} \omega_{i_1 \cdots i_p} \chi_{j_1 \cdots j_p} \frac{1}{(n-p)!} \epsilon^{j_1 \cdots j_p}_{\quad j_{p+1} \cdots j_n} \epsilon^{i_1 \cdots i_p j_{p+1} \cdots j_n} dx^1 \wedge \cdots \wedge dx^n$$

$$= p! \int_{\mathcal{M}} \omega_{i_1 \cdots i_p} \chi_{j_1 \cdots j_p} \delta^{i_1 j_1} \cdots \delta^{i_p j_p} dx^1 \wedge \cdots \wedge dx^n,$$

$$(12.18)$$

The inner product is symmetric, $(\omega_p, \chi_p) = (\chi_p, \omega_p)$.

The next structure we will define is the adjoint of the exterior derivative, which we call δ. The inner product

$$\begin{aligned}
(\omega_p, d\chi_{p-1}) &= \int_{\mathcal{M}} \omega_p \wedge *d\chi_{p-1} = \int_{\mathcal{M}} (d\chi_{p-1}) \wedge *\omega_p \\
&= \int_{\mathcal{M}} d(\chi_{p-1} \wedge *\omega_p) - (-1)^{p-1} \chi_{p-1} \wedge d * \omega_p \\
&= \int_{\mathcal{M}} -(-1)^{p-1} \chi_{p-1} \wedge (-1)^{(n-p+1)(n-n+p-1)} * *d * \omega_p \\
&= \int_{\mathcal{M}} ((-1)^{np+n+1} *d * \omega_p) \wedge *\chi_{p-1} \\
&\equiv (\delta\omega_p, \chi_{p-1}),
\end{aligned} \tag{12.19}$$

where we have used trivial identities such as $(-1)^{2n} = 1$. Thus $\delta = (-1)^{np+n+1} * d*$, and note for n even the sign is always -1 and $\delta = - *d*$, while for n odd we get $\delta = (-1)^p * d*$. It is also easy to see $\delta\delta\omega_p = 0$.

The exterior algebra naturally gives rise to a Stokes theorem for manifolds with boundaries. If $\partial\mathcal{M}$ is the boundary of a p-dimensional manifold $\mathcal{M}$ and ω_{p-1} is an arbitrary $(p-1)$-form, then Stokes theorem states

$$\int_{\mathcal{M}} d\omega_{p-1} = \int_{\partial\mathcal{M}} \omega_{p-1}. \tag{12.20}$$

This theorem contains and generalizes all three of the usual Green, Gauss and Stokes theorems that are taught in an elementary course on vector calculus.

12.2.4 The Laplacian and the Hodge Decomposition

The Laplacian is now defined as

$$\nabla^2 = (d+\delta)^2 = d\delta + \delta d. \tag{12.21}$$

The Laplacian does not change the degree of the form. The Laplacian is a positive operator as

$$(\omega_p, \nabla^2 \omega_p) = (\omega_p, d\delta\omega_p + \delta d\omega_p) = (\delta\omega_p, \delta\omega_p) + (d\omega_p, d\omega_p) \geq 0, \tag{12.22}$$

assuming there are no boundaries. Therefore, $\nabla^2 \omega_p = 0$, and then ω is called a harmonic p-form, if and only if both $d\omega_p = 0$ (we say ω_p is closed) and $\delta\omega_p = 0$ (we say ω_p is co-closed).

A p-form that can be globally written as the exterior derivative of a $p-1$-form, *i.e.*

$$\omega_p = d\zeta_{p-1} \tag{12.23}$$

is called an exact p-form while if ω can be globally written as

$$\omega_p = \delta\xi_{p+1} \tag{12.24}$$

then it is called a co-exact p-form. The Hodge theorem states that on a compact manifold without boundary any p-form ω_p can be uniquely decomposed as the sum of an exact form, a co-exact form and a harmonic form

$$\omega_p = d\zeta_{p-1} + \delta\xi_{p+1} + \rho_p, \tag{12.25}$$

where ρ_p is a harmonic form, meaning that $\nabla^2\rho_p = 0$.

12.2.5 Homology

The homology of a manifold is the set of equivalence classes of sub-manifolds called cycles, boundaryless collections of sub-manifolds of dimension p, which differ only by boundaries. We start with our initial n-dimensional manifold $\mathcal{M}$. Then we define a p-chain as a formal sum of p-dimensional, smooth, oriented, sub-manifolds, $\mathcal{N}_i^p$, the formal, finite sum being written as $\alpha_p = \sum_i c_i \mathcal{N}_i^p$, where c_i are real, complex or integer, or even in the group $\mathbb{Z}_2$, giving rise to the corresponding p-chain. We continue to use the symbol ∂ as the operator that corresponds to taking the boundary, $\partial\alpha_p = \sum_i c_i \partial\mathcal{N}_i^p$, which is evidently a $(p-1)$-chain. Let Z_p be the set of boundaryless p-chains, which are called p-cycles. This means $\alpha_p \in Z_p \Rightarrow \partial\alpha_p = \varnothing$. Let X_p be the set of p-chains that are boundaries, *i.e.* $\alpha_p \in X_p \Rightarrow \alpha_p = \partial\alpha_{p+1}$. Since the boundary of a boundary is always empty, $X_p \subseteq Z_p$. Then the simplicial homology of $\mathcal{M}$ is defined as the set of equivalence classes H_p

$$H_p = Z_p/X_p, \tag{12.26}$$

i.e. the set of p-cycles that only differ from each other by boundaries are considered equivalent, $\alpha_p \sim \alpha_p' \Rightarrow \alpha_p = \alpha_p' + \partial\alpha_{p+1}$. H_p is obviously a group under the formal addition. The formal sum of two p-cycles commutes with the process of making equivalence classes with respect to p-cycles which are boundaries. The integral homology groups are the most fundamental, we can get the real, complex or $\mathbb{Z}_2$ homologies from them. We will write the homology groups as $H_p(\mathcal{M}, G)$, where $G = \mathbb{C}, \mathbf{R}, \mathbb{Z}, \mathbb{Z}_2$. $H_p(\mathcal{M}, G) = \varnothing$ for $p > n$.

$H_0(\mathcal{M}, G) = G$ if $\mathcal{M}$ is connected, since 0-cycles are just collections of points, the boundary of a point is empty. We can reduce any finite collection of points with arbitrary coefficients to a 0-cycle consisting of single point, $P \in \mathcal{M}$. Any 0-cycle, $\alpha_0 = \sum_i c_i P_i$, can be reduced to single point P with a coefficient $\sum_i c_i \in G$, using

$$\alpha_0 = \sum_i c_i P_i = \sum_i (c_i P_i - c_i P + c_i P) \sim \left(\sum_i c_i\right) P \tag{12.27}$$

as every pair of points with alternating coefficient, as appears above $c_i P_i - c_i P$, is the boundary of a 1-cycle corresponding to any curve joining the two points. However, cP is not equivalent to $\tilde{c}P$ for $c \neq \tilde{c} \in G$, thus the elements of $H_0(\mathcal{M}, G)$ are in a one-to-one correspondence with G. Obviously $H_n(\mathcal{M}, G) = G$ also, since there is only one sub-manifold of dimension n, $\mathcal{M}$ itself, in $\mathcal{M}$, and we see that $H_0(\mathcal{M}, G) = H_n(\mathcal{M}, G)$. This generalizes to what is called Poincaré duality, when $G = \mathbf{R}, \mathbb{C}$ or $\mathbb{Z}_2$ (all fields) we have $H_p(\mathcal{M}, G) = H_{n-p}(\mathcal{M}, G)$.

Finally, for $G = \mathbf{R}, \mathbb{C}$ or $\mathbb{Z}_2$ the homology group $H_p(\mathcal{M}, G)$ is clearly a vector space over G. We define the cohomology group $H^p(\mathcal{M}, G)$ simply as the dual vector space to $H_p(\mathcal{M}, G)$.

12.2.6 De Rham Cohomology

We define the de Rham cohomology group with respect to differential forms for $G = \mathbf{R}, \mathbb{C}$. With the definitions Z^p as the set of closed p-forms and X^p as the set of exact p-forms, the de Rham cohomolgy group is defined as

$$H^p_{dR}(\mathcal{M}, G) = Z^p / X^p, \tag{12.28}$$

i.e. the equivalence classes of closed modulo exact p-forms, $\omega_p \sim \omega'_p \Rightarrow \omega_p = \omega'_p + d\alpha_{p-1}$. For the special case of $H^0_{dR}(\mathcal{M}, G)$ we define this as the space of constant functions, as their exterior derivative vanishes. A zero-form cannot be the exterior derivative of any "-1" form, as these do not exist. The spectacular conclusion of the de Rham theorem asserts that these cohomology groups are in fact identical to the simplicial cohomology groups and hence dual to the simplicial homology groups.

We define the inner product of a p-cycle $\alpha_p \in Z_p$ with a closed p-form $\omega_p \in Z^p$ through the integral

$$\pi(\alpha_p, \omega_p) = \int_{\alpha_p} \omega_p. \tag{12.29}$$

It is easy to see that this inner product only depends on the equivalence class of α_p and of ω_p. Indeed,

$$\int_{\alpha_p} (\omega_p + d\chi_{p-1}) = \int_{\alpha_p} \omega_p + \int_{\alpha_p} d\chi_{p-1} = \int_{\alpha_p} \omega_p + \int_{\partial \alpha_p} \chi_{p-1} = \int_{\alpha_p} \omega_p \tag{12.30}$$

and

$$\int_{\alpha_p + \partial \beta_{p+1}} \omega_p = \int_{\alpha_p} \omega_p + \int_{\partial \beta_{p+1}} \omega_p = \int_{\alpha_p} \omega_p + \int_{\beta_{p+1}} d\omega_p = \int_{\alpha_p} \omega_p \tag{12.31}$$

as $\partial \alpha_p = \varnothing$ and $d\omega_p = 0$. Thus π gives a mapping

$$\pi : H_p(\mathcal{M}, G) \otimes H^p_{dR}(\mathcal{M}, G) \to G. \tag{12.32}$$

De Rham proved the following theorems. Let $\{c_i\}, i = 1, \cdots, dim(H_p(\mathcal{M}, R)$, be a set of independent p-cycles that form a basis of $H_p(\mathcal{M}, R)$. Then

1. For any given set of periods $\nu_i, i = 1, \cdots, dim(H_p(\mathcal{M}, R)$ there exists a closed p-form ω_p such that

$$\nu_i = \pi(c_i, \omega_p) = \int_{c_i} \omega_p. \tag{12.33}$$

2. If all the periods vanish for a give p-form ω_p, then ω_p is exact, *i.e.* $\omega_p = d\chi_{p-1}$.

This means that if $\{\omega_j\}$ is a basis of p-forms of $H^p_{dR}(\mathcal{M}, \mathbf{R})$ then the period matrix $\pi_{ij} = \pi(c_i, \omega_j)$ is invertible. This is equivalent to saying that $H^p_{dR}(\mathcal{M}, \mathbf{R})$ is dual to $H_p(\mathcal{M}, \mathbf{R})$. Consequently, the de Rham cohomology and the simplicial cohomology are naturally isomorphic and can be identified.

The Hodge theorem asserts that for each de Rham cohomology class there is an essentially unique harmonic form that can be taken as the representative of the class. Indeed, we have from the Hodge decomposition

$$\omega_p = d\zeta_{p-1} + \delta\xi_{p+1} + \rho_p. \tag{12.34}$$

Then evidently, the exact form $d\zeta_{p-1}$ is irrelevant in determining the equivalence class. ω_p being closed and ρ_p being harmonic, thus $d\omega_p = d\rho_p = 0$ which implies that $d\delta\xi_{p+1} = 0$, but then $0 = (\xi_{p+1}, d\delta\xi_{p+1}) = (\delta\xi_{p+1}, \delta\xi_{p+1})$ requires $\delta\xi_{p+1} = 0$. Thus $\omega_p = d\zeta_{p-1} + \rho_p$ and the de Rham cohomology class of ω_p is determined by the unique harmonic form ρ_p in its Hodge decomposition. This fact will be very important in the supersymmetric quantum mechanics that we will analyse in the later sections.

We define the Betti numbers as the dimension of the homology groups and consequently also the cohomology groups

$$B_p = dim(H_p(\mathcal{M}, \mathbf{R})) = dim(H^p_{dR}(\mathcal{M}, \mathbf{R})) = dim(H^p(\mathcal{M}, \mathbf{R})), \tag{12.35}$$

where B_p is the pth Betti number. The alternating sum of the Betti number is the Euler characteristic

$$\chi(\mathcal{M}) = \sum_{p=0}^{n} (-1)^p B_p \tag{12.36}$$

and we will see it is a topological invariant of the manifold. Morse theory relates the critical points of functions defined on a manifold to its Betti numbers.

12.3 Supersymmetric Quantum Mechanics

After this brief, condensed exposition of manifolds, structures defined on them and of the de Rham cohomology we can now move on to show how supersymmetry and instantons can be used to prove the global topological results framed in the Morse inequalities [62].

12.3.1 The Supersymmetry Algebra

In any quantum theory we can separate the Hilbert space $\mathbb{H}$ into $\mathbb{H} = \mathbb{H}^+ \oplus \mathbb{H}^-$, where $\mathbb{H}^+$ and $\mathbb{H}^-$ are the subspaces of bosonic and fermionic states, respectively. A supersymmetry corresponds to a transformation generated by conserved hermitean operators $Q_i, i = 1, \cdots, N$ that maps $\mathbb{H}^+$ to $\mathbb{H}^-$ and vice versa. We also define the operator $(-1)^F$, where F is the fermion number. Then $(-1)^F|\psi\rangle = |\psi\rangle$ for $|\psi\rangle \in \mathbb{H}^+$ while $(-1)^F|\psi\rangle = -|\psi\rangle$ for $|\psi\rangle \in \mathbb{H}^-$. The supersymmetry generators must anti-commute with $(-1)^F$, $\{Q_i, (-1)^F\} = 0$, which means that they are fermionic operators. On the other hand, they commute with the Hamiltonian $\mathcal{H}$, $[Q_i, \mathcal{H}] = 0$, which means that they are conserved. Finally, to define a supersymmetric theory we also impose $Q_i^2 = \mathcal{H}$ for any i and $\{Q_i.Q_j\} = 0$ for $i \neq j$, together giving

$$\{Q_i.Q_j\} = 2\delta_{ij}\mathcal{H}. \tag{12.37}$$

This definition of supersymmetry does not allow for Lorentz-invariant theories. This is because Lorentz transformations combine the Hamiltonian to the momentum generators. In $1 + 1$ dimensions we have only one momentum generator, P. The simplest algebra preserving Lorentz symmetry requires two supersymmetry operators, Q_1 and Q_2 and the algebra

$$Q_1^2 = \mathcal{H} + P, \quad Q_2^2 = \mathcal{H} - P, \quad \{Q_1, Q_2\} = 0. \tag{12.38}$$

This is compatible with the idea that $(\mathcal{H}, P)$ transform as a vector and (Q_1, Q_2) transform as a spinor. There is just one generator of Lorentz transformation M, taken hermitean, and

$$[M, \mathcal{H}] = iP, \quad [M, P] = i\mathcal{H}, \quad [M, Q_1] = i\frac{1}{2}Q_1, \quad [M, Q_2] = -i\frac{1}{2}Q_2. \tag{12.39}$$

Then, for example, $[M, \mathcal{H} + P] = i(\mathcal{H} + P)$, which is compatible with

$$[M, Q_1^2] = [M, Q_1]Q_1 + Q_1[M, Q_1] = i\frac{1}{2}Q_1Q_1 + Q_1 i\frac{1}{2}Q_1 = iQ_1^2 = i(\mathcal{H} + P). \tag{12.40}$$

From Equation (12.38) we easily find

$$\mathcal{H} = \frac{1}{2}(Q_1^2 + Q_2^2) \tag{12.41}$$

and therefore the Hamiltonian is positive semi-definite. Also, $[\mathcal{H}, (-1)^F] = [P, (-1)^F] = 0$ as they are quadratic in the supercharges, hence the Hamiltonian and the momentum generator are bosonic operators.

If there exists a single state $|0\rangle$ in the Hilbert space that is annihilated by the supercharges

$$Q_i|0\rangle = 0 \quad i = 1, 2, \tag{12.42}$$

then the supersymmetry is unbroken. Such a state obviously has zero energy and, since the Hamiltonian is positive semi-definite, $|0\rangle$ is the vaccum state. If there are many solutions to Equation (12.42), then the supersymmetry is also unbroken, but presumably the Hilbert space separates into superselection sectors of states constructed over each vacuum. If there are no states that satisfy Equation (12.42), then the supersymmetry is spontaneously broken. It is generally quite difficult to directly prove the existence of solutions to

Equation (12.42), or the lack thereof. However, the following indirect method sheds light on the question: one computes the index of one of the supersymmetry generators.

We are looking for states that are annihilated by both supersymmetry generators, $Q_i|0\rangle = 0$. Then with the algebra (12.38) it is easy to see that $P|0\rangle = 0$, thus we can restrict to the subspace $\mathbb{H}_{P=0}$, which is all states annihilated by P. This subspace also splits into a bosonic and a fermionic subspace, $\mathbb{H}_{P=0} = \mathbb{H}^+_{P=0} \oplus \mathbb{H}^-_{P=0}$. Within this subspace, $Q_1^2 = Q_2^2 = \mathcal{H}$, restricted to $\mathbb{H}_{P=0}$, we can look for states that are annihilated by one of the supercharges, call it $\tilde{Q}$, where $\tilde{Q}$ could be Q_1 or Q_2 or a linear combination of the two. $\tilde{Q}$ necessarily can only take a state in $\mathbb{H}^+_{P=0} \to \mathbb{H}^-_{P=0}$ and a state in $\mathbb{H}^-_{P=0} \to \mathbb{H}^+_{P=0}$. $\tilde{Q}$ has no other action. This fact then allows for the decomposition $\tilde{Q} = Q_+ + Q_-$, where Q_+ acts only on and maps $\mathbb{H}^+_{P=0} \to \mathbb{H}^-_{P=0}$ while Q_- acts only on and maps $\mathbb{H}^-_{P=0} \to \mathbb{H}^+_{P=0}$. Q_- is the adjoint of Q_+. The index of $\tilde{Q}$ restricted to $\mathbb{H}_{P=0}$ is then defined as

$$index(\tilde{Q}) = dim(Ker(Q_+)) - dim(Ker(Q_-)), \qquad (12.43)$$

where $Ker(Q_\pm)$ is the subspace of $\mathbb{H}^\pm_{P=0}$ that is annihilated by $Q_\pm$. If the index is non-zero then we know for sure that there are states that are annihilated by $\tilde{Q}$ and hence supersymmetry is unbroken. The $index(\tilde{Q})$ can be written as

$$index(\tilde{Q}) = Tr(-1)^F = n_B(E=0) - n_F(E=0) \qquad (12.44)$$

as the bosonic zero modes in $\mathbb{H}^+_{P=0}$ count as $+1$ for each mode and the fermionic zero modes in $\mathbb{H}^-_{P=0}$ count as -1 for each mode. The non-zero energy modes are necessarily paired because of the supersymmetry, and hence cancel pairwise in their contribution to the trace. The index being non-zero requires necessarily that there exists at least one zero energy state and hence we can conclude that in this case the supersymmetry is unbroken. In the sequel we will drop the subscript $P=0$ and take as given that we are working in the subspace with $P=0$.

12.3.2 Supersymmetric Cohomology

The Hamiltonian is given by

$$\mathcal{H} = QQ^\dagger + Q^\dagger Q \qquad (12.45)$$

with the superalgebra

$$Q^2 = Q^{\dagger 2} = 0, \qquad (12.46)$$

with consequently

$$[\mathcal{H}, Q] = [\mathcal{H}, Q^\dagger] = 0. \qquad (12.47)$$

There also exists the operator $(-1)^F$ and usually the fermion number operator F, which both commute with the Hamiltonian. The states in the Hilbert spaces are graded by the eigenvalue of $(-1)^F$. Bosonic states, a subspace denoted by

$\mathbb{H}^+$, take eigenvalue $+1$ while fermionic states, a subspace denoted by $\mathbb{H}^-$, take the eigenvalue -1. The fermion number operator is integer-valued, with bosonic states having an even number of fermions and fermionic states having an odd number. The Hamiltonian maps bosonic states to bosonic states and fermionic states to fermionic states, while the supercharges switch the two, mapping bosonic states to fermionic states and fermionic states to bosonic states.

$$\mathcal{H} : \mathbb{H}^+ \to \mathbb{H}^+, \quad \mathbb{H}^- \to \mathbb{H}^-$$
$$Q, Q^\dagger : \mathbb{H}^+ \to \mathbb{H}^-, \quad \mathbb{H}^- \to \mathbb{H}^+. \tag{12.48}$$

If we write the energy levels in an ordered list $E_0 < E_1 < \cdots$ then the Hamiltonian preserves the energy-level subspace and the Hilbert space can be decomposed in terms of subspaces $\mathbb{H}_m$ of fixed energy levels

$$\mathbb{H} = \bigoplus_m \mathbb{H}_m \tag{12.49}$$

with the action of the Hamiltonian, the supercharges and $(-1)^F$ satisfying

$$\mathcal{H}|_{\mathbb{H}_m} = E_m, \quad Q, Q^\dagger, (-1)^F : \mathbb{H}_m \to \mathbb{H}_m. \tag{12.50}$$

The energy-level subspace further decomposes into bosonic and fermionic subspaces $\mathbb{H}_m = \mathbb{H}_m^+ \oplus \mathbb{H}_m^-$ and while the Hamiltonian preserves the bosonic and fermionic subspaces (they are indeed eigensubspaces of the Hamiltonian) the supercharges exchange the two

$$Q, Q^\dagger : \mathbb{H}_m^+ \to \mathbb{H}_m^-, \quad \mathbb{H}_m^- \to \mathbb{H}_m^+. \tag{12.51}$$

The action of the operator Q twice, vanishes, $Q^2 = 0$. Thus we have the exact sequence:

$$\mathbb{H}^- \xrightarrow{\ Q\ } \mathbb{H}^+ \xrightarrow{\ Q\ } \mathbb{H}^- \xrightarrow{\ Q\ } \mathbb{H}^+ \tag{12.52}$$

An exact sequence means that the image of a given map in the sequence is the kernel of the subsequent map. This is called a $\mathbb{Z}_2$-graded complex of vector spaces as the fermionic and bosonic Hilbert spaces are graded with the $\mathbb{Z}_2$ charge with respect to the operator $(-1)^F$. This gives rise to the cohomology groups:

$$H^+(Q) = Kernel\left\{Q : \mathbb{H}^+ \to \mathbb{H}^-\right\} / Image\left\{Q : \mathbb{H}^- \to \mathbb{H}^+\right\}$$
$$H^-(Q) = Kernel\left\{Q : \mathbb{H}^- \to \mathbb{H}^+\right\} / Image\left\{Q : \mathbb{H}^+ \to \mathbb{H}^-\right\}. \tag{12.53}$$

We can further refine this complex by noting that at energy level $E_m \neq 0$, the action of Q does not take you out of the energy sector, since Q commutes with the Hamiltonian. If a vector $|E_m\rangle$ is Q closed, $Q|E_m\rangle = 0$, *i.e.* in the kernel of Q, then it is necessarily exact, *i.e.* in the image of the previous map, since

$$|E_m\rangle = \mathcal{H}|E_m\rangle / E_m = \left(QQ^\dagger + Q^\dagger Q\right)|E_m\rangle / E_m = Q\left(Q^\dagger |E_m\rangle / E_m\right). \tag{12.54}$$

Hence all states that are closed are also exact for all the non-zero energy levels, and thus cohomology groups are just determined by the states in the zero energy sector. For a state of zero energy $|E_0\rangle$ we have

$$0 = \langle E_0 | \mathcal{H} | E_0 \rangle = \langle E_0 | \left(QQ^\dagger + Q^\dagger Q \right) | E_0 \rangle = \left| Q^\dagger | E_0 \rangle \right|^2 + |Q|E_0\rangle|^2, \qquad (12.55)$$

which is only possible if both $Q|E_0\rangle = 0$ and $Q^\dagger |E_0\rangle = 0$. Thus the zero energy states are annihilated by Q and hence closed. But none of them are exact, $|E_0\rangle \neq Q|\alpha\rangle$, since, if this were true, $Q^\dagger |E_0\rangle = 0 = Q^\dagger Q|\alpha\rangle$, which implies $\langle \alpha | Q^\dagger Q | \alpha \rangle = |Q|\alpha\rangle|^2 = 0$, which is only possible if $Q|\alpha\rangle = 0$. Thus the cohomology groups can be identified with the set of zero energy states:

$$H^+(Q) = H_0^+$$
$$H^-(Q) = H_0^- \qquad (12.56)$$

where $H_0^\pm$ are the states of zero energy.

Q takes states of p fermions to states of $p+1$ fermions. It is reasonable to assign vanishing fermion number to states without fermions, and the action of Q an even number of times always gives back a bosonic subspace, while an odd number of times give us a fermionic subspace, hence with the notation that $\mathbb{H}^p$ is the subspace of states of p fermions, we have:

$$\mathbb{H}^+ = \oplus_{p \text{ even}} \mathbb{H}^p$$
$$\mathbb{H}^- = \oplus_{p \text{ odd}} \mathbb{H}^p. \qquad (12.57)$$

Then the $\mathbb{Z}_2$-graded exact sequence in Equation (12.52) becomes a $\mathbb{Z}$-graded exact sequence

$$\cdots \xrightarrow{Q} \mathbb{H}^{p-1} \xrightarrow{Q} \mathbb{H}^p \xrightarrow{Q} \mathbb{H}^{p+1} \xrightarrow{Q} \cdots \qquad (12.58)$$

and we can define the cohomology group at each p:

$$H^p(Q) = Kernel\left\{ Q : \mathbb{H}^p \to \mathbb{H}^{p+1} \right\} / Image\left\{ Q : \mathbb{H}^{p-1} \to \mathbb{H}^p \right\}. \qquad (12.59)$$

The Witten index then becomes the "Euler" characteristic of the complex

$$Tr(-1)^F = \sum_{p=0,1,\cdots} (-1)^p dim\left(H^p(Q) \right). \qquad (12.60)$$

12.3.3 1-d Supersymmetric Quantum Mechanics

Consider the action

$$S = \int dt L(t) = \int dt \left(\frac{1}{2} \dot{x}^2 - \frac{1}{2} \left(h'(x) \right)^2 + \frac{i}{2} \left(\psi^\dagger \dot{\psi} - \dot{\psi}^\dagger \psi \right) - h''(x) \psi^\dagger \psi \right).$$
$$(12.61)$$

The variables ψ and $\psi^\dagger$ are anti-commuting variables which will eventually be realized by the exterior derivative or some deformation of the exterior derivative. For the moment we just impose

$$\{\psi, \psi^\dagger\} = 0. \qquad (12.62)$$

The supersymmetric transformation is

$$\delta x = \epsilon \psi^\dagger - \epsilon^\dagger \psi$$
$$\delta \psi = \epsilon \left(i\dot{x} + h'(x)\right)$$
$$\delta \psi^\dagger = \epsilon^\dagger \left(-i\dot{x} + h'(x)\right), \qquad (12.63)$$

where $\epsilon = \epsilon_1 + i\epsilon_2$ is a complex fermionic parameter. It is reasonably easy to see that the action is invariant under the supersymmetry transformation. The conserved supercharges can be obtained by Noether's theorem

$$Q = \psi^\dagger \left(i\dot{x} + h'(x)\right)$$
$$Q^\dagger = \psi \left(-i\dot{x} + h'(x)\right). \qquad (12.64)$$

Quantizing the system corresponds to imposing the canonical commutation and anti-commutation relations

$$[x, p] = i$$
$$\{\psi, \psi^\dagger\} = 1 \qquad (12.65)$$

as the canonically conjugate momenta are $p = \partial L / \partial \dot{x}$ and $\pi_\psi = \partial L / \partial \dot{\psi} = i\psi^\dagger$, with $\{\psi, \pi_\psi\} = i$. (The convention taken with Grassmann derivatives is action from the left, $\partial \psi_1 \psi_2 / \partial \psi_1 = -\psi_2$, in the final analysis, it is just the algebra of the operators that counts.) The Hamiltonian is given by

$$\mathcal{H} = \frac{1}{2} p^2 + \frac{1}{2} \left(h'(x)\right)^2 + \frac{1}{2} h''(x) \left(\psi^\dagger \psi - \psi \psi^\dagger\right). \qquad (12.66)$$

The fermion number operator is $F = \psi^\dagger \psi$ and satisfies the commutation relations

$$[F, \psi] = -\psi, \quad [F, \psi^\dagger] = \psi^\dagger. \qquad (12.67)$$

As $\{\psi, \psi\} = 0 = \{\psi^\dagger, \psi^\dagger\}$ the fermionic fields satisfy the algebra of fermionic annihilation and creation operators and, if there exists the state $|0\rangle$ that is annihilated by ψ, which we assume $\psi|0\rangle = 0$, then the state $\psi^\dagger|0\rangle$ is the only other independent state in the theory. Evidently $\psi \psi^\dagger |0\rangle = |0\rangle$ and $\psi^\dagger \psi^\dagger |0\rangle = 0$. Thus we can write the fermionic operators as

$$\psi = \begin{pmatrix} 0 & 1 \\ 0 & 0 \end{pmatrix}, \quad \psi^\dagger = \begin{pmatrix} 0 & 0 \\ 1 & 0 \end{pmatrix}. \qquad (12.68)$$

The full Hilbert space of the theory will be the Hilbert space of the bosonic variable x, which is the space of complex-valued square-integrable functions of the variable x denoted by $L^2(\mathbf{R}, \mathbb{C})$, multiplying (tensored with the states $|0\rangle$ and the state $\psi^\dagger|0\rangle$),

$$\mathbb{H} = L^2(\mathbf{R}, \mathbb{C})|0\rangle \oplus L^2(\mathbf{R}, \mathbb{C})\psi^\dagger|0\rangle \qquad (12.69)$$

the first component is identified with the bosonic subspace and the second with the fermionic subspace. The supercharges remain form-invariant from their classical expressions

$$Q = \psi^\dagger(ip + h'(x))$$
$$Q^\dagger = \psi(-ip + h'(x)), \qquad (12.70)$$

and commute with the Hamiltonian. We can compute, with a little straightforward algebra, that indeed

$$\{Q, Q^\dagger\} = 2\mathcal{H}, \qquad (12.71)$$

hence the supersymmetry algebra is satisfied.

The supersymmetric ground states are determined by the two conditions:

$$Q|E_0\rangle = \begin{pmatrix} 0 & 0 \\ d/dx + h'(x) & 0 \end{pmatrix}|E_0\rangle = 0$$

$$Q^\dagger|E_0\rangle = \begin{pmatrix} 0 & -d/dx + h'(x) \\ 0 & 0 \end{pmatrix}|E_0\rangle = 0. \qquad (12.72)$$

Expanding $|E_0\rangle = \xi_1(x)|0\rangle + \xi_2(x)\psi^\dagger|0\rangle$ gives

$$\left(\frac{d}{dx} + h'(x)\right)\xi_1(x) = 0$$

$$\left(-\frac{d}{dx} + h'(x)\right)\xi_2(x) = 0, \qquad (12.73)$$

which are trivially solved as

$$\xi_1(x) = c_1 e^{-h(x)}$$
$$\xi_2(x) = c_2 e^{h(x)}. \qquad (12.74)$$

Obviously these solutions cannot both be square-integrable and the square-integrability depends on the behaviour of $h(x)$ as $x \to \pm\infty$. The four cases are $\lim_{x\to\pm\infty} h(x) = \pm\infty$, $\lim_{x\to\pm\infty} h(x) = \mp\infty$, $\lim_{x\to\pm\infty} h(x) = +\infty$ and $\lim_{x\to\pm\infty} h(x) = -\infty$, the first case being equivalent to the second. The first two cases yield no square-integrable solution and hence there are no supersymmetric ground states and $Tr(-1)^F = 0$. The latter two yield a solution with either $c_2 = 0$ or $c_1 = 0$; in each case there is one supersymmetric ground state, bosonic if $c_2 = 0$ yielding $Tr(-1)^F = 1$ and fermionic if $c_1 = 0$ yielding $Tr(-1)^F = -1$. Thus we know exactly that in the first two cases there are no supersymmetric ground states, while in the latter two cases there is exactly one, which is bosonic if the potential rises to $+\infty$ as $x \to \pm\infty$ and fermionic if the potential falls to $-\infty$ as $x \to \pm\infty$. We underline that these are exact results.

12.3.3.1 Supersymmetric harmonic oscillator The example of a harmonic oscillator is particularly simple; here we take $h(x) = \frac{\omega}{2}x^2$. Then the potential in our Hamiltonian is $\frac{1}{2}(h'(x))^2 = \frac{\omega^2}{2}x^2$ while the coefficient of the fermionic term is $h''(x) = \omega$. Thus the Hamiltonian is given by

$$\mathcal{H} = \frac{1}{2}p^2 + \frac{\omega^2}{2}x^2 + \frac{1}{2}\omega\left(\psi^\dagger\psi - \psi\psi^\dagger\right). \tag{12.75}$$

The harmonic oscillator has spectrum

$$\mathcal{E}_n = \left(n + \frac{1}{2}\right)\omega \quad n = 0, 1, 2, \cdots \tag{12.76}$$

for eigenstate $\phi_n(x)$, which are known to be Hermite polynomials multiplied by a Gaussian. The fermionic part yields the matrix

$$\frac{\omega}{2}\begin{pmatrix} -1 & 0 \\ 0 & 1 \end{pmatrix}, \tag{12.77}$$

which commutes with the harmonic oscillator and has the spectrum $\tilde{\mathcal{E}} = \left(-\frac{\omega}{2}, \frac{\omega}{2}\right)$. Thus the spectrum of the Hamiltonian is for $\omega > 0$,

$$E_n = \begin{cases} n\omega & \text{for } \phi_n(x)|0\rangle \\ (n+1)\omega & \text{for } \phi_n(x)\psi^\dagger|0\rangle \end{cases} \quad n = 0, 1, 2, \cdots \tag{12.78}$$

and for $\omega < 0$

$$E_n = \begin{cases} (n+1)|\omega| & \text{for } \phi_n(x)|0\rangle \\ n|\omega| & \text{for } \phi_n(x)\psi^\dagger|0\rangle \end{cases} \quad n = 0, 1, 2, \cdots. \tag{12.79}$$

We notice that for positive ω we have a bosonic zero mode but for negative ω the supersymmetric zero mode is fermionic.

12.3.4 A Useful Deformation

We will next consider a deformation where the supersymmetric harmonic oscillator corresponds to the lowest-level approximation. Consider the theory with $h(x)$ replaced with $th(x)$, where t is just a parameter (in no sense the time).

$$h(x) \to th(x). \tag{12.80}$$

Then the Hamiltonian becomes

$$\mathcal{H}_t = \frac{1}{2}p^2 + \frac{t^2}{2}\left(h'(x)\right)^2 + \frac{t}{2}h''(x)\left(\psi^\dagger\psi - \psi\psi^\dagger\right) \tag{12.81}$$

and we are interested in what happens as $t \to \infty$. In this limit, the potential $\frac{t^2}{2}(h'(x))^2$ becomes very large for most values of x, and the wave function is pushed into regions where the potential is small. The potential is small only at the places x_i where $h'(x_i) = 0$, *i.e.* critical points of the function $h(x)$. Around critical points, the potential can be approximated in the lowest approximation as a quadratic polynomial $\sim (x - x_i)^2$, which brings us back to the harmonic oscillator that we have just analysed. The frequency of the harmonic oscillator becomes $t\omega$, where $\omega = h''(x_i)$ at the critical point, and then the energy levels are linear in t. The fermionic term now has coefficient $th''(x_i) = t\omega$, and thus also gives a linear contribution in t to the energy, which exactly cancels the oscillator ground-state zero-point energy for the bosonic case if $\omega > 0$ and for the fermionic case if $\omega < 0$, just as we have seen explicitly above for the exact harmonic oscillator.

Thus we are left with exactly one energy level at each critical point whose energy does not scale linearly with t. The energy of the state is zero in the approximation that we have employed. It may well be exactly zero, but this is not yet determined. However, we do know that without any approximations there is only one or no exact supersymmetric ground state in the theory, depending on the asymptotic behaviour of $h(x)$. Thus all or all but one of the zero-energy levels that we have found approximately must in fact have non-zero energy. What will be clear is that the exact energy levels of the corresponding exact eigenstates, which are concentrated about the critical points of $h(x)$ (as we have found to be approximately the case), will not scale linearly with t. To first order in the approximation, they are zero-energy modes. Perturbatively, they will actually remain zero-energy modes to all orders. Their energy can only become non-zero through non-perturbative corrections. These non-perturbative corrections are just instanton corrections, corresponding to tunnelling transitions between the perturbative zero-energy modes.

Expanding the function $h(x)$ about a critical point x_i where $h'(x_i) = 0$, and assuming $h''(x_i) \neq 0$, which simply means that the critical points are non-degenerate, we have

$$h(x) = h(x_i) + \frac{1}{2}h''(x_i)(x - x_i)^2 + \frac{1}{6}h'''(x_i)(x - x_i)^3 + \cdots \qquad (12.82)$$

and evidently

$$h'(x) = h''(x_i)(x - x_i) + \frac{1}{2}h'''(x_i)(x - x_i)^2 + \cdots . \qquad (12.83)$$

Scaling $x - x_i \to \tilde{x} - \tilde{x}_i = (x - x_i)/\sqrt{t}$ and correspondingly $p \to \tilde{p} = \sqrt{t}p$ gives

$$h(x) = h(x_i) + \frac{1}{2t}h''(x_i)(\tilde{x} - \tilde{x}_i)^2 + \frac{1}{6t^{3/2}}h'''(x_i)(\tilde{x} - \tilde{x}_i)^3 + o\left(\frac{1}{t^2}\right) \qquad (12.84)$$

and for the Hamiltonian

$$\mathcal{H}_t = t\left(\frac{1}{2}\tilde{p}^2 + \frac{1}{2}\left(h''(x_i)\right)^2 (\tilde{x} - \tilde{x}_i)^2 + \frac{1}{2}h''(x_i)\left(\psi^\dagger\psi - \psi\psi^\dagger\right)\right)$$

$$+ o(\sqrt{t}) + o(1) + o\left(\frac{1}{\sqrt{t}}\right) + \cdots . \qquad (12.85)$$

Thus we can imagine computing perturbatively in $1/\sqrt{t}$ where the leading term is given by

$$
\begin{aligned}
\mathcal{H}_{local} &= t\left(\frac{1}{2}\tilde{p}^2 + \frac{1}{2}\left(h''(x_i)\right)^2(\tilde{x} - \tilde{x}_i)^2 + \frac{1}{2}h''(x_i)\left(\psi^\dagger\psi - \psi\psi^\dagger\right)\right) \\
&= \frac{1}{2}p^2 + t^2\frac{1}{2}\left(h'(x_i)\right)^2 + t\frac{1}{2}h''(x_i)\left(\psi^\dagger\psi - \psi\psi^\dagger\right)
\end{aligned}
\tag{12.86}
$$

where in the last equality we have put back $\tilde{x} \to x = \sqrt{t}\tilde{x}$ after shifting so that the critical point occurs at $x = 0$. Obviously this is the supersymmetric harmonic oscillator that we have just treated and completely understand. There will be, in this approximation, one bosonic supersymmetric ground state of zero energy, as in Equation (12.78), for each critical point with $h''(x_i) > 0$ and one fermionic supersymmetric ground state of zero energy, as in Equation (12.79), for each critical point with $h''(x_i) < 0$. The eigenstate, say if bosonic, will be of the form (unnormalized)

$$
|E_0\rangle \approx e^{-\frac{t}{2}h''(x_i)(x-x_i)^2}|0\rangle,
\tag{12.87}
$$

which is the first approximation to the exact zero-energy state (unnormalized) which in this case is

$$
|E_0\rangle = e^{-th(x)}|0\rangle
\tag{12.88}
$$

but with $h(x)$ expanded about x_i with the constant value of $h(x_i)$ absorbed into the normalization. Evidently, if we compute the perturbative corrections to the energy state in Equation (12.87), using the perturbatively (in $1/\sqrt{t}$) expanded Hamiltonian (12.85), we will simply rebuild the exact zero-energy state given in Equation (12.88) from a Taylor expansion of $h(x)$. However, at each stage of the perturbative calculation the wave function will be concentrated around $x = x_i$, a Gaussian multiplied by polynomial corrections corresponding to the higher levels of the harmonic oscillator. The energy admits an expansion in even powers of $1/\sqrt{t}$ since the contribution from odd powers vanishes due to parity. However, the energy must actually remain zero at all stages of the perturbation, since we know that the energy of the exact wave function is exactly zero. Perturbative contributions at a higher order cannot correct a non-zero contribution to the energy at a lower order, hence the correction to the energy must be absent at each order. We can do this calculation around each critical point and, hence, perturbatively we will construct as many zero-energy modes as there are critical points.

Since we know that in fact there is at most only one exact zero-energy mode, all but one combination of these perturbatively found zero modes must be non-perturbatively corrected to finite energy. The Witten index will be given as

$$
Tr(-1)^F = \sum_{i=1}^{N}\mathrm{sign}(h''(x_i))
\tag{12.89}
$$

as each bosonic zero mode for $h''(x_i) > 0$ contributes $+1$ and each fermionic zero mode for $h''(x_i) < 0$ contributes -1. Evidently, this sum must equal ± 1 or 0, as we have found, dependent on the asymptotic behaviour of $h(x)$. This makes perfect sense as the number of concave and convex critical points can only change in equal numbers if we deform $h(x)$ locally, as long as the asymptotic behaviour of $h(x)$ is kept invariant. As we have said, these perturbatively found zero modes must not be exact zero modes, thus they must lift away from zero energy due to non-perturbative corrections. But then supersymmetry imposes that for each bosonic mode lifting away from zero energy there must be a corresponding fermionic one that is exactly degenerate in (non-zero) energy. Thus the non-perturbative corrections must simultaneously lift the bosonic and fermionic perturbatively found zero modes away from zero energy in pairs.

The generalization of this theory to n dimensions and on a Riemannian manifold will bring us to Morse theory in the next section.

12.4 Morse Theory

There is a connection between the Betti numbers and critical points of real-valued functions defined on a manifold [94, 10, 89, 125, 62, 19, 64]. We do not consider arbitrary real-valued functions, but an essentially generic class of functions that are called Morse functions. Morse functions, for which we will use the notation $h(x)$, are defined to be those real-valued functions that have a finite number of non-degenerate, isolated critical points. A critical point is where the first derivative of the function vanishes, which evidently is independent of the system of coordinates that are used. Thus the critical points of a Morse function occur at a finite number of discrete points, P_a, and the condition that they be non-degenerate means that the determinant of the matrix of second derivatives in any system of local coordinates containing P_a, the so-called Hessian matrix, has a non-zero determinant. This means the eigenvalues of the Hessian are non-zero. We can diagonalize the Hessian, a real symmetric matrix, by an orthogonal transformation of the coordinates, and shift the coordinates so that the critical point occurs at the origin of the coordinates. We can also rescale the resulting coordinates so that the positive eigenvalues are $+1$ and the negative eigenvalues are -1. Then around a critical point P_a with p negative directions, there exists a coordinate system in which a Morse function appears as

$$h(x) = c_a - \sum_{i=1}^{p} x_i^2 + \sum_{i=p+1}^{n} x_i^2, \tag{12.90}$$

where c_a is the value of the Morse function at the critical point. This rather reasonable fact corresponds to the Morse Lemma. It is clear that with an infinitesimal deformation of the Morse function, all values of the function c_a at the critical points can be taken to be distinct. Furthermore, we can assume that the values of the Morse function at the critical points are labelled in a monotone, ascending order, $c_l < c_{l+1}$, $l = 0, \cdots, N$. p is called the Morse index of a critical point, and the number of critical points with Morse index p is called the M_p.

A surprising fact corresponds to the understanding that the manifold can be reconstructed out of any Morse function that is defined on it. One considers the inverse map defining the submanifold (not including its boundary)

$$\mathcal{M}_c = \{x \in \mathcal{M} \ni h(x) < c\}. \tag{12.91}$$

Clearly for $c < c_0$, where c_0 is the global minimum of the Morse function, $\mathcal{M}_{c<c_0} = \varnothing$. The global minimum must exist as the manifold is assumed to be compact. As we increase c, when we pass c_0, but stay below the next critical point where the value of the Morse function is c_1, the manifold $\mathcal{M}_{c_1>c>c_0}$ is topologically always the same and what is called a 0-cell. The nomenclature, 0-cell, corresponds to the fact that the critical point which is the global minimum has 0 negative directions. A 0-cell is in fact topologically an n-dimensional ball, without its boundary. It is evident that the topology of $\mathcal{M}_{c>c_0}$ does not change as we increase c, until we come to the value of c_1, the next critical point of the Morse function. At the critical point c_1, there are p negative directions and $n-p$ positive directions. The manifold must attach a p-cell to the 0-cell that rises from the global minimum and the topology of the manifold must change as c passes from below c_1 to above c_1 by the attachment of a p-cell. A p-cell corresponds to a topological manifold that has p negative directions and $n-p$ positive directions, such a manifold is sometimes called a p-handle.

This construction will continue at each critical point of the Morse function. The topology of the set $\mathcal{M}_c$ will be invariant for $c_l < c < c_{l+1}$, the topology change occurring exactly and only at the critical points of the Morse function with values c_l. At each critical point of the Morse index p we will have to attach a p-cell. Finally, for $c > c_N$, where c_N is the global maximum of the Morse function,

$$\mathcal{M}_{c>c_N} = \mathcal{M} \tag{12.92}$$

and at this point we will have reconstructed the entire manifold. As we approach the final critical point, we must attach an n-cell, as the global maximum has n negative directions. An n-cell is also, topologically, an n-dimensional ball, as was the 0-cell at the global minimum, except that it now has n negative directions. Nothing precludes the attachment of n-cells, 0-cells or in general any number of p-cells at intermediate critical points; if there are local critical points with p negative directions, that is what is required. Indeed, in principle, for a critical point of the Morse index p, we must add a p-cell. The detailed description of this attachment of p-cells, or p-handles as they are sometimes called, is rather straightforward and unremarkable. We will not describe it in any more detail. The reader can consult the literature cited above for the full details.

Obviously, the reconstruction of the manifold based on a given Morse function must obey some constraints imposed on it due to the actual global topology of the manifold. The actual global topology of the manifold cannot arbitrarily change by its reconstruction based on a given Morse function. The actual topology of the manifold specifically constrains how many p-cells exist in the manifold. Hence the reconstruction based on a Morse function must be in some sense redundant. This gives the first hint that the number of critical points with Morse index p must be restricted by the global topology of the manifold.

The crudest example of such a restriction is, for example, the condition that there must exist only one global maximum and one global minimum for the Morse function. The topology of the manifold, that it is compact, imposes this condition. As any Morse function on the manifold can be interpreted as a height function, with a corresponding topology preserving deformation of the manifold, we can easily see that it is possible to eliminate pairwise, for example, a local maximum and a local minimum by simply deforming the Morse function or equivalently the manifold. Indeed, we will be able to show that the number of critical points, M_p, of the Morse index p is bounded below by exactly the topological properties of the manifold expressed in the Betti number B_p,

$$M_p \geq B_p. \tag{12.93}$$

These correspond to the weak Morse inequalities. There are also strong Morse inequalities, which we will introduce when appropriate in the sequel.

12.4.1 Supersymmetry and the Exterior Algebra

The realization of supersymmetry that we will use corresponds to the following identification in the exterior algebra of a Riemannian manifold, $\mathcal{M}$, of dimension, n, where we will further assume that it is equipped with a smooth metric g_{ij}. Let $Q = d$, $Q^\dagger = \delta$, and

$$Q_1 = d + \delta \quad Q_2 = i(d - \delta), \quad \mathcal{H} = d\delta + \delta d. \tag{12.94}$$

Then

$$\mathcal{H} = Q_1^2 = Q_2^2 \quad \text{and} \quad \{Q_1, Q_2\} = 0, \tag{12.95}$$

i.e. the supersymmetry algebra is satisfied. p-forms are bosonic or fermionic depending on whether p is even (bosonic) or odd (fermionic). The Q_i map bosonic states to fermionic states.

What are the supersymmetric ground states for this quantum-mechanical theory? Evidently they are the zero modes of the Laplacian, those p-forms that are annihilated by the Laplacian, $\mathcal{H} = d\delta + \delta d = \nabla^2$. But these are just the harmonic forms. The harmonic forms satisfy exactly

$$\nabla^2 \rho_p = 0. \tag{12.96}$$

The number of harmonic p-forms is exactly the dimension of the pth homology group, $dim H_p(\mathcal{M}, \mathbf{R})$. Hence the number of supersymmetric ground states, $dim H^p(Q)$, of a supersymmetric quantum mechanics defined on a Riemannian manifold, $\mathcal{M}$, is exactly equal to the Betti numbers, B_p, of the manifold. Interestingly, supersymmetry has some relation to the global topology of the manifold as defined by the Betti numbers.

The Witten index is obviously a topological invariant, the number of supersymmetric ground states can only change by pairs of bosonic–fermionic states lifting away from zero energy or coming down to zero energy. Therefore, we see that the Euler characteristic

$$\chi(\mathcal{M}) = \sum_{p=0}^{n} (-1)^p B_p = \sum_{p=0}^{n} (-1)^p dim H^p(Q) = Tr\,(-1)^F \qquad (12.97)$$

is in fact a topological invariant of the manifold.

12.4.2 The Witten Deformation

We deform the exterior algebra with an additional real parameter, t, and an arbitrary smooth real-valued function, $h(x)$, defined on $\mathcal{M}$, which will be the appropriate Morse function, and then we let

$$d_t = e^{-ht} d e^{ht} \quad \delta_t = e^{ht} \delta e^{-ht}. \qquad (12.98)$$

These operators continue to satisfy $d_t^2 = 0 = \delta_t^2$, and so we define

$$Q_{1t} = d_t + \delta_t, \quad Q_{2t} = i(d_t - \delta_t), \quad \mathcal{H}_t = d_t \delta_t + \delta_t d_t \qquad (12.99)$$

and the supersymmetry algebra is satisfied for each t

$$Q_{1t}^2 = Q_{2t}^2 = \mathcal{H}_t, \quad \{Q_{1t}, Q_{2,t}\} = 0. \qquad (12.100)$$

Then with $Q_t = (Q_{1t} - iQ_{2t})/2$, deformed supercharges are given by

$$Q_t = d + t\,dh \wedge \quad Q_t^\dagger = \delta + t(dh \wedge)^\dagger. \qquad (12.101)$$

As before, the exact supersymmetric ground states are those that are exactly annihilated by Q_t and by $Q_t^\dagger$. These would be the analogue of the harmonic forms. In the local coordinate system these are easily determined; for example, for states annihilated by Q_t we need to find p-forms that satisfy

$$Q_t \omega_p = (d + t\,dh)\omega_p = 0. \qquad (12.102)$$

Writing $\omega_p = \omega_{i_1 \cdots i_p} dx^{i_1} \wedge \cdots \wedge dx^{i_p}$ we get

$$(\partial_i \omega_{i_1 \cdots i_p} + t\partial_i h \omega_{i_1 \cdots i_p}) dx^i \wedge dx^{i_1} \wedge \cdots \wedge dx^{i_p} = 0. \qquad (12.103)$$

This has an evident solution

$$\omega_{i_1 \cdots i_p} = e^{-th} c_{i_1 \cdots i_p}, \qquad (12.104)$$

where $c_{i_1 \cdots i_p}$ is constant, and similarly for $Q_t^\dagger$. However, this does not mean that we have actually found a harmonic form, the coordinate system is in principle only a patch on the manifold. To find the set of harmonic forms is, in general, a complicated exercise. The set of exact supersymmetric ground states does exist and their numbers are given by the corresponding Betti numbers.

We define the Betti numbers, B_p, analogous to the definition of the de Rham cohomology, as the number of linearly independent p-forms that satisfy $d_t \omega_p = 0$, *i.e.* closed with respect to d_t, but which cannot be written as the exterior derivative of a $p-1$-form, *i.e.* $\omega_p \neq d_t \chi_{p-1}$, *i.e.* that are not exact with respect to d_t. The point is that this definition of the Betti numbers is actually independent of the parameter t, the Betti numbers so defined must be equal to their usual values at $t = 0$. d_t differs from d by conjugation with an invertible operator e^{ht}, thus the mapping $\omega_p \to e^{-ht} \omega_p$ is an invertible mapping of closed but not exact p-forms in the sense of d, mapped to closed but not exact p-forms in the sense of d_t. The dimensions of these spaces are independent of t.

At each point, P, of the manifold, $\mathcal{M}$, choose a basis, $\{a_k\}$, of the tangent space, T_P. We will also consider the dual basis $\{a^{*k}\}$ of the cotangent space T_P^*. The tangent space basis vectors and the dual space basis vectors can be thought of as operators on the exterior algebra, acting through what is called interior product for the $\{a_k\}$ and through the usual exterior product for the $\{a^{*k}\}$. Thus explicitly we have

$$a^{*i} = dx^i \wedge$$
$$a_i = \iota_{\partial/\partial_i}, \tag{12.105}$$

where the interior product ι_V is defined as

$$\iota_V(\omega_p) = \chi_{p-1} \ni \chi_{p-1}(V_1, \cdots, V_{p-1}) = \omega_p(V, V_1, \cdots, V_{p-1}). \tag{12.106}$$

This is just a fancy way of saying that we contract the vector index on the first index of the differential form. Thus for the present case $V = \delta_k^i a_i = a_k$ and $\omega_p = \omega_{i_1 \cdots i_p} a^{*i_1} \wedge \cdots \wedge a^{*i_p}$ then $\iota_{a_k}(\omega_p) = \omega_{k, i_2 \cdots i_p} a^{*i_2} \wedge \cdots \wedge a^{*i_p}$. Even more explicitly

$$a_k(a^{*i_1} \wedge \cdots \wedge a^{*i_p}) = \sum_{l=1}^{p} (-1)^{l-1} \delta_k^{i_l} a^{*i_1} \wedge \cdots \wedge a^{*i_{l-1}} \wedge a^{*i_{l+1}} \wedge \cdots a^{*i_p}. \tag{12.107}$$

The operators $\{a^{*k}\}$ are dual to the $\{a_k\}$, and their action on the exterior algebra corresponds simply to exterior multiplication. Explicitly, the action on a given p-form is simply given by $a^{*k}(\omega_p) = a^{*k} \wedge \omega_p$. These operators play the role of fermion creation and annihilation operators. The supercharges can be written in this notation as

$$Q_t = d + t \partial_i h a^{*i} \wedge \quad \text{and} \quad Q_t^\dagger = \delta + t g^{ij} \partial_i h a_j. \tag{12.108}$$

The function $h(x)$ can be differentiated in the coordinate system, then one can calculate in a straightforward, but somewhat tedious, manner,

$$\mathcal{H}_t = d\delta + \delta d + t^2 g^{ij} \partial_i h \partial_j h + t g^{jk} D_i D_j h \left[a^{*i}, a_k \right], \tag{12.109}$$

where g^{ij} is the assumed Riemannian metric on the manifold $\mathcal{M}$ and D_i is the covariant derivative with respect to the Levi–Civita connection associated to the metric, explicitly, $D_i D_j h = D_i \partial_j h = \partial_i \partial_j h - \Gamma_{ij}^l \partial_l h$ with $\Gamma_{jk}^i = \frac{1}{2} g^{il} \left(\partial_i g_{lj} + \partial_j g_{li} - \partial_l g_{ij} \right)$. For large t, the potential $t^2 g^{ij} \partial_i h \partial_j h$ dominates, and the wave function concentrates about the minima (critical points) of this potential. Corrections can be computed as an expansion in powers of $1/\sqrt{t}$, exactly as in the one-dimensional case.

12.4.3 The Weak Morse Inequalities

$h(x)$ will be called the Morse function, and we will assume it is non-degenerate, meaning that it only has isolated critical points at coordinates x^a, at which $\partial_j h(x^a) = 0$. Therefore, at each critical point the matrix of second derivatives, $D_i D_j h$, must be non-singular, *i.e.* it does not have any vanishing eigenvalues. We define M_p to be the number of critical points with p negative eigenvalues. The first Morse inequality states that $M_p \geq B_p$, which we will be able to prove with our supersymmetric quantum mechanical model.

Let $\lambda_p^{(n)}(t)$ be the nth smallest eigenvalue of $\mathcal{H}_t$ acting on p-forms. We will see that

$$\lambda_p^{(n)}(t) = t A_p^{(n)} + o(1) + o(1/t), \tag{12.110}$$

which admits an expansion in powers of $1/t$ due to parity. The Betti number, B_p, is equal to the number of exactly zero eigenvalues. For large t, the number of the eigenvalues that vanish can be no larger than the number of $A_p^{(n)}$ that vanish, simply because a vanishing eigenvalue requires $A_p^{(n)} = 0$. We will show that the number of $A_p^{(n)}$ that vanish is equal to the number of critical points of the Morse function with p negative eigenvalues, which means that $M_p \geq B_p$.

At each critical point we can use Gaussian normal coordinates x^i, coordinates in which the metric is simply δ_{ij} and shift the origin so that they are chosen to vanish at the position of the critical point. The Morse function can be expanded in a Taylor series; in general, this gives

$$h(x) = h(0) + \frac{1}{2} \sum_{i,j=0}^{n} \left(\frac{\partial^2}{\partial_i \partial_j} h(0) \right) x^i x^j + \cdots. \tag{12.111}$$

A further orthogonal rotation of the coordinates keeps the metric δ_{ij}; however, the real symmetric matrix of second partial derivatives can be diagonalized, with eigenvalues $\lambda_i = \partial_i \partial_i h(0)$ in the new coordinates. Then we get

$$h(x) = h(0) + \frac{1}{2} \sum_i \lambda_i (x^i)^2 + \cdots \tag{12.112}$$

and

$$\partial_i h(x) = \lambda_i x^i + \cdots. \tag{12.113}$$

The Hamiltonian then also admits a local expansion about each critical point, using the general expression Equation (12.109) and noting that the metric is δ_{ij}, the Levi–Civita connection vanishes so that covariant derivatives are ordinary derivatives and $\partial_i h(x^i) = \lambda_i x^i + \cdots$ using Equation (12.112)

$$\mathcal{H}_t = \sum_{i=1}^{n} \left(-\frac{\partial^2}{\partial x^i \partial x^i} + t^2 \lambda_i^2 (x^i)^2 + t\lambda_i [a^{*i}, a_i] \right) + \cdots. \tag{12.114}$$

The explicitly written term, although an approximation to the full Hamiltonian, is sufficient to compute the $A_p^{(n)}$. To compute the expansion of the eigenvalues in powers of $1/t$ requires calculating the higher-order terms in the Hamiltonian and continuing the perturbative expansion.

As the operators a_i and a^{*i} are also simply linear operators on the exterior algebra by exterior or interior multiplication, they commute with the simple harmonic oscillator part and hence the local Hamiltonian in lowest approximation can be written as two commuting terms

$$\mathcal{H}_{local} = \sum_{i=1}^{n} H_i + t\lambda_i K_i \tag{12.115}$$

with

$$H_i = -\frac{\partial^2}{\partial x^i \partial x^i} + t^2 \lambda_i^2 (x^i)^2 \tag{12.116}$$

while

$$K_i = [a^{*i}, a_i]. \tag{12.117}$$

H_i is the Hamiltonian of the simple harmonic oscillator, with the well-known spectrum $E_i(N_i) = t|\lambda_i|(1+2N_i)$, where $N_i = 0, 1, 2, \cdots$, taking into account that the λ_i are not necessarily positive. The corresponding eigenfunctions are Hermite polynomials multiplied by Gaussians centred at the origin, and hence rapidly fall off for $|\lambda_i x^i| \gg 1/\sqrt{t}$.

The eigenvalues of K_i are simply ± 1. The action of K_i on a p-form ω is $K_i \omega = [a^{*i}, a_i]\omega = a^{*i} a_i \omega - a_i a^{*i} \omega = 2a^{*i} a_i \omega - \{a_i, a^{*i}\}\omega = (2a^{*i} a_i - 1)\omega$. The first operator is simply the fermionic Hamiltonian for one degree of freedom for each i, which has eigenspectrum 0 or 2, acting on the fermionic vacuum state or the one fermion state, which yields the eigenspectrum ± 1 for K_i. Another way to see this is to realize that the action of K_i on a p-form $\omega = \omega_{i_1 \cdots i_p} a^{*i_1} \wedge \cdots \wedge a^{*i_p}$ obviously gives back ω if $i \in (i_1 \cdots i_p)$ but gives back $-\omega$ if $i \notin (i_1 \cdots i_p)$.

As K_i and H_i commute, the eigenvalues simply add; thus, the spectrum of $\mathcal{H}_{local}$ is

$$E_t(N_i, n_i) = t \sum_{i=1}^{n} (|\lambda_i|(1+2N_i) + n_i \lambda_i), \quad N_i = 0, 1, 2, \cdots, n_i = \pm 1. \tag{12.118}$$

If we restrict the action of $\mathcal{H}_{local}$ to p-forms, then the sum over K_i in the Hamiltonian contains p terms for which the eigenvalue of K_i is $+1$; thus, the number of n_i that equal $+1$ must be equal to p. The remaining K_i will have eigenvalue -1, thus the number of these will be $n - p$, where n is the dimension of $\mathcal{M}$.

The only way it is possible for the energy $E_t(N_i, n_i)$ to vanish is if all $N_i = 0$, $n_i = 1$ for each negative λ_i and $n_i = -1$ for each positive λ_i. We can solve this constraint if we choose the p-form to consist of the p-fold exterior product of coordinate differentials of exactly those coordinate directions which correspond to the negative eigenvalues. Thus the energy eigenvalue is (allowing for a minor relabelling of the independent directions in the manifold)

$$E_t(N_i, n_i) = t \sum_{i=1}^{p} (|\lambda_i|(1) + \lambda_i) + t \sum_{i=p+1}^{n} (|\lambda_i|(1) - \lambda_i) = 0 \qquad (12.119)$$

as $n_i = +1$ for the first p directions with negative eigenvalues and $n_i = -1$ for the $n - p$ remaining directions for which the eigenvalues are positive.

Thus for a critical point with Morse index equal to p, *i.e.* with p negative directions at the critical point of the Morse function h, it is possible to satisfy these conditions. We choose a p-form with a coefficient function given by the ground state of the harmonic oscillator (which puts all the $N_i = 0$), and which consists of exactly those coordinate differentials which correspond to the p negative directions, λ_i, which gives the desired $n_i = +1$. Thus at a critical point of Morse index p, we can construct exactly one eigenfunction which could have a zero eigenvalue. These are zero-energy eigenfunctions of the approximate Hamiltonian given in Equation (12.115). We could, in principle, compute the corrections that are brought to these approximate zero-energy levels, but we can be assured that they will remain low-lying levels even as $t \to \infty$, the key point being that $A_p^{(n)}$ vanishes for all of these levels. The dimension of the subspace spanned by these levels is M_p, the number of critical points with Morse index p.

For an actual vanishing eigenvalue of the full Hamiltonian (12.109), all higher perturbative and non-perturbative corrections must also vanish. This will happen for the exact supersymmetric ground states. The number of exact supersymmetric ground states is given by the Betti number, B_p, which is equal to the number of p-forms with zero eigenvalues of the Laplacian, $d\delta + \delta d$, or the deformed Laplacian, $d_t \delta_t + \delta_t d_t$. For each actual zero eigenvalue, we know that the $A_p^{(n)}$ must also vanish, as the computation of $A_p^{(n)}$ is the first step of computing the exact eigenvalue in perturbation. We have determined that the number of approximate states corresponding to $A_p^{(n)} = 0$ is M_p, the number of critical points of Morse index p. Hence the number of actual zero eigenvalue states must be less than or equal to M_p. Thus we obtain the result

$$M_p \geq B_p. \qquad (12.120)$$

These are called the weak Morse inequalities.

12.4.4 Polynomial Morse Inequalities

We actually wish to prove something stronger, that the Morse numbers always dominate the Betti numbers as encapsulated in the polynomial Morse inequality which states that there exists a set of non-negative integers Q_p such that

$$\sum_{p=0}^{n} M_p t^p - \sum_{p=0}^{n} B_p t^p = (1+t)\sum_{p=0}^{n-1} Q_p t^p. \tag{12.121}$$

This is an inequality in the sense that $Q_p \geq 0$. As the weak Morse inequalities give us that $M_p \geq B_p$, it is clear that the coefficient of t^p on the left-hand side is necessarily positive semi-definite. The right-hand side has the coefficient $Q_p + Q_{p-1}$ (with $Q_n = Q_{-1} = 0$) for t^p, which then must be positive semi-definite.

The polynomial Morse inequality is equivalent to the following two assertions, called the strong Morse inequalities (as originally proven by Morse):

$$\sum_{p=0}^{m}(-1)^{p+m} M_p \geq \sum_{p=0}^{m}(-1)^{p+m} B_p \quad \text{for} \quad m = 0, 1, \cdots, n \tag{12.122}$$

$$\sum_{p=0}^{n}(-1)^p M_p = \sum_{p=0}^{n}(-1)^p B_p. \tag{12.123}$$

We can prove the equivalence as follows. If we take the second equality, Equation (12.123), we have

$$\left(\sum_{p=0}^{n} M_p t^p - \sum_{p=0}^{n} B_p t^p\right)\Bigg|_{t=-1} = 0, \tag{12.124}$$

i.e. $t = -1$ is a root of the polynomial $\sum_{p=0}^{n} M_p t^p - \sum_{p=0}^{n} B_p t^p$ and hence it is divisible by $1+t$. Thus we have immediately and trivially

$$\sum_{p=0}^{n} M_p t^p - \sum_{p=0}^{n} B_p t^p = (1+t)\sum_{p=0}^{n-1} Q_p t^p. \tag{12.125}$$

Since the coefficients are integers on the left-hand side, the Q_n must also be integers. It remains to show that $Q_n \geq 0$. To see this we analyse the identity power by power in t. We start with the t^0 term. This term gives

$$M_0 - B_0 = Q_0, \tag{12.126}$$

which the first inequality, Equation (12.122), for $m = 0$ requires $Q_0 \geq 0$. Next, for the t term we have

$$M_1 - B_1 = Q_1 + Q_0 \tag{12.127}$$

or replacing for Q_0 from above

$$M_1 - M_0 - (B_1 - B_0) = Q_1, \tag{12.128}$$

which the first inequality, Equation (12.122), for $m = 1$ then requires $Q_1 \geq 0$. Doing one more step, before concluding the general relation, we have for the coefficient of t^2

$$M_2 - B_2 = Q_2 + Q_1 \tag{12.129}$$

replacing for Q_1 from above

$$M_2 - M_1 + M_0 - (B_2 - B_1 + B_0) = Q_2. \tag{12.130}$$

Again from the first inequality, Equation (12.122), for $m = 2$ then requires $Q_2 \geq 0$. We see then that in general

$$\sum_{p=0}^{m}(-1)^{p+m}M_p - \sum_{p=0}^{m}(-1)^{p+m}B_p = Q_m \quad \text{for} \quad m = 0, 1, \cdots, n-1 \tag{12.131}$$

and hence we can conclude that $Q_m \geq 0$ for all $m = 0, 1, 2, \cdots n - 1$.

To prove the converse, the polynomial Morse inequality, Equation (12.121), by comparing powers of t, as we have just seen, implies

$$\sum_{p=0}^{n}(-1)^{p+m}M_p - \sum_{p=0}^{n}(-1)^{p+m}B_p = Q_m \quad \text{for} \quad m = 0, 1, \cdots, n-1 \tag{12.132}$$

but now we assume that the $Q_m \geq 0$. Hence we recover the first inequalities in Equation (12.122) trivially. To recover the second equality, Equation (12.123), we simply put $t = -1$ in Equation (12.121).

The second equality, Equation (12.123), is related to the Euler characteristic of the manifold. This is defined as the alternating sum of the Betti numbers

$$\chi(\mathcal{M}) = \sum_{p=0}^{n}(-1)^p B_p = \sum_{p=0}^{n}(-1)^p M_p. \tag{12.133}$$

From the weak Morse inequalities, we know that $M_p \geq B_p$. Thus the number of critical points of Morse index, p, could be greater than the Betti number, B_p, but then there must be exactly the same surplus of critical points with opposite value of $(-1)^p$, *i.e.* each additional critical point of Morse index, p, must pair with another critical point of Morse index of opposite parity. As p determines if the state is fermionic or bosonic, we identify these pairs of critical points with the approximate, supersymmetric, bosonic and fermionic zero energy pairs of states associated with each critical point, but those which must actually lift away from exact zero energy when non-perturbative corrections are taken into account, as the actual number of supersymmetric zero energy states is strictly given by the Betti numbers.

It is straightforward [64] to prove the strong Morse inequalities using the simple ideas of supersymmetry and what we have understood about the spectrum. We know that the eigenvalues and corresponding eigenstates of $\mathcal{H}_t$ separate into two subsets as $t \to \infty$; those whose energies diverge linearly as t gets large and a finite number whose energies do not. There are M_p states for each p whose energies do not diverge with t. These further split into two subsets, the first B_p states whose energies are exactly zero and the remaining $M_p - B_p$ states whose energies are of $o(1)$. We will call these latter $M_p - B_p$ states the low-lying states. But now we recall that, since the low-lying states have non-zero energy, supersymmetry requires that they come in bosonic–fermionic pairs. The fermionic states correspond to odd p and the bosonic states correspond to even p, hence we must have

$$\sum_{p \text{ odd}} (M_p - B_p) = \sum_{p \text{ even}} (M_p - B_p). \tag{12.134}$$

This immediately implies the second of the strong Morse inequalities, Equation (12.123)

$$\sum_{p=0}^{n} (-1)^p M_p = \sum_{p=0}^{n} (-1)^p B_p. \tag{12.135}$$

To obtain the first strong Morse inequality we consider the mapping that Q_{1t} induces on the fermionic and bosonic subspaces of low-lying levels. As $Q_{1t}^2 = \mathcal{H}_t$ and evidently Q_{1t} commutes with the Hamiltonian, it must preserve the eigensubspaces of $\mathcal{H}_t$. Q_{1t}, being a fermionic operator, maps the eigensubspace of p-forms to the eigensubspace of $p+1$-forms and $p-1$ forms. Let Λ_t^p denote the subspace of low-lying eigenstates of p-forms, clearly of dimension $M_p - B_p$. For any state $|\psi\rangle$ in this subspace $Q_{1t}|\psi\rangle \neq 0$ and $\mathcal{H}_t Q_{1t}|\psi\rangle = Q_{1t}^3|\psi\rangle = Q_{1t}\mathcal{H}_t|\psi\rangle = EQ_{1t}|\psi\rangle$, where $\mathcal{H}_t|\psi\rangle = E|\psi\rangle$. If Q_{1t} maps two distinct states to the same state, then it must annihilate their difference, which is not possible as this does not preserve the eigenspace. Thus the mapping $Q_{1t} : \Lambda_t^p \to \Lambda_t^{p-1} \oplus \Lambda_t^{p+1}$ must be one-to-one, into (injective). Hence we can conclude

$$Q_{1t}: \bigoplus_{p \text{ odd } p=1}^{2j-1} \Lambda_t^p \to \bigoplus_{p \text{ even } p=0}^{2j} \Lambda_t^p$$

$$Q_{1t}: \bigoplus_{p \text{ even } p=0}^{2j} \Lambda_t^p \to \bigoplus_{p \text{ odd } p=1}^{2j+1} \Lambda_t^p \tag{12.136}$$

for each $j \ni 0 \leq 2j < n$ and $0 \leq 2j+1 < n$. But since the mappings are injective, the dimension of the domain must be less than or equal to the dimension of the image. This yields:

$$(M_1 - B_1) + \cdots + (M_{2j-1} - B_{2j-1}) \leq (M_0 - B_0) + \cdots + (M_{2j} - B_{2j})$$
$$(M_0 - B_0) + \cdots + (M_{2j} - B_{2j}) \leq (M_1 - B_1) + \cdots + (M_{2j+1} - B_{2j+1}). \tag{12.137}$$

These inequalities are identical to the first strong Morse inequality, Equation (12.122), if we bring the Ms and Bs to opposite sides.

12.4.5 Witten's Coboundary Operator

The polynomial Morse inequality is equivalent to the understanding that the critical points of a Morse function form a model for the cohomology of the manifold $\mathcal{M}$. We define X_p to be a vector space of dimension M_p for each $p \in 0, 1, 2, \cdots, n$. X_p can be thought of as a vector space spanned by the critical points of Morse index p. The polynomial Morse inequality, Equation (12.121), means that there exists a coboundary operator $\delta_W : X_p \to X_{p+1}$ (we add a subscript W to honour Witten), where $\delta_W^2 = 0$ and the corresponding Betti numbers, the dimension of the cohomology groups associated to δ_W, are identical to the Betti numbers of the manifold $\mathcal{M}$. The homotopy classes in this cohomology are elements of X_p, which are closed under the action of δ_W, but differ only by elements which are obtained by the action of δ_W on some element of X_{p-1}, the analogue of the standard notion of closed modulo exact forms, or cycles, etc. The explicit expression for δ_W is not given in the original work of Morse or others; however, Witten found an appropriate expression for it.

Witten proposed the following construction. First, consider possible zero modes of the Laplacian. The number of independent such p-forms gives the Betti numbers, B_p. We have an upper bound on the Betti numbers, $M_p \geq B_p$ in the Morse inequalities. However, although perturbation theory might suggest a given mode is a zero mode, tunnelling effects can lift the degeneracy. Exact instanton effects can give energies of the order of $\sim e^{-tS}$ where S is the action of the instanton, which for large t is smaller than any perturbative correction. Thus Witten was led to consider instanton configurations that tunnel from one zero mode to another. In fact, tunnelling from putative zero modes which are p-forms to putative zero modes which are $p+1$-forms are exactly the instanton modes that are required. However, as we have seen, the $p+1$-form chosen at a given critical point of Morse index, $p+1$, requires a choice of the exterior product of all the coordinate differentials that correspond to the $p+1$ negative directions. The orientation or order of the differentials remains arbitrary. Thus a tunnelling transition from a state at a critical point of Morse index, p, to a state at a critical point of Morse index, $p+1$, must also fix a sign. We determine the sign with the following construction.

Consider instanton paths, Γ, that pass from a critical point, B, of Morse index, $p+1$, to a critical point, A, of Morse index, p. The instanton path has initial tangent vector v within V_B, the $p+1$-dimensional vector space of negative directions at B. Let $|b\rangle$, a $p+1$-form, be the state of zero energy at the critical point B. Then $|b\rangle$ chooses an orientation of V_B, and we can choose an orientation of the p-dimensional subspace, $\tilde{V}_B$, corresponding to the orthogonal complement of v within $|b\rangle$, as we can generate a p-form from $|b\rangle$ by contracting it (by interior multiplication) with v. Then the instanton path from B to A gives a mapping of $\tilde{V}_B$ to V_A, the p-dimensional vector space of negative directions at A. This mapping induces an orientation of V_A. However, the state $|a\rangle$ corresponding to the perturbative zero mode at A already gave an orientation of V_A. We define

$$n_\Gamma = \begin{cases} +1 & \text{if the induced orientation agrees with that fixed by } |a\rangle \\ -1 & \text{if the induced orientation disagrees with that fixed by } |a\rangle \end{cases}$$

(12.138)

and

$$n(a,b) = \sum_\Gamma n_\Gamma, \tag{12.139}$$

where the sum runs over all instantons paths (paths of steepest descent) from B to A. Then we can define the coboundary operator, for any basis element $|a\rangle$ of X_p at A

$$\delta_W |a\rangle = \sum_b n(a,b)|b\rangle, \tag{12.140}$$

where the sum runs over all basis elements of X_{p+1} (in other words, this is a set of perturbative zero modes that are $p+1$-forms that are concentrated at the critical points of Morse index $p+1$ of the Morse function). The effect of the instantons is to non-perturbatively correct the energy of some of the perturbative zero modes, their energy behaves as $\sim e^{-tS}$, for large t. Thus all states in X_p are not annihilated by the Laplacian $\delta_W \delta_W^* + \delta_W^* \delta_W$.

Denoting Y_p as the number of actual zero eigenvalues of $\delta_W \delta_W^* + \delta_W^* \delta_W$ acting on X_p, then Y_p also give upper bounds on the Betti numbers, and the strong Morse inequality, Equation (12.121), remains valid with M_p replaced with Y_p. Witten conjectures that, in fact, $Y_p = B_p$.

12.4.6 Supersymmetric Sigma Model

To demonstrate that δ_W as defined in Equation (12.140) provides the appropriate coboundary operator, Witten considered the Lagrangian version of the supersymmetric quantum-mechanical model that we have been considering, that for which the supercharge is given explicitly by d_t. Canonical quantization of the model defined by the action, in Minkowski time

$$\int d\tau \mathcal{L} = \frac{1}{2} \int d\tau \left(g_{ij} \left(\frac{dx^i}{d\tau} \frac{dx^j}{d\tau} + \bar\psi^i i \frac{D\psi^j}{D\tau} \right) + \frac{1}{4} R_{ijkl} \bar\psi^i \psi^k \bar\psi^j \psi^l \right.$$
$$\left. - t^2 g^{ij} \frac{dh}{dx^i} \frac{dh}{dx^j} - t \frac{D^2 h}{Dx^i Dx^j} \bar\psi^i \psi^j \right)$$

(12.141)

where a sum over all repeated indices is understood, gives the required algebraic symmetries and explicitly the supercharge. This is the Lagrangian of the 1+1-dimensional supersymmetric sigma model restricted to 0+1 dimensions. Here the ψ and $\bar\psi$ (the complex conjugate field to ψ) are anti-commuting fermionic fields, x^i are local coordinates, g_{ij} is the metric tensor and R_{ijkl} is the corresponding Riemann curvature tensor on $\mathcal{M}$, and D/Dx^i is the covariant derivative with the Levi–Civita connection of the metric while $D/D\tau$ is the covariant derivative along the direction tangent to the time trajectory. Specifically, acting on the fermions we have

$$\frac{D}{D\tau}\psi^i = \partial_\tau\psi^i + \Gamma^i_{jk}\partial_\tau x^j\psi^k. \tag{12.142}$$

Under the supersymmetry transformations

$$\delta x^i = \epsilon\bar\psi^i - \bar\epsilon\psi^i$$
$$\delta\psi^i = \epsilon\left(i\dot x^i - \Gamma^i_{jk}\bar\psi^j\psi^k + tg^{ij}\partial_j h\right)$$
$$\delta\bar\psi^i = \bar\epsilon\left(-i\dot x^i - \Gamma^i_{jk}\bar\psi^j\psi^k + tg^{ij}\partial_j h\right) \tag{12.143}$$

for infinitesimal anti-commuting parameters ϵ and $\bar\epsilon$, the action is invariant, $\delta\int d\tau\mathcal{L} = 0$. The corresponding supercharges are as required

$$Q_t = \bar\psi^i(ig_{ij}\dot x^j + t\partial_i h)$$
$$\bar Q_t = \psi^i(-ig_{ij}\dot x^j + t\partial_i h). \tag{12.144}$$

There is also a symmetry-conserving fermion number, $\psi^i \to e^{-i\theta}\psi^i$, $\bar\psi^i \to e^{i\theta}\bar\psi^i$, which gives the conserved charge, the fermion number

$$F = g_{ij}\bar\psi^i\psi^j. \tag{12.145}$$

In quantizing the system we will first consider the system at $t = 0$ (all the supersymmetry and other symmetries are equally valid at $t = 0$). We impose the canonical commutation and anti-commutation relations

$$[x^i,p_j] = i\delta^i_j$$
$$\{\psi^i,\bar\psi^j\} = g^{ij}, \tag{12.146}$$

then the conserved supercharges are simply $Q = i\bar\psi^i p_i$ and $\bar Q = -i\psi^i p_i$. The supercharges have the opposite fermion number

$$[F,Q] = Q, \ [F,\bar Q] = -\bar Q. \tag{12.147}$$

We impose that the Hamiltonian is given by the supersymmetry algebra

$$\{Q,\bar Q\} = 2\mathcal{H}_0 \tag{12.148}$$

and consequently the fermion number is conserved, $[F,\mathcal{H}_0] = 0$. The natural realization of this algebra is, as we have been using, provided by the exterior algebra of differential forms, $\Lambda^*(\mathcal{M}) \otimes \mathbb{C}$ equipped with its hermitian inner product from Equation (12.17)

$$(\omega,\chi) = \int_{\mathcal{M}} \bar\omega \wedge *\chi \tag{12.149}$$

for two p-forms, ω and χ. Then the observables in this realization of the algebra on this Hilbert space, when acting explicitly on a p-form ω, are:

$$x^i : x^i \omega$$
$$p_i : -i\partial_i \omega$$
$$\bar{\psi} : dx^i \wedge \omega$$
$$\psi^i : g^{ij} \iota_{\partial/\partial x^j} \omega \qquad (12.150)$$

(ι_V is the interior multiplication defined in Equation (12.106)). Then with the state $|0\rangle$ denoting the form annihilated by all of the ψ^i we have the schema:

$$|0\rangle = 1$$
$$\bar{\psi}^i |0\rangle = dx^i$$
$$\bar{\psi}^i \bar{\psi}^j |0\rangle = dx^i \wedge dx^j$$
$$\cdots$$
$$\bar{\psi}^1 \cdots \bar{\psi}^n |0\rangle = dx^1 \wedge \cdots \wedge dx^n. \qquad (12.151)$$

The fermion number of a state that is a p-form is simply equal to p, thus the Hilbert space separates into bosonic and fermionic subspaces depending on whether p is even or odd, respectively. Thus the canonically quantized system reproduced with complete fidelity the supersymmetric system of the exterior algebra that we studied in subsection (12.4.1).

Recall then that the supersymmetric states are just the zero-energy states, those annihilated by the Laplacian, the so-called harmonic forms. We underline that the set of harmonic forms of the manifold characterize the de Rham cohomology of the manifold. Equally well, the space of supersymmetric ground states characterize the cohomology of the Q-operator. As there is the conserved fermion number which satisfies $[F, Q] = Q$, the Q-cohomology is graded by the fermion number and equal to the degree p of the form. As Q is identified with the exterior derivative d, the graded Q-cohomology and the de Rham cohomology must be equal

$$H^p(Q) = H^p_{dR}(\mathcal{M}). \qquad (12.152)$$

The Witten index, $(-1)^F$, can be evaluated and we find

$$Tr\left((-1)^F\right) = \sum_{p=0}^{n} (-1)^p dim(H^p(Q)) = \sum_{p=0}^{n} (-1)^p dim(H^p_{dR}(\mathcal{M})) = \chi(\mathcal{M}), \quad (12.153)$$

where $\chi(\mathcal{M})$ is the Euler characteristic of the manifold. The Witten index only receives contributions from the supersymmetric ground states; as we have seen, the non-zero energy modes are all paired in fermionic–bosonic pairs and their contributions cancel. Thus the calculation of the topological invariant, the Euler characteristic, can be done by studying the zero-energy modes of this supersymmetric quantum mechanical system. Witten's magical trick was to add an external field to this system, which causes a separation of the zero- and low-energy modes from the finite-energy modes, and in the limiting case makes the calculation of the zero mode sector very simple.

Now adding in the deformation by th, the supercharges are then given by

$$Q_t = \bar{\psi}^i(ig_{ij}\dot{x}^j + t\partial_i h) = dx^i \wedge (\frac{\partial}{\partial x^i} + t\partial_i h) = d + tdh\wedge = e^{-th}de^{th} = d_t \quad (12.154)$$

and

$$Q_t^\dagger = \delta + t(dh\wedge)^* = e^{th}\delta e^{-th} = \delta_t, \quad (12.155)$$

where $*$ denotes the adjoint. The Hamiltonian then is as before

$$\mathcal{H} = \frac{1}{2}\{Q_t, \bar{Q}_t\} = \frac{1}{2}(d_t d_t^* + d_t^* d_t), \quad (12.156)$$

chosen to satisfy the supersymmetry algebra. The supersymmetric ground states again define the Q_t-cohomology. However, since the th deformation is obtained by a similarity transformation

$$Q_t = e^{-th}Qe^{th} \quad (12.157)$$

the cohomology is isomorphic to the undeformed case. As the cohomology of the undeformed Q is isomorphic to the de Rham cohomology, we can compute the de Rham cohomology with the deformed operator Q_t.

The perturbative approximation to the Hamiltonian around a critical point is given by

$$\mathcal{H}_t = \sum_{i=1}^n \left(-\frac{\partial^2}{\partial x^i \partial x^i} + t^2\lambda_i^2(x^i)^2 + t\lambda_i[a^{*i}, a_i]\right) \quad (12.158)$$

with exact, zero-energy ground-state wave functions, which we will label $|\phi\omega\rangle$, corresponding to the harmonic oscillator ground state, ϕ, multiplied by an appropriate p-form, ω, where p is the Morse index of the critical point, as discussed previously. Indeed, perturbative corrections to the energy of these wave functions must vanish to all orders: the energy remains exactly zero to all orders in perturbation theory. One can find the modification of the wave function, order by order, so that its energy remains zero in each order in perturbation theory. This is because the corrections are calculated in terms of local data at the critical point. From local data it is not possible to know which critical points are actually necessary because of the global topology of the manifold and which critical points are removable by deformations. States that have zero energy to lowest order have zero energy to all orders. The same reasoning applies to the calculation of tunnelling in the double-well potential. In perturbation theory we can never get a non-zero tunnelling amplitude, these amplitudes are non-perturbative in the coupling and are not seen at any order in perturbation theory.

However, the wave functions, $|\phi_i\omega_i\rangle$, are not necessarily exact ground states, where we have added a label i to denote different critical points. The perturbative zero-energy states are not necessarily exact ground states. Hence the number of exact, supersymmetric actual ground states are clearly less than or equal to the number of critical points. An exact supersymmetric ground state is annihilated by the supercharge. For the case of the perturbative ground states, although we will find, if calculated perturbatively,

$$\mathcal{H}_t|\phi_i\omega_i\rangle = (Q_t Q_t^\dagger + Q_t^\dagger Q_t)|\phi_i\omega_i\rangle = 0, \tag{12.159}$$

which also requires

$$Q_t|\phi_i\omega_i\rangle = 0 \tag{12.160}$$

to all orders in perturbation theory, we can in fact have non-perturbative corrections

$$Q_t|\phi_i\omega_i\rangle = \sum_{j=1}^{N} |\phi_j\omega_j\rangle\langle\phi_j\omega_j|Q_t|\phi_i\omega_i\rangle + \cdots, \tag{12.161}$$

where the $+\cdots$ corresponds to amplitudes to non-zero-energy states (which are suppressed by large energy denominators as $t \to \infty$). The explicit mixing that can be important is between the perturbative zero-energy states. Thus we want to compute

$$\langle\phi_j\omega_j|Q_t|\phi_i\omega_i\rangle = \int_\mathcal{M} \phi_j\omega_j \wedge *(d + tdh\wedge)\phi_j\omega_i. \tag{12.162}$$

But such an amplitude is exactly what we are looking for with the coboundary operator δ_W between zero modes localized at different critical points. If ω_j is a q-form and ω_i is a p-form, this matrix element can only be non-zero if $q = p+1$, *i.e.* transitions between perturbative zero-energy modes correspond to critical points of Morse indices that differ by one negative direction. This can also be seen from fermion number conservation, the action of Q_t on the state $|\phi_i\omega_i\rangle$ changes its fermion number by one unit. It also should not be surprising that the eventual δ_W that we will be able to define will satisfy $\delta_W^2 = 0$, since it is obtained from the action of $Q_t = d_t$. Clearly $Q_t^2 = 0$, hence we can expect $\delta_W^2 = 0$.

We will return below to the notation of subsection 12.4.5 with $|a\rangle$ for $|\phi_i\omega_i\rangle$ and $\langle b|$ for $\langle\phi_j\omega_j|$ and the understanding that if $|a\rangle$ corresponds to a p-form then $\langle b|$ corresponds to a $p+1$-form. It is also clear that the action of Q_t on the low-lying states annihilates any exact, supersymmetric ground state that is a p-form as these are harmonic with respect to Q_t. Thus only the $M_p - B_p$ low-lying but not exact supersymmetric ground states will be mixed with low-lying $p+1$-forms. But additionally, none of these states can be the exact supersymmetric ground states that are $p+1$-forms, since the inner product

$$\langle b|Q_t|a\rangle = (Q_t|b\rangle)^\dagger|a\rangle = 0 \tag{12.163}$$

if $|b\rangle$ corresponds to an exact supersymmetric ground state, as these are also harmonic with respect to Q_t. Thus the action of Q_t on the set of states $|a\rangle$ only mixes the $M_p - B_p$ not exact ground states but low-lying states with the $M_{p+1} -$

B_{p+1} corresponding low-lying states $|b\rangle$. This is as it should be; since mixing causes the energies to go up, this cannot happen to any exact supersymmetric ground state.

12.4.7 The Instanton Calculation

We will use the path integral to compute this amplitude, since we know that it is exactly through the path integral that we can uncover tunnelling amplitudes through the path integral, and from the amplitude we will extract the coboundary operator, δ_W. The bosonic sector of the model is governed by the Lagrangian, in Euclidean time

$$\int d\tau \mathcal{L}_b = \frac{1}{2}\int d\tau \left(g_{ij}\frac{dx^i}{d\tau}\frac{dx^j}{d\tau} + t^2 g^{ij}\frac{dh}{dx^i}\frac{dh}{dx^j}\right). \tag{12.164}$$

We can show that the stationary points of the corresponding action are the paths of steepest descent using a Bogomolny-type identity [17]. Indeed,

$$\int d\tau \mathcal{L}_b = \frac{1}{2}\int d\tau g_{ij}\left(\frac{dx^i}{d\tau}\pm tg^{ik}\frac{dh}{dx^k}\right)\left(\frac{dx^j}{d\tau}\pm tg^{jl}\frac{dh}{dx^l}\right)\mp t\int d\tau\frac{dh}{d\tau}. \tag{12.165}$$

The first integral is positive semi-definite, while the second integral is equal to $t\Delta h$. Therefore, if $t\Delta h \geq 0$, we choose the plus (lower) sign, while if $t\Delta h \leq 0$, we choose the minus (upper) sign. Then the second term is always positive, and thus

$$\int d\tau \mathcal{L}_b \geq t|\Delta h| \tag{12.166}$$

with equality for (assuming Δh is positive)

$$\frac{dx^i}{d\tau} - tg^{ij}\frac{dh}{dx^j} = 0. \tag{12.167}$$

This is exactly the equation of steepest descent, physically stating that the tangent vector to the curve is parallel to the gradient, up to reparametrization. Also it should be noted that this equation is not the same as the usual instanton equation which we have seen can be interpreted as ordinary, conservative, Newtonian cinematic motion of a particle in the reversed potential. Such a motion would never follow a path of steepest descent and stop at a lower value of the potential. Here the equation of steepest descent is first order in the "time" coordinate, and thus allows such motion. The solution to Equation (12.167) obviously exists, which then implies $S_E = \int d\tau \mathcal{L}_b = t|\Delta h|$. Then for the operator d_t whose matrix elements we want to compute, they will then be proportional to $e^{-t|\Delta h|}$. If we want to compute matrix elements of the Hamiltonian $dd^* + d^*d$ then we get two factors of S_E and hence the amplitude is proportional to $e^{-2t|\Delta h|}$.

The next step in the calculation is to compute the determinant of the fluctuations in Gaussian approximation about the instanton configuration. It is in this stage that the calculation dramatically simplifies due to supersymmetry. The non-zero eigenvalues are all paired in bosonic and fermionic multiplets. The fermionic determinant is exactly cancelled by the bosonic square root of the determinant. The bosonic zero mode corresponding to Euclidean time translation invariance, which would normally give rise to a diverging factor of β, is also exactly cancelled by a corresponding fermionic zero mode which would normally give rise to a vanishing determinant. These zero modes can be explicitly obtained first for the bosonic case in the usual way, the bosonic zero mode corresponds to the Euclidean time derivative of the instanton. Then the fermionic zero mode is obtained by a supersymmetry transformation of the bosonic zero mode. To show the cancellation of the contribution of the zero modes requires some care, we refer the reader to the detailed calculation in [62]. Finally the amplitude is given by the factor

$$\langle b|d_t|a\rangle = e^{-t|\Delta h|}. \tag{12.168}$$

However, we still have not determined the sign of the amplitude; the functional integral always gives rise to an ambiguous sign due to the fermions. To determine the sign, we go back to the calculation of the amplitude in the usual WKB method of Schrödinger quantum mechanics. Here we know that the states $|a\rangle$ at the critical point A and $|b\rangle$ at the critical point B rapidly die off, away from their respective critical points. Any overlap is greatest along the paths that connect the two critical points that are the semi-classical solutions to the equations of motion, the paths that keep the Euclidean action stationary. These paths are the instantons, the paths of steepest descent or ascent between the critical points. Thus the behaviour of the states along the paths of steepest descent are enough to determine the sign of the matrix element $\langle b|d_t|a\rangle$. The quantum mechanical problem becomes effectively one-dimensional along the path of steepest descent, and we find that the state $|a\rangle$ drops off as e^{-th} along the instanton that ascends from A to B. It must ascend, as $|a\rangle$ was a p-form, hence A was a critical point of p negative directions while $|b\rangle$ was a $p+1$-form, hence B was a critical point of $p+1$ negative directions. If we descend from A we can only reach other critical points with fewer negative directions, we can never reach B.

To determine the sign, we start at $|b\rangle$ at B and the orientation of the space of negative directions at $|b\rangle$, which we called V_B. Calculating $|b\rangle$ along the path of steepest descent in the WKB approximation we find the wave function of the state $|b\rangle$ at A, but it is still a $p+1$-form. However, with the limiting direction of the tangent vector as we arrive at A, we can induce an orientation of V_A, the space of negative directions at A. Then the sign of the matrix element $\langle b|d_t|a\rangle$ is $+1$ if this induced orientation of V_A matches that furnished by the state $|a\rangle$, otherwise it is -1. This is exactly the construction of the sign $n(a,b)$ that was described in section 12.4.5, but appended with the explicit transport afforded by the WKB calculation of the wave function along the path of steepest descent.

Hence the Witten coboundary operator is given by

$$\delta_W |a\rangle = \sum_b e^{-t(h(B)-h(A))} n(a,b)|b\rangle. \tag{12.169}$$

Since the path descends from B to A, the exponent has the right sign. This factor can be removed by rescaling the wave functions by

$$|a\rangle \to e^{th(A)}|a\rangle, \tag{12.170}$$

which corresponds to undoing the conjugation by e^{th} which transformed d to d_t. Hence the Witten coboundary operator is given by

$$\delta_W |a\rangle = \sum_b n(a,b)|b\rangle \tag{12.171}$$

and the notion that the set of critical points of a Morse function form a model of the cohomology of the manifold $\mathcal{M}$ is verified.

Appendix A

An Aside on $O(4)$

$O(4)$ is the group defined by the multiplication properties of the set of orthogonal matrices which keep the quadratic form

$$\sum_{i=1}^{4} x_i^2 \tag{A.1}$$

invariant. If

$$\vec{x}' = \mathcal{O}\vec{x} \tag{A.2}$$

then

$$\vec{x}' \cdot \vec{x}' = \vec{x} \cdot \mathcal{O}^{\mathrm{T}}\mathcal{O} \cdot \vec{x} = \vec{x} \cdot \vec{x}$$
$$\Rightarrow \mathcal{O}^{\mathrm{T}}\mathcal{O} = 1 \tag{A.3}$$

where $\vec{x}$ is a four-dimensional vector. Looking in the neighbourhood of the identity, we find, with $\mathcal{O} = 1 + \delta$, then

$$\mathcal{O}^{\mathrm{T}}\mathcal{O} = \left(1 + \delta^{\mathrm{T}}\right)\left(1 + \delta\right) = 1 + \delta + \delta^{\mathrm{T}} + o(\delta^2) = 1$$
$$\Rightarrow \delta + \delta^{\mathrm{T}} = 0. \tag{A.4}$$

This means that δ must be an anti-symmetric, 4×4 matrix. This defines the Lie algebra of $O(4)$. The complete set of anti-symmetric 4×4 matrices is given by

$$(M_{\mu\nu})_{\sigma\tau} = \delta_{\mu\sigma}\delta_{\nu\tau} - \delta_{\mu\tau}\delta_{\nu\sigma}$$
$$= \frac{1}{2}\epsilon_{\mu\nu\lambda\rho}\epsilon_{\lambda\rho\sigma\tau}. \tag{A.5}$$

It is easy to calculate

$$[M_{\mu\nu}, M_{\sigma\tau}] = \frac{1}{4}\left(\epsilon_{\mu\nu\lambda\rho}\epsilon_{\lambda\rho\gamma\delta}\epsilon_{\sigma\tau\delta\beta}\epsilon_{\sigma\tau\alpha\beta} - \epsilon_{\alpha\beta\sigma\tau}\epsilon_{\sigma\tau\gamma\delta}\epsilon_{\mu\nu\lambda\rho}\epsilon_{\lambda\rho\delta\omega}\right). \tag{A.6}$$

We can expand this further; it is easy to do some of the sums over dummy indices, but it is more illuminating to define

$$J_i = \frac{1}{2}\epsilon_{ijk}(M_{jk})_{lm} = \frac{1}{2}\epsilon_{ijk}(\delta_{jl}\delta_{km} - \delta_{jm}\delta_{kl}) = \epsilon_{ilm} \tag{A.7}$$

and

$$K_i = (M_{0i})_{lm}.$$ (A.8)

Then the commutators

$$[J_i, J_j] = \epsilon_{ijk} J_k$$
$$[J_i, K_j] = \epsilon_{ijk} K_k$$ (A.9)

follow directly, with $J_1 = M_{23}$, $J_2 = M_{31}$ and $J_3 = M_{12}$. To calculate $[K_i, K_j]$ we consider the generators

$$\tilde{J}_1 = M_{12}, \quad \tilde{J}_2 = M_{20}, \quad \tilde{J}_3 = M_{01},$$ (A.10)

which generate the subgroup that leaves the form $x_0^2 + x_1^2 + x_2^2$ invariant. Then because of rotational symmetry we must have

$$\left[\tilde{J}_i, \tilde{J}_j\right] = \epsilon_{ijk} \tilde{J}_k.$$ (A.11)

(We can check this, for example, with $\left[\tilde{J}_1, \tilde{J}_2\right] = [M_{12}, M_{20}] = [M_{12}, M_{20}] = [J_3, -K_2] = -\epsilon_{321} K_1 = K_1 = M_{01} = \tilde{J}_3$.) Thus

$$\left[\tilde{J}_2, \tilde{J}_3\right] = [M_{20}, M_{01}] = [-K_2, K_1] = \tilde{J}_1 = M_{12} = J_3$$ (A.12)

thus

$$[K_1, K_2] = J_3$$ (A.13)

hence rotational covariance dictates the general relation

$$[K_i, K_j] = \epsilon_{ijk} J_k.$$ (A.14)

The combinations

$$M_i^{\pm} = \frac{1}{2}(J_i \pm K_i)$$ (A.15)

satisfy the commutators

$$\left[M_i^{\pm}, M_j^{\pm}\right] = \epsilon_{ijk} M_k^{\pm}$$ (A.16)

while

$$\left[M_i^{+}, M_j^{-}\right] = 0.$$ (A.17)

Appendix B

Asymptotic Analysis

Asymptotic analysis concerns the notion of the behaviour of functions, $f(x)$, as certain parameters go to their limiting values, usually zero or infinity. For convenience and without loss of generality, we will consider functions as their arguments go to infinity. Obviously the limit to any finite value x_0 can be obtained by taking $y = \frac{1}{(x-x_0)}$ to infinity.

We define

$$f(x) \sim g(x) \tag{B.1}$$

if and only if

$$\lim_{n \to \infty} \frac{f(n)}{g(n)} \to 1. \tag{B.2}$$

The binary relation of equivalence satisfies many obvious properties: for any smooth function $F(y)$, if $f(x) \sim g(x)$ then

$$F(f(x)) \sim F(g(x)). \tag{B.3}$$

This specifically is useful when applied to powers, $f \sim g$ implies

$$f^r \sim g^r \tag{B.4}$$

for any real number r. If $f(x) \sim g(x)$ and $a(x) \sim b(x)$ then

$$a(x)f \sim b(x)g(x). \tag{B.5}$$

Asymptotic analysis is most useful in the application of asymptotic expansions of functions. An asymptotic expansion of a function $f(x)$ is a series representation of a function that does not necessarily converge, and hence must be truncated at the expense of adding a remainder term. A very famous example of an asymptotic expansion is the Stirling approximation for the factorial, $N!$. The Stirling approximation is given by

$$\ln(\Gamma(z)) = z \ln z - z - \frac{1}{2} \ln z + \ln 2\pi + \sum_{n=1}^{N-1} \frac{B_{2n}}{2n(2n-1)x^{2n-1}} + R_N(z), \tag{B.6}$$

where B_n are the Bernoulli numbers with

$$R_N \leq \frac{|B_{2N}|}{2N(2N-1)|z|^{2N-1}} \tag{B.7}$$

for real z. The Bernoulli numbers behave as

$$B_{2N} = (-1)^{N+1} \frac{2(2N)!}{(2\pi)^{2N}} \zeta(2N). \tag{B.8}$$

The zeta function being bounded, we see that the Stirling approximation diverges.

A series expansion can be obtained for the factorial of a positive integer N by expanding the Γ function. Typically, the series expansion gives a very accurate approximation for the function that becomes maximally accurate after a certain number of terms in the expansion. At any finite truncation of the series, the remainder can be understood to be smaller than the subsequent term that has been dropped. Thus if we have a function $f(x)$ and its asymptotic series $g_1(x) + g_2(x) + \cdots$ then

$$f(x) - (g_1(x) + g_2(x) + \cdots + g_{k-1}(x)) \sim g_k(x) \tag{B.9}$$

for each k up to a maximum $k_{\max}$ which depends on x. For larger values of x, $k_{\max}$ increases. But after this term, the expansion starts to diverge, and it is not a good approximation to the original function. Thus for the Stirling approximation, for a given N, we should sum a finite number of terms to obtain a good approximation to $N!$, that number fixed by the value of N. However, if we look at the subsequent terms in the expansion, we find that they start to increase, and eventually they increase so much that the series fails to converge. Truncating the series at a given term $k_{\max}$ gives an approximation that is as small as the first term neglected, which can be very good approximation even though the asymptotic series does not converge.

We use the notation

$$f(x) - (g_1(x) + g_2(x) + \cdots + g_{k-1}(x)) = o(g_k(x)), \tag{B.10}$$

which generally in physics is translated as the difference

$$f(x) - (g_1(x) + g_2(x) + \cdots + g_{k-1}(x)) \tag{B.11}$$

is of the order of $g_k(x)$. However, there is a precise mathematical sense to this relation, it means that for every positive ϵ there exists a positive real number X such that, for $x \geq X$,

$$f(x) - (g_1(x) + g_2(x) + \cdots + g_{k-1}(x)) \leq \epsilon g_k(x). \tag{B.12}$$

If $f(x) = o(g(x))$ and $g(x) \neq 0$, then

$$\lim_{x \to \infty} \frac{f(x)}{g(x)} = 0. \tag{B.13}$$

Bibliography

[1] S. A. Abel, C.-S. Chu, J. Jaeckel and V. V. Khoze. "SUSY breaking by a metastable ground state: Why the early universe preferred the non-supersymmetric vacuum". *JHEP*, **01** (2007), p. 089. DOI: 10.1088/1126-6708/2007/01/089. arXiv: hep-th/0610334 [hep-th].

[2] M. Abramowitz and I. A. Stegun. *Handbook of Mathematical Functions with Formulas, Graphs, and Mathematical Tables* (Dover, 1965).

[3] A. A. Abrikosov. "On the magnetic properties of superconductors of the second group". *Sov. Phys. JETP*, **5** (1957). [Zh. Eksp. Teor. Fiz.32,1442(1957)], pp. 1174–1182.

[4] I. K. Affleck and N. S. Manton. "Monopole pair production in a magnetic field". *Nucl. Phys.*, **B194** (1982), pp. 38–64. DOI: 10.1016/0550-3213(82)90511-9.

[5] A. Altland and B. D. Simons. *Condensed Matter Field Theory* (Cambridge University Press, 2010).

[6] P. W. Anderson. "An approximate quantum theory of the antiferromagnetic ground state". *Phys. Rev.*, **86** (5 June 1952), pp. 694–701. DOI: 10.1103/PhysRev.86.694. URL: 10.1103/PhysRev.86.694.

[7] N. W. Ashcroft and N. D. Mermin. *Solid State Physics*. HRW International Editions (Holt, Rinehart and Winston, 1976).

[8] M. F. Atiyah and I. M. Singer. "The index of elliptic operators on compact manifolds". *Bull. Am. Math. Soc.*, **69** (1969), pp. 422–433. DOI: 10.1090/S0002-9904-1963-10957-X.

[9] L. Balents. "Spin liquids in frustrated magnets". *Nature*, **464.7286** (Mar. 2010), pp. 199–208. DOI: 10.1038/nature08917.

[10] A. Banyaga and D. Hurtubise. *Lectures on Morse Homology*. Texts in the Mathematical Sciences (Springer, 2013).

[11] A. Barone. *Superconductive Particle Detectors: Advances in the Physics of Condensed Matter* (World Scientific Pub. Co. Inc., 1987).

[12] A. A. Belavin and A. M. Polyakov. "Quantum fluctuations of pseudoparticles". *Nucl. Phys.* **B123** (1977), pp. 429–444. DOI: 10.1016/0550-3213(77)90175-4.

[13] F. A. Berezin. "The method of second quantization". *Pure Appl. Phys.*, **24** (1966), pp. 1–228.

[14] H. Bethe. "On the theory of metals. 1. Eigenvalues and eigenfunctions for the linear atomic chain". *Z. Phys.*, **71** (1931), pp. 205–226. DOI: 10.1007/BF01341708.

[15] K. Binder and A. P. Young. "Spin glasses: Experimental facts, theoretical concepts, and open questions". *Rev. Mod. Phys.*, **58** (4 Oct. 1986), pp. 801–976. DOI: 10.1103/RevModPhys.58.801.

[16] M. Blasone and P. Jizba. "Nambu–Goldstone dynamics and generalized coherent-state functional integrals". *Journal of Physics A: Mathematical and Theoretical*, **45.24** (2012), p. 244009.

[17] E. B. Bogomolny. "Stability of classical solutions". *Sov. J. Nucl. Phys.*, **24** (1976) [Yad. Fiz.24,861(1976)], p. 449.

[18] R. Bott. "An application of the Morse theory to the topology of Liegroups." English. *Bull. Soc. Math. Fr.*, **84** (1956), pp. 251–281. ISSN: 0037-9484.

[19] R. Bott. "Morse theory indomitable". English. *Publications Mathématiques de l'IHÉS*, **68** (1988), pp. 99–114. URL: http://eudml.org/doc/104046.

[20] H.-B. Braun and D. Loss. "Chiral quantum spin solitons". *Journal of Applied Physics*, **79.8** (1996), pp. 6107–6109. DOI: 10.1063/1.362102.

[21] E. Brézin, J. C. Le Guillou and J. Zinn-Justin, *Phys. Rev.* D15 (1977), 1544.

[22] L. Brillouin. "La mécanique ondulatoire de Schrödinger: une méthode générale de resolution par approximations successives". *Comptes Rendus de l'Académie des Sciences*, **183** (Oct. 1926), pp. 24–26.

[23] C. G. Callan Jr. and S. R. Coleman. "The fate of the false vacuum. 2. First quantum corrections". *Phys. Rev.*, **D16** (1977), pp. 1762–1768. DOI: 10.1103/PhysRevD.16.1762.

[24] S. Chadha and P. Di Vecchia "Zeta function regularization of the quantum fluctuations around the Yang–Mills pseudoparticle". *Phys. Lett.*, **B72** (1977), pp. 103–108. DOI: 10.1016/0370-2693(77)90073-9.

[25] W. Chen, K. Hida and B. C. Sanctuary. "Ground-state phase diagram of $S = 1$ XXZ chains with uniaxial single-ion-type anisotropy". *Phys. Rev. B*, **67** (10 Mar. 2003), p. 104401. DOI: 10.1103/PhysRevB.67.104401.

[26] Y. Choquet-Bruhat, C. DeWitt-Morette and M. Dillard-Bleick. *Analysis, Manifolds, and Physics*. Analysis, Manifolds, and Physics pt. 1 (North-Holland Publishing Company, 1982).

[27] E. M. Chudnovsky and L. Gunther. "Quantum theory of nucleation in ferromagnets". *Phys. Rev. B*, **37** (16 June 1988), pp. 9455–9459. DOI: 10.1103/PhysRevB.37.9455.

[28] E. M. Chudnovsky and L. Gunther. "Quantum tunneling of magnetization in small ferromagnetic particles". *Phys. Rev. Lett.*, **60** (8 Feb. 1988), pp. 661– 664. DOI: 10.1103/PhysRevLett.60.661.

[29] E. M. Chudnovsky and J. Tejada. *Lectures on Magnetism*. Lectures on Magnetism: With 128 Problems (Rinton Press, 2006).

[30] E. M. Chudnovsky, J. Tejada, C. Calero and F. Macia. *Problem Solutions to Lectures on Magnetism by Chudnovsky and Tejada* (Rinton Press, 2007).

[31] S. Coleman. *Aspects of Symmetry: Selected Erice Lectures* (Cambridge University Press, 1988).

[32] S. R. Coleman. "The fate of the false vacuum. 1. Semiclassical theory". *Phys. Rev.*, **D15** (1977). [Erratum: Phys. Rev.D16,1248(1977)], pp. 2929–2936. DOI: 10.1103/PhysRevD.15.2929, DOI: 10.1103/PhysRevD.16.1248.

[33] S. R. Coleman and F. De Luccia. "Gravitational effects on and of vacuum decay". *Phys. Rev.*, **D21** (1980), p. 3305. DOI: 10.1103/PhysRevD.21.3305.

[34] S. R. Coleman, V. Glaser and A. Martin. "Action minima among solutions to a class of Euclidean scalar field equations". *Commun. Math. Phys.*, **58** (1978), p. 211. DOI: 10.1007/BF01609421.

[35] J. C. Collins and D. E. Soper. "Large order expansion in perturbation theory". *Annals Phys.*, **112** (1978), pp. 209–234. DOI: 10.1016/0003-4916(78)90084-2.

[36] R. F. Dashen, B. Hasslacher and A. Neveu. "Nonperturbative methods and extended hadron models in field theory. 1. Semiclassical functional methods". *Phys. Rev.*, **D10** (1974), p. 4114. DOI: 10.1103/PhysRevD.10.4114.

[37] P. J. Davis. *Circulant Matrices*. Pure and Applied Mathematics (Wiley, 1979).

[38] J. von Delft and C. L. Henley. "Destructive quantum interference in spin tunnelling problems". *Phys. Rev. Lett.*, **69** (22 Nov. 1992), pp. 3236– 3239. DOI: 10.1103/ PhysRevLett.69.3236.

[39] F. Devreux and J. P. Boucher. "Solitons in Ising-like quantum spin chains in a magnetic field: a second quantization approach". *J. Phys. France*, **48.10** (1987), pp. 1663–1670. DOI: 10.1051/jphys:0198700480100166300.

[40] P. A. M. Dirac. "The Lagrangian in quantum mechanics". *Phys. Z. Sowjetunion*, **3** (1933), pp. 64–72.

[41] A. J. Dolgert, S. J. Di Bartolo and A. T. Dorsey. "Superheating fields of superconductors: Asymptotic analysis and numerical results". *Phys. Rev. B*, **53** (9 Mar. 1996), pp. 5650–5660. DOI: 10.1103/PhysRevB.53.5650.

[42] T. Eguchi, P. B. Gilkey and A. J. Hanson. "Gravitation, gauge theories and differential geometry". *Phys. Rept.*, **66** (1980), p. 213. DOI: 10.1016/0370-1573(80)90130-1.

[43] M. Enz and R. Schilling. "Magnetic field dependence of the tunnelling splitting of quantum spins". *Journal of Physics C: Solid State Physics*, **19.30** (1986), p. L711.

[44] L. D. Faddeev and V. N. Popov. "Feynman diagrams for the Yang–Mills field". *Phys. Lett.*, **25B** (1967), pp. 29–30. DOI: 10.1016/0370-2693(67)90067-6.

[45] R. P. Feynman. "Space-time approach to nonrelativistic quantum mechanics". *Rev. Mod. Phys.*, **20** (1948), pp. 367–387. DOI: 10.1103/RevModPhys.20.367.

[46] R. P. Feynman and A. R. Hibbs. *Quantum Mechanics and Path Integrals.* International Series in Pure and Applied Physics (McGraw-Hill, 1965).

[47] W. Fischler, V. Kaplunovsky, C. Krishnan, L. Mannelli and M. Torres. "Metastable supersymmetry breaking in a cooling universe". *JHEP*, **03** (2007), p. 107. DOI: 10.1088/1126-6708/2007/03/107. arXiv: hep-th/0611018 [hep-th].

[48] E. Fradkin. *Field Theories of Condensed Matter Physics* (Cambridge University Press, 2013).

[49] E. Fradkin and M. Stone. "Topological terms in one- and twodimensional quantum Heisenberg antiferromagnets". *Phys. Rev. B*, **38** (10 Oct. 1988), pp. 7215–7218. DOI: 10.1103/PhysRevB.38.7215.

[50] K. Fujikawa. "Path integral measure for gauge invariant fermion theories". *Phys. Rev. Lett.*, **42** (1979), pp. 1195–1198. DOI: 10.1103/PhysRevLett.42.1195.

[51] D. B. Fuks. "Spheres, homotopy groups of the". In *Encyclopedia of Mathematics* (2001).

[52] D. A. Garanin. "Spin tunnelling: a perturbative approach". *J. Phys. A-Math. Gen.*, **24.2** (1991), p. L61.

[53] A. Garg and G.-H. Kim. "Macroscopic magnetization tunneling and coherence: Calculation of tunneling-rate prefactors". *Phys. Rev. B*, **45** (22 June 1992), pp. 12921–12929. DOI: 10.1103/PhysRevB.45.12921.

[54] H. Georgi and S. L. Glashow. "Unified weak and electromagnetic interactions without neutral currents". *Phys. Rev. Lett.*, **28** (1972), p. 1494. DOI: 10.1103/ PhysRevLett.28.1494.

[55] J. Glimm and A. Jaffe. *Quantum Physics: A Functional Integral Point of View* (Springer 2012). DOI: 10.1007/BF02812722.

[56] J. Goldstone. "Field theories with superconductor solutions". *Nuovo Cim.*, **19** (1961), pp. 154–164. DOI: 10.1007/BF02812722.

[57] D. J. Gross and F. Wilczek. "Ultraviolet behavior of nonabelian gauge theories". *Phys. Rev. Lett.*, **30** (1973), pp. 1343–1346. DOI: 10.1103/PhysRevLett.30.1343.

[58] D. Haldane. "Large-D, and intermediate-D states in an $S = 2$ quantum spin chain with on-site and XXZ anisotropies". *Phys. Soc. Jn.*, **80.4** (2011), p. 043001. DOI: 10.1143/JPSJ.80.043001.

[59] F. D. M. Haldane. "Nonlinear field theory of large-spin Heisenberg antiferromagnets: semiclassically quantized solitons of the one-dimensional easy-axis Néel state". *Phys. Rev. Lett.*, **50** (15 Apr. 1983), pp. 1153–1156. DOI: 10.1103/PhysRevLett.50.1153.

[60] S. W. Hawking and G. F. R. Ellis. *The Large Scale Structure of Space-Time*. (Cambridge University Press, 2011). DOI: 10.1017/CBO9780511524646.

[61] P. W. Higgs. "Broken symmetries and the masses of gauge bosons". *Phys. Rev. Lett.*, **13** (1964), pp. 508–509. DOI: 10.1103/PhysRevLett.13.508.

[62] K. Hori, S. Katz, A. Klemm et al. *Mirror Symmetry*. Vol. 1. Clay Mathematics Monographs (AMS, 2003). URL: www.claymath.org/library/monographs/cmim01.pdf.

[63] L. Hulthén. "Uber has Austauschproblem eines Kristalls". *Arkiv Mat. Astron. Fysik*, **26A** (1938), pp. 1–10.

[64] K. Jain Rohit. *Supersymmetric Schrodinger operators with applications to Morse theory*. 2017. arXiv: 1703.06943v2.

[65] A. Jevicki. "Treatment of zero frequency modes in perturbation expansion about classical field configurations". *Nucl. Phys.*, **B117** (1976), pp. 365–376. DOI: 10.1016/0550-3213(76)90403-X.

[66] B. Julia and A. Zee. "Poles with both magnetic and electric charges in nonabelian gauge theory". *Phys. Rev.*, **D11** (1975), pp. 2227–2232. DOI: 10.1103/PhysRevD.11.2227.

[67] S. Kachru, R. Kallosh, A. Linde and S. P. Trivedi. "De Sitter vacua in string theory". *Phys. Rev.*, **D68** (2003), p. 046005. DOI: 10.1103/PhysRevD.68.046005. arXiv: hep-th/0301240[hep-th].

[68] A. Khare and M. B. Paranjape. "Suppression of quantum tunneling for all spins for easy-axis systems". *Phys. Rev. B*, **83** (17 May 2011), p. 172401. DOI: 10.1103/PhysRevB.83.172401.

[69] T. W. B. Kibble. "Some implications of a cosmological phase transition". *Phys. Rept.*, **67** (1980), p. 183. DOI: 10.1016/0370-1573(80)90091-5.

[70] T. W. B. Kibble. "Topology of cosmic domains and strings". *J. Phys.*, **A9** (1976), pp. 1387–1398. DOI: 10.1088/0305-4470/9/8/029.

[71] G.-H. Kim. "Level splittings in exchange-biased spin tunneling". *Phys. Rev. B*, **67** (2 Jan. 2003), p. 024421. DOI: 10.1103/PhysRevB.67.024421.

[72] G.-H. Kim. "Tunneling in a single-molecule magnet via anisotropic exchange interactions". *Phys. Rev. B*, **68** (14 Oct. 2003), p. 144423. DOI: 10.1103/PhysRevB.68.144423.

[73] A. Kitaev. "Anyons in an exactly solved model and beyond". *Ann. Phys.*, **321.1** (2006), pp. 2–111. DOI: 10.1016/j.aop.2005.10.005.

[74] J. A. Kjäll, M. Zalatel, R. Mong, J. Bardarson and F. Pollmann. "Phase diagram of the anisotropic spin-2 XXZ model: infinite-system density matrix renormalization group study". *Phys. Rev. B*, **87** (23 June 2013), p. 235106. DOI: 10.1103/PhysRevB.87.235106.

[75] J. R. Klauder. "Path integrals and stationary-phase approximations". *Phys. Rev. D*, **19** (8 Apr. 1979), pp. 2349–2356. DOI: 10.1103/PhysRevD.19.2349.

[76] H. Kleinert. *Path Integrals in Quantum Mechanics, Statistics, Polymer Physics, and Financial Markets*, 5th Edition (World Scientific Publishing Co., 2009). DOI: 10.1142/7305.

[77] H. A. Kramers. "Wellenmechanik und halbzahlige Quantisierung". *Z. Phys.*, **39** (Oct. 1926), pp. 828–840. DOI: 10.1007/BF01451751.

[78] H. A. Kramers, "Théorie générale de la rotation paramagnétique dans les cristaux". *Proc. Acad. Sci. Amsterdam*, **33** (1930), p. 959.

[79] B. Kumar, M. B. Paranjape and U. A. Yajnik. "Fate of the false monopoles: induced vacuum decay". *Phys. Rev.*, **D82** (2010), p. 025022. DOI: 10.1103/PhysRevD.82.025022. arXiv: 1006.0693[hep-th].

[80] B. Kumar and U. Yajnik. "Graceful exit via monopoles in a theory with OfRaifeartaigh type supersymmetry breaking". *Nucl. Phys.*, **B831** (2010), pp. 162–177. DOI: 10.1016/j.nuclphysb.2010.01.011. arXiv: 0908.3949[hep-th].

[81] B. Kumar and U. A. Yajnik. "Stability of false vacuum in supersymmetric theories with cosmic strings". *Phys. Rev.*, **D79** (2009), p. 065001. DOI: 10.1103/PhysRevD.79.065001. arXiv: 0807.3254[hep-th].

[82] Laurascudder. *Baryon Decuplet ó Wikipedia, The Free Encyclopedia*. 2007.

[83] Laurascudder. *Baryon Octet ó Wikipedia, The Free Encyclopedia*. 2007.

[84] Laurascudder. *Meson Octet ó Wikipedia, The Free Encyclopedia*. 2007.

[85] B.-H. Lee, W. Lee, R. MacKenzie et al. "Tunneling decay of false vortices". *Phys. Rev.*, **D88** (2013), p. 085031. DOI: 10.1103/PhysRevD.88.085031. arXiv: 1308.3501[hep-th].

[86] A. J. Leggett. "A theoretical description of the new phases of liquid 3He". *Rev. Mod. Phys.*, **47** (2 Apr. 1975), pp. 331–414. DOI: 10.1103/RevModPhys.47.331.

[87] E. H. Lieb. "The classical limit of quantum spin systems". *Comm. Math. Phys.*, **31.4** (1973), pp. 327–340. URL: http://projecteuclid.org/euclid.cmp/1103859040.

[88] D. Loss, D. P. DiVincenzo and G. Grinstein. "Suppression of tunneling by interference in half-integer-spin particles". *Phys. Rev. Lett.*, **69** (22 Nov. 1992), pp. 3232–3235. DOI: 10.1103/PhysRevLett.69.3232.

[89] Y. Matsumoto. *An Introduction to Morse Theory*. Trans. Kiki Hudson and Masahico Saito (American Mathematical Society, 2002).

[90] F. Meier, J. Levy and D. Loss. "Quantum computing with antiferromagnetic spin clusters". *Phys. Rev. B*, **68** (13 Oct. 2003), p. 134417. DOI: 10.1103/PhysRevB.68.134417.

[91] F. Meier and D. Loss. "Electron and nuclear spin dynamics in antiferromagnetic molecular rings". *Phys. Rev. Lett.*, **86** (23 June 2001), pp. 5373–5376. DOI: 10.1103/PhysRevLett.86.5373.

[92] F. Meier and D. Loss. "Thermodynamics and spin-tunneling dynamics in ferric wheels with excess spin". *Phys. Rev. B*, **64** (22 Nov. 2001), p. 224411. DOI: 10.1103/PhysRevB.64.224411.

[93] H.-J. Mikeska and M. Steiner. "Solitary excitations in one-dimensional magnets". *Adv. Phys.*, **40.3** (1991), pp. 191–356. DOI: 10.1080/00018739100101492.

[94] J. Milnor. *Morse Theory (AM-51)*. Annals of Mathematics Studies (Princeton University Press, 2016).

[95] S. E. Nagler, W. J. L. Buyers, R. L. Armstrong and B. Briat. "Propagating domain walls in CsCoBr3". *Phys. Rev. Lett.*, **49** (8 Aug. 1982), pp. 590–592. DOI: 10.1103/PhysRevLett.49.590.

[96] H. B. Nielsen and P. Olesen. "Vortex line models for dual strings". *Nucl. Phys. B*, **61** (1973), pp. 45–61. DOI: 10.1016/0550-3213(73)90350-7.

[97] S. P. Novikov. "The Hamiltonian formalism and a many valued analog of Morse theory". *Usp. Mat. Nauk*, **37N5.5** (1982). [Russ. Math. Surveys(1982),37(5):1], pp. 3–49. DOI: 10.1070/RM1982v037n05ABEH004020.

[98] F. R. Ore Jr. "Quantum field theory about a Yang–Mills pseudoparticle". *Phys. Rev. D*, **15** (1977), p. 470. DOI: 10.1103/PhysRevD.15.470.

[99]   S. A. Owerre and M. B. Paranjape. "Macroscopic quantum spin tunneling with two interacting spins". *Phys. Rev. B*, **88** (22 Dec. 2013), p. 220403. DOI: 10.1103/PhysRevB.88.220403.

[100] A. Perelomov. *Generalized Coherent States and their Applications* (Springer-Verlag New York Inc., Jan. 1986).

[101] M. E. Peskin and D. V. Schroeder. *An Introduction to Quantum Field Theory* (Addison-Wesley, 1995).

[102] H. D. Politzer. "Reliable perturbative results for strong interactions?" *Phys. Rev. Lett.*, 30 (26 June 1973), pp. 1346–1349. DOI: 10.1103/PhysRevLett.30.1346.

[103] A. M. Polyakov. "Quark confinement and topology of gauge groups". *Nucl. Phys. B*, **120** (1977), pp. 429–458. DOI: 10.1016/0550-3213(77)90086-4.

[104] M. K. Prasad and C. M. Sommerfield. "An exact classical solution for the ít Hooft monopole and the Julia-Zee dyon". *Phys. Rev. Lett.*, **35** (1975), pp. 760–762. DOI: 10.1103/PhysRevLett.35.760.

[105] K. Pretzl. "Superconducting granule detectors". *J. Low Temp. Phys.*, **93.3** (1993), pp. 439–448. ISSN: 1573-7357. DOI: 10.1007/BF00693458.

[106] J. M. Radcliffe. "Some properties of coherent spin states". *J. Phys. A: General Physics*, **4.3** (1971), p. 313. DOI: 10.1088/0305-4470/4/3/009.

[107] M. Reed and B. Simon. *I: Functional Analysis*. Methods of Modern Mathematical Physics (Elsevier Science, 1981).

[108] H. J. Schulz. "Phase diagrams and correlation exponents for quantum spin chains of arbitrary spin quantum number". *Phys. Rev. B*, **34** (9 Nov. 1986), pp. 6372–6385. DOI: 10.1103/PhysRevB.34.6372.

[109] J. Schwinger. "On gauge invariance and vacuum polarization". *Phys. Rev.*, **82** (5 June 1951), pp. 664–679. DOI: 10.1103/PhysRev.82.664.

[110] J. Simon, W. Bakr and R. Ma. "Quantum simulation of antiferromagnetic spin chains in an optical lattice". *Nature*, **472.7343** (Apr. 2011), pp. 307–312. DOI: 10.1038/nature09994.

[111] P. J. Steinhardt. "Monopole dissociation in the early universe". *Phys. Rev. D*, **24** (1981), p. 842. DOI: 10.1103/PhysRevD.24.842.

[112] G. 't Hooft. "Computation of the quantum effects due to a four-dimensional psuedoparticle". *Phys. Rev. D*, **14** (12 Dec. 1976), pp. 3432–3450. DOI: 10.1103/PhysRevD.14.3432.

[113] J. Ummethum, J. Nehrkorn, S. Mukherjee et al. "Discrete antiferromagnetic spin-wave excitations in the giant ferric wheel Fe18". *Phys. Rev. B*, **86** (10 Sept. 2012), p. 104403. DOI: 10.1103/PhysRevB.86.104403.

[114] J. H. Van Vleck. "On sigma-type doubling and electron spin in the spectra of diatomic molecules". *Phys. Rev.*, **33** (1929), pp. 467–506. DOI: 10.1103/PhysRev.33.467.

[115] J. Villain. "Propagative spin relaxation in the Ising-like antiferromagnetic linear chain". *Physica B+C*, **79.1** (1975), pp. 1–12. ISSN: 0378-4363. DOI: 10.1016/0378-4363(75)90101-1.

[116] O. Waldmann, C. Dobe, H. Güdel and H. Mutka. "Quantum dynamics of the Néel vector in the antiferromagnetic molecular wheel CsFe8". *Phys. Rev. B*, **74** (5 Aug. 2006), p. 054429. DOI: 10.1103/PhysRevB.74.054429.

[117] S. Weinberg. "Dynamical approach to current algebra". *Phys. Rev. Lett.*, **18** (1967), pp. 188–191. DOI: 10.1103/PhysRevLett.18.188.

[118] S. Weinberg. "The U(1) problem". *Phys. Rev. D*, **11** (1975), pp. 3583–3593. DOI: 10.1103/PhysRevD.11.3583.

[119] G. Wentzel. "Eine Verallgemeinerung der Quantenbedingungen für die Zwecke der Wellenmechanik". *Z. Phys.*, **38** (June 1926), pp. 518–529. DOI: 10.1007/BF01397171.

[120] J. Wess and B. Zumino. "Consequences of anomalous Ward identities". *Phys. Lett. B*, **37** (1971), pp. 95–97. DOI: 10.1016/0370-2693(71)90582-X.

[121] N. Wiener. "Differential space". *J. Math. and Phys.*, **2** (1923), pp. 132–174.

[122] E. Witten. "Baryons in the 1/n Expansion". *Nucl. Phys. B*, **160** (1979), pp. 57–115. DOI: 10.1016/0550-3213(79)90232-3.

[123] E. Witten. "Constraints on supersymmetry breaking". *Nucl. Phys. B*, **202** (1982), p. 253. DOI: 10.1016/0550-3213(82)90071-2.

[124] E. Witten. "Dynamical breaking of supersymmetry". *Nucl. Phys. B*, **188** (1981), p. 513. DOI: 10.1016/0550-3213(81)90006-7.

[125] E. Witten. "Supersymmetry and Morse theory". *J. Diff. Geom.*, **17.4** (1982), pp. 661–692.

[126] U. A. Yajnik. "Phase transitions induced by cosmic strings". *Phys. Rev. D*, **34** (1986), pp. 1237–1240. DOI: 10.1103/PhysRevD.34.1237.

[127] W.-M. Zhang, D. H. Feng and R. Gilmore. "Coherent states: Theory and some applications". *Rev. Mod. Phys.*, **62** (4 Oct. 1990), pp. 867–927. DOI: 10.1103/RevModPhys.62.867.

[128] J. Zinn-Justin. "Perturbation series at large orders in quantum mechanics and field theories: Application to the problem of resummation". *Physics Reports*, **70.2** (1981), pp. 109–167. ISSN: 0370-1573. DOI: 10.1016/0370-1573(81)90016-8.

[129] W. H. Zurek. "Cosmic strings in laboratory superfluids and the topological remnants of other phase transitions". *Acta Phys. Pol. B*, **24** (1993), pp. 1301–1311.

[130] W. H. Zurek. "Cosmological experiments in superfluid helium?" *Nature* **317** (1985), pp. 505–508. DOI: 10.1038/317505a0.

Index

Printed in the United States
by Baker & Taylor Publisher Services